Biotechnology and Bioengineering Symposium   No. 9

Computer Applications in Fermentation Technology

Proceedings of the Second International Conference on
Computer Applications in Fermentation Technology
Held at Philadelphia, Pennsylvania, August 28–30, 1978
Sponsored by
The National Science Foundation
and the University of Pennsylvania

Editor
William B. Armiger
BioChem Technology, Inc.

an Interscience® Publication
published by John Wiley & Sons

BIOTECHNOLOGY AND BIOENGINEERING

SYMPOSIUM NO. 9

E. L. GADEN, JR., *Editor*

William B. Armiger has been appointed Editor for this Symposium by the Editorial Board of *Biotechnology and Bioengineering*.

This book constitutes a part of the annual subscription to *Biotechnology and Bioengineering,* Vol. XXI, and as such is supplied without additional charge to subscribers. Single copies can be purchased from the Subscription Department, John Wiley & Sons, Inc.

Subscription price, *Biotechnology and Bioengineering,* Vol. XXII, 1980: $150.00 per volume. Postage and handling outside U.S.A.: $19.00.

Copyright © 1979 by John Wiley & Sons, Inc., 605 Third Avenue, New York, New York 10016. All rights reserved. Reproduction or translation of any part of this work work beyond that permitted by Sections 107 or 108 of the United States Copyright Law without the permission of the copyright owner is unlawful. Second-class postage paid at New York, New York, and at additional mailing offices.

The code and the copyright notice appearing at the bottom of the first page of an article in this journal indicate the copyright owner's consent that copies of the article may be made for personal or internal use, or for the personal or internal use of specific clients, on the condition that the copier pay for copying beyond that permitted by Sections 107 or 108 of the United States Copyright Law. The per-copy fee for each article appears after the dollar sign and is to be paid through the Copyright Clearance Center, Inc. This consent does not extend to other kinds of copying, such as copying for general distribution, for advertising or promotional purposes, for creating new collective works, or for resale. Such permission requests and other permission inquiries should be addressed to the publisher.

Printed in the United States of America.

Contents

Computer Application in Fermentation Technology—A Perspective
  C. L. Cooney 1

On-Line Gas Analysis for Material Balances and Control
  H. Y. Wang, C. L. Cooney, and D. I. C. Wang 13

Analysis and Control of Mixed Cultures
  R. T. Hatch, C. Wilder, and T. W. Cadman 25

IMPAC/FORTRAN Schemes for Fermentation Data
  D. D. Dobry and J. L. Jost 39

Applications of Mass-Energy Balances in On-Line Data Analysis
  L. E. Erickson 49

Instrumentation for Fermentation Monitoring and Control
  Å. Undén 61

Affinity Sensors for Individual Metabolites
  J. S. Schultz and G. Sims 65

Cultivation of Microorganisms with a DO-Stat and a Silicone Tubing Sensor
  T. Kobayashi, T. Yano, H. Mori, and S. Shimizu 73

Redox Potential as a State Variable in Fermentation Systems
  L. Kjaergaard and B. B. Joergensen 85

Indirect Fermentation Measurements as a Basis for Control
  J. R. Swartz and C. L. Cooney 95

Sensors and Instrumentation: Steam-Sterilizable Dissolved Oxygen Sensor and Cell Mass Sensor for On-Line Fermentation System Control
  M. Ohashi, T. Watabe, T. Ishikawa, Y. Watanabe, K. Miwa, M. Shoda,
  Y. Ishikawa, T. Ando, T. Shibata, T. Kitsunai, N. Kamiyama, and Y. Oikawa 103

Use of Culture Fluorescence for Monitoring of Fermentation Systems
  D. W. Zabriskie 117

Use of an Immobilized Penicillinase Electrode in the Monitoring of the Penicillin Fermentation
  J. W. Hewetson, T. H. Jong, and P. P. Gray 125

Mathematical Models and Computer Simulations of Dialysis and Nondialysis Continuous Processes for Ammonium-Lactate Fermentation
  R. W. Stieber and P. Gerhardt 137

Modeling and Optimal Control of an SCP Fermentation Process
  J. Alvarez and J. Ricaño 149

Growth Kinetics and Antibiotic Synthesis during the Repeated Fed-Batch Culture of *Streptomycetes*
  M. Bošnjak, V. Topolovec, and V. Johanides 155

Kinetic Model for the Control of Wine Fermentations
  R. Boulton 167

Process Modeling Based on Biochemical Mechanisms of Microbial Growth
  A. R. Moreira, G. Van Dedem, and M. Moo-Young 179

Modeling and Optimal Control of Bakers' Yeast Production in Repeated Fed-Batch Culture
  P. Peringer and H. T. Blachere 205

CONTENTS

Review of Alternatives and Rationale for Computer Interfacing and System Configuration
 W. B. Armiger and D. M. Moran ... 215

Microprocessor in Fermentation Control
 M. Cordonnier, J. P. Kernevez, J. M. Lebeault, and J. Kryze ... 227

Single-Board Microcomputer for Fermentation Control
 R. P. Jefferis III, S. S. Klein, and J. Drakeford ... 231

Fermentation Control Systems—A Time for Change
 M. C. Beaverstock and G. P. Trearchis ... 241

Development of a Computerized Fermentation System Having Complete Feedback Capabilities for Use in a Research Environment
 R. D. Mohler, P. J. Hennigan, H. C. Lim, G. T. Tsao, and W. A. Weigand ... 257

Review of Process Control and Optimization in Fermentation
 S. Aiba ... 269

Application of Modern Control Theories to a Fermentation Process
 T. Takamatsu, S. Shioya, M. Shiota, and T. Kitabata ... 283

Optimization of Erythromycin Biosynthesis by Controlling pH and Temperature: Theoretical Aspects and Practical Application
 A. Cheruy and A. Durand ... 303

Optimal Strategy for Batch Alcoholic Fermentation
 I. Endo, T. Nagamune, and I. Inoue ... 321

Optimization of a Repeated Fed-Batch Reactor for Maximum Cell Productivity
 W. A. Weigand, H. C. Lim, C. C. Creagan, and R. D. Mohler ... 335

Control of the Quasi-Steady-State in Fed-Batch Fermentation
 T. J. Boyle ... 349

Computer Control of Glucose Feed to a Continuous Aerobic Culture of *Saccharomyces cerevisiae* Using the Respiratory Quotient
 R. Spruytenburg, I. J. Dunn, and J. R. Bourne ... 359

Computer Control of the Pekilo Protein Process
 A. Halme ... 369

Computer Control of Fermentation Plants
 R. Lundell ... 381

Author Index ... 395

Subject Index ... 397

Preface

With the expanding role of microbiological systems in the production of industrial chemicals, the conversion of biomass to fuels, and the treatment of industrial and municipal wastes, the need for improved techniques for fermentation monitoring and control has arisen. Computer control and on-line optimization of the fermentation step can have a major impact on the economics of these processes. To date, fermentation control has barely advanced beyond the point of single, independent, closed-loop feedback control of culture conditions such as temperature, pH, and dissolved oxygen. The field is becoming ready for utilizing more sophisticated strategies of adaptive or interactive control. Eventually, the on-line optimization of a fermentation based upon model reference control of the process will be realized.

Computer applications in the chemical process industry began in the late 1950s. By the late 1960s, publications on computer applications in batch process control were fairly common. With the widespread availability of minicomputers in the 1970s, there has been an abundance of control applications in the chemical process industry. However, the fermentation industry has lagged behind in the application of computers largely due to the more complex control problems that are involved. It has only been in the last 10 years that computers have begun to be applied to fermentation processes.

In response to the growing amount of activity in this field, The Second International Conference on Computer Applications in Fermentation Technology was held at the University of Pennsylvania from August 27–30, 1978. The objective of the conference was to bring together the leading experts from around the world to discuss the recent advances in and the necessary future developments for computer applications in biotechnology. The first conference was held at Dijon, France, in September 1973 (First European Conference on Computer Control of Fermentation). This meeting was followed by a workshop held in July, 1976, at Braunschweig–Stockheim, Germany (Workshop, Computer Application in Fermentation).

The conference was sponsored by the National Science Foundation and the University of Pennsylvania. It was organized and directed by:

Chairman:
   William B. Armiger, BioChem Technology, Inc.

Organizing Committee:
   Arthur E. Humphrey, University of Pennsylvania
   Dane W. Zabriskie, BioChem Technology, Inc.

There were six technical sessions organized by the following cochairmen:

Session I. Techniques for On-Line Data Analysis
    D. I. C. Wang, MIT
    H. Topiwala, Shell Research, Great Britain
Session II. Sensors and Instrumentation
    D. W. Zabriskie, BioChem Technology, Inc.
    G. W. Undén, AB Fermenta, Sweden
Session III. Process Modeling
    A. G. Frederickson, University of Minnesota
    R. Tanner, Vanderbilt University
Session IV. Interfacing and System Configuration
    J. L. Jost, The Upjohn Company
    D. E. F. Harrison, University of Kent, Great Britain
Session V. Process Control and Optimization
    A. Constantinides, Rutgers University
    S. Aiba, Osaka University, Japan
Session VI. Round Table—Future Developments
    Moderator: C. L. Cooney, MIT

We are pleased to have the proceedings of the symposium published as a supplement to *Biotechnology and Bioengineering*. A few of the original papers have not been published as part of the proceedings due to the unavailability of the manuscripts. Suggestions, comments, or discussion regarding the format of succeeding symposia on this topic are encouraged and should be addressed to the conference chairman.

WILLIAM B. ARMIGER

Computer Application in Fermentation Technology—A Perspective

CHARLES L. COONEY

Department of Nutrition and Food Science, Massachusetts Institute of Technology, Cambridge, Massachusetts 02139

INTRODUCTION

Considerable excitement has been aroused and substantial progress has been made since the First International Conference on Computer Applications to Fermentation Technology [1] was held in Dijon in 1973. It is in this context that it is interesting to examine the broad topic of computer-aided fermentation processes and ask where is this technology today? Where is it going? And what is the resistance to progress? The answers to these three questions are the essence of this Second International Conference on Computers in Fermentation Technology. Before proceeding, however, to an examination of these questions, it is useful to review the historical perspective leading to the status of computer-aided fermentations today. A summary of some of the major developments leading to the current state-of-the-art is summarized in Table I. It is difficult to truly pinpoint the starting point of modern fermentation technology. However, I have chosen 1929 with Fleming's discovery of penicillin as one of the first of a chain of events leading to interest in applying computers to fermentation processes. The historic discovery of penicillin (see Wilson [2]) represents the introduction into fermentation technology of products other than foods for human and animal consumption. It is also an interesting point in time because it was in 1932 that Nyquist first presented the basis for feedback control theory [3]. In the 1940s there was the advent of digital computers, the forerunners of the computational machines that are in such widespread use today [4]. Then, in 1945, there was simultaneously the development of submerged culture for use in antibiotics and especially penicillin production and the beginning applications of pH probes and gas analyzers for monitoring chemical processes. Although these events are isolated developments, they represent the foundation for the application of sensors, computers, and control theory to the development of modern fermentation technology. By 1960 there was beginning to appear knowledge of how controlled addition of sugar feeds could be used to improve, in particular, antibiotic fermentations [5,6]. At this same time, the application of the development of the theory for feed forward control and direct digital control was introduced [4]. Then, in 1965 Ajinamoto described the first application of computer control to a fermentation for the production of monosodium glutamate (MSG) [7]. The

TABLE I

Historical Perspective

1929	Fleming - Discovery of penicillin
1932	Nyquist - Feedback control theory
1940's	Advent of digital computers
1945	Submerged culture for antibiotics
	Applications of pH probes and gas analyzers
1960	Controlled sugar feed
1960	Feedforward control and DDC
1965	Ajinamoto - Computer control of MSG fermentation
1966	Merck - Computerized chemical pilot plant
1969	Dista - Computerized penicillin plant
1970's	Wide application of computer-aided fermentation processes

following year Bacher and Kaufman from Merck [8] described the extensive computerization of a chemical pilot plant, which in principle is very similar to a fermentation pilot plant. Later, in 1969, Grayson [9] of Dista Corporation described the use of computer control in a fermentation plant with its application not only in the pilot plant but also large-scale fermentors. And in the 1970s we are seeing a rapid proliferation of computer-coupled fermentation systems many of which are used in research and development programs and some utilized at plant scale level [10,11]. Looking back over the past 50 years, one sees the development of biotechnology: modern control theory, and computer technology, and while each of these areas had its own humble beginning, it is the integration of these topics that brings us together at this meeting.

It is interesting to examine the accumulated experience of computer process control in the chemical industry. As shown in Figure 1, the number of computer-control applications in the petroleum and petrochemical industries has increased exponentially during the past decade (with a specific growth rate of 0.43 yr^{-1}) and there is no indication of saturation in this particular market [4]. The total number of on-line applications of computers is estimated to be greater than 200,000, yet it is only within the past couple of years that substantial progress has been made in the application of computers to fermentation processes. Therefore, the major question becomes, why? What are the major bottlenecks that limit progress in fermentation? Are there problems unique to fermentation relative to other industries that make the application of computer control more difficult or is it a conflict between the art and science of fermen-

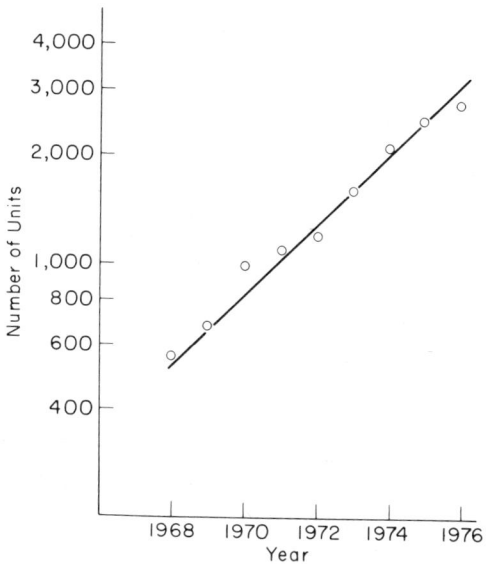

Fig. 1. Total number (cumulative) of computer-control applications in the petroleum and petrochemical industry [4].

tation technology and a general industrial conservatism that provides the major bottleneck?

COMPUTER-AIDED FERMENTATION TECHNOLOGY

The essential components in a computer-aided fermentation system are shown schematically in Figure 2. A fermentor monitored with available sensors is interfaced to a computer through a multiplexer and analog-to-digital conversion. The computer becomes the tool for data acquisition and analysis and its software is used to direct the fermentation itself. The technological bottleneck in such a system is not the availability of sophisticated computers. In fact, the cost of computer hardware is one of the few commodities whose price has decreased in recent years. The availability of sophisticated fermentation equipment also is not a bottleneck; substantial progress has been made in the design and construction of convenient, multipurpose, flexible fermentation pilot plants. But rather, the bottlenecks are twofold. First, there is a major limitation in the availability of sensors to measure those biological and physical–chemical parameters that we need in order to follow the fermentation. Second, there are major limitations in our ability to interpret the available information in the context of the biological system so that the information obtained can be used as a rational basis for closed loop control. These two factors, sensors and interpretation of data, are the major restrictions to the wide-scale application of computers to fermentation technology today.

A summary of the available sensors suitable for monitoring the culture en-

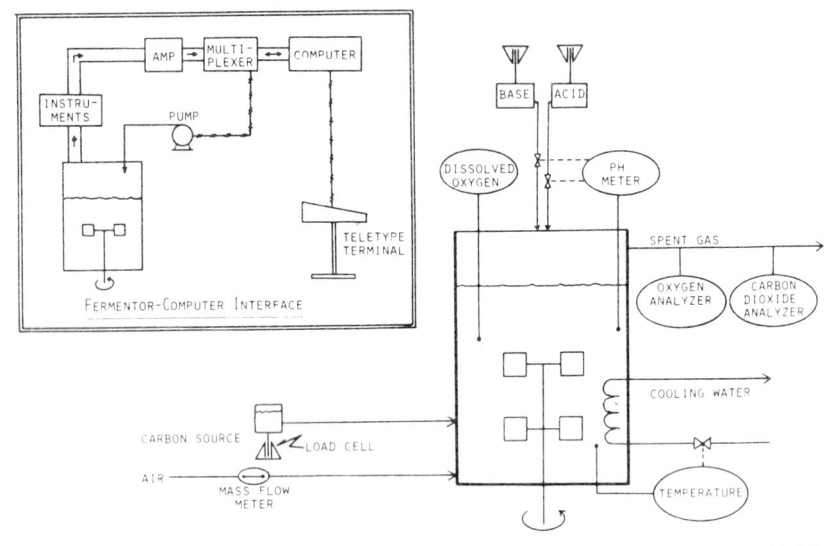

Fig. 2. Fermentation system and process sensors: major components in a computer-aided fermentation system.

vironment and cell properties are presented in Tables II and III. One can see from this summary that the sensors available today do not differ greatly from the sensors available a decade ago, particularly when you ask which of the available sensors are suitable for use in plant-scale operation. The restrictions are great, particularly in the availability of reliable sensors that will directly measure biological properties at the plant-scale level. Most notable in these lists is the absence of sensors to measure cell mass concentration, substrate and

TABLE II
Biological Sensors: Environmental

SENSOR	LABORATORY	PLANT
PH	YES	YES
TEMPERATURE	YES	YES
DISSOLVED O_2	YES	YES
REDOX POT.	YES	YES
DISSOLVED CO_2	YES	NO
SUBSTRATE	(YES)*	NO
PRODUCT	(YES)*	NO
MASS FLOW RATES	YES	YES

* Specialized analytical schemes for laboratory use are available for only a limited number of substrates and products.

TABLE III
Biological Sensors: Cellular

SENSOR	LABORATORY	PLANT
GASEOUS O_2	YES	YES
GASEOUS CO_2	YES	YES
HEAT EVOLUTION	YES	(YES)
ATP	YES	NO
NAD/NADH	YES	NO
PRODUCT	(YES)*	NO
CELL MASS	(YES)*	NO
ENZYME LEVEL	(YES)*	NO

* Several specialized sensors are under development.

product concentration, and dissolved carbon dioxide. Certainly, the problems of sensor availability are not unique to the fermentation industry. However, the problem itself is considerably greater because of the physical heterogeneity of fermentation processes as well as the dynamic nature of microorganisms in changing themselves and their environment with time. The problems of physical heterogeneity make sampling and measurement very difficult. The problem of fermentation dynamics makes interpretation of data very difficult. Yet microorganisms do not violate the fundamental laws of thermodynamics or conservation of mass and energy. Therefore, it should be possible to take advantage of many of the developments in the chemical process industry and apply them to fermentation processes. For instance, by the careful use of material and energy balancing one can hope to use available information from sensors such as gaseous carbon dioxide, gaseous oxygen, mass flow rates, and load cells to indirectly assess the state of the fermentation [12]. There is, however, a need to learn to better use the sensors and instruments available today.

OBJECTIVES OF COMPUTER-AIDED FERMENTATION

When discussing the potential for the application of computers to fermentation processes, the obvious question and focal point of discussion becomes, what are the objectives of fermentation process control? What, in fact, can be achieved by the application of computer technology to fermentation processes? Certainly a computer is capable of logging data through sophisticated acquisition systems, of analyzing data through the use of sophisticated computer algorithms, and of using this information to perform control functions for process optimization. Some more specific objectives are summarized in Table IV.

Our interest in closer monitoring of fermentations and the increasing use of

TABLE IV
Objectives of Fermentation Process Control

Data Management

Raw Materials Management

Energy Conservation

Process Optimization
    Productivity
    Titer
    Conversion Yield
    Reproducibility

---

available sensors for monitoring fermentations makes data management a time-consuming problem in the operation of a fermentation pilot plant or plant. Clearly, computers have a major advantage in this operation and in the preparation of reports. Beyond this, it is possible to use the computer for managing the use of raw materials and energy in hopes of achieving tighter control and more efficient use of both of these costly resources. The importance of this point becomes apparent when you examine the economics for fermentation processes where energy costs can represent 10–20% of the manufacturing cost and raw materials can account for 25–75% of operating costs [13].

However, the use of the computer in the above manner requires little knowledge of the biological system itself. The major objective for the future is the use of computers to achieve process optimization—i.e., the maximization of volumetric productivity, product concentration, conversion yield of substrate to

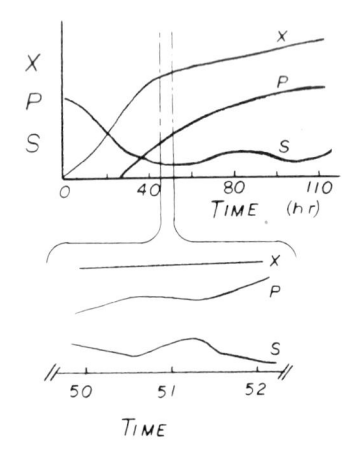

Fig. 3. Computer-aided monitoring of fermentation processes allows events in time to be magnified so that short-term changes are readily seen. Shown here is the amplification of a 2-hr interval of time in a typical fermentation process.

product, and process reproducibility. All of these parameters directly affect the economics of operation. Improvements in one or all of these areas can be used to directly calculate the payoff time required for the application of computer control. It is in these applications that the real excitement today is in the use of computers in fermentation systems.

In asking the question, how can a computer help us to achieve these goals, one needs to ask, what does the computer actually allow us to do? To answer this, it is instructive to draw an analogy with a microscope. When van Leeuwenhoek developed the microscope, the only thing that the microscope would allow us to do that could not be done before was to resolve distances or elements in space with much greater precision and accuracy than ever before. As a consequence, one could see details that were previously inaccessible. In the application of computers to fermentation processes the major benefit is that is is possible to resolve events in time much better than can be done by manual means. This is illustrated in Figure 3. For instance, in a long fermentation lasting 100 hr or more one typically takes samples at 1, 2, or 4 hr intervals. As a consequence, events happening with shorter time frequencies are invisible to the investigator. However, by continuous monitoring, high-frequency events can be seen and hopefully interpreted. Herein lies the power of the computer, because now one can see and resolve events in time in a way that was previously inaccessible. In the remainder of this conference many examples will be examined that demonstrate the power of the computer in allowing us to see phenomena that have not been seen before.

COMPUTER CONTROL

The ultimate objective of our efforts in the use of computer-aided fermentation is the application of computers to achieve optimal process control. This objective is only realized, however, after careful monitoring of the process in order to develop an understanding of the dynamics as a basis for process control. In this context, I would like to examine the use of a computer for on-line monitoring and control in the production of the antibiotic penicillin by *Penicillium chrysogenum*. For this example I will draw upon the recent work of Mou [17].

Process control in the penicillin fermentation is not a new concept. It was discovered almost 30 years ago that controlled addition of the carbon source could be used to prevent catabolite repression of penicillin production [14]. Although the biological basis for this effect was not understood until the mid-1960s, the use of controlled carbon source addition became a standard operating procedure in most penicillin fermentations. Control was achieved in a number of ways. Squires [15] developed a method utilizing monitoring of dissolved oxygen as a basis for controlling carbon source addition. Pan et al. [16] developed a control strategy based on changes in broth pH. More commonly, preset feed schedules which change with time are used to strive for reproducible operation of a successful penicillin process. All of these methods suffer from a lack of flexibility in that they are not based upon the supply of the carbon source to meet

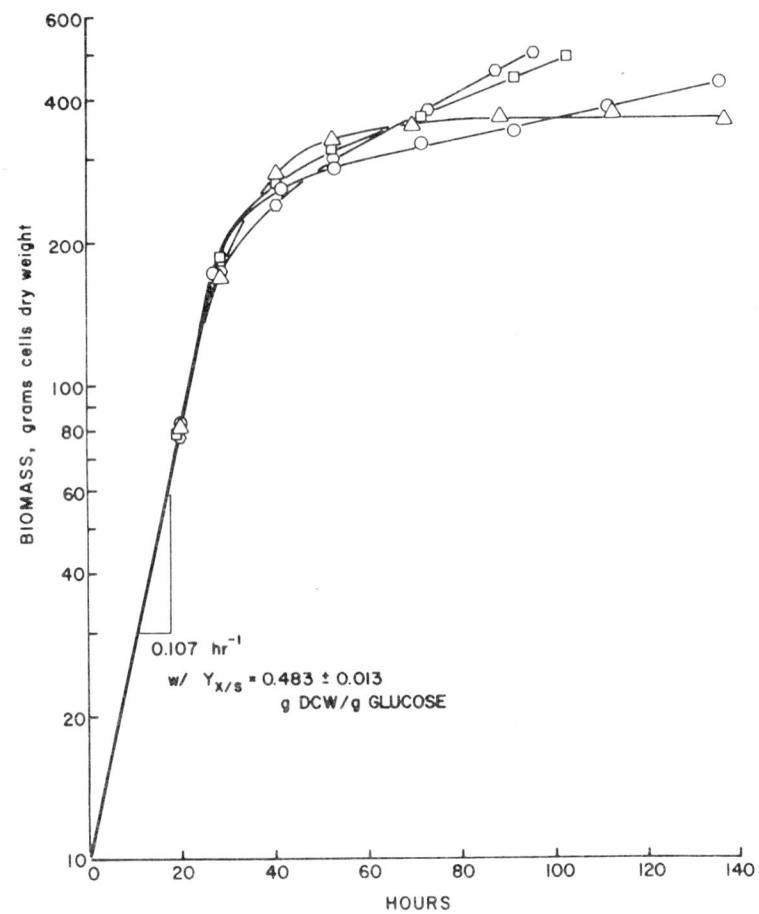

Fig. 4. Growth of *P. chrysogenum* under computer control of a fed-batch fermentation. Objective function was to maximize growth rate during the primary growth phase and control the growth rate at constant, preset values (as shown) during production phase [17]. (△) 1: $\mu_{pr}^* = 0.000$ hr^{-1}; μ_{pr} = 0.000 hr^{-1}. (○) 2: $\mu_{pr}^* = 0.006$ hr^{-1}; $\mu_{pr} = 0.006$ hr^{-1}. (□) 3: $\mu_{pr}^* = 0.010$ hr^{-1}; $\mu_{pr} = 0.009$ hr^{-1}. (○) 4: $\mu_{pr}^* = 0.015$ hr^{-1}, $\mu_{pr} = 0.014$ hr^{-1}.

the biological needs at each point in time. Therefore, we set out to develop a control strategy that would balance the supply with the demand for the carbon source throughout the penicillin fermentation. For this purpose we used a high-producing *P. chrysogenum* P2 strain for penicillin production in a simple medium.

The first goal was to develop a method for monitoring the amount of cell mass and the rate of growth during the growth phase of the penicillin fermentation. It was found that CO_2 production was a most valuable parameter, the rate of CO_2 evolution was indicative of the cell growth rate, and integration of the total CO_2 evolved provided an accurate measurement of the amount of mycelium present. With this information, it was possible to develop an adaptive control

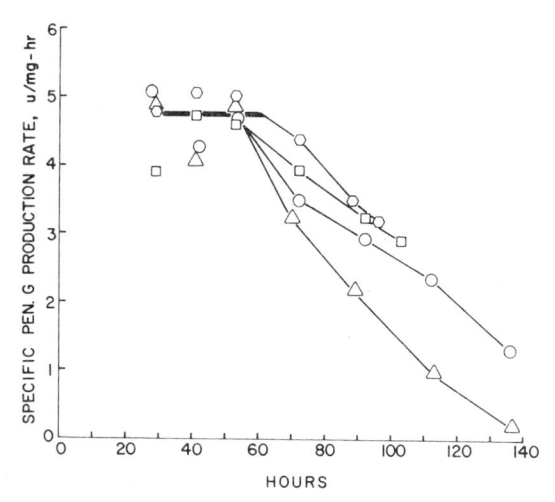

Fig. 5. Specific rate of penicillin production by *P. chrysogenum* during a computer-controlled fed-batch fermentation (see also Fig. 4) where the specific growth rate (μ_{pr}) was controlled at preset values during the production phase [17]. μ_{pr} (hr^{-1}): (O) 0.014; (□) 0.009; (O) 0.006; (△) 0.000.

strategy for continuous addition of glucose in order to meet the demands of the growing mycelium. This control strategy would add sufficient carbon source to maintain the maximum or any desired growth rate, yet not overfeed the glucose. Overfeeding would lead to wasteful side-product formation as well as repression of antibiotic synthesis. Once a desired cell concentration was obtained, the computer could be used to change the control strategy from rapid growth to slow controlled growth required for the production of penicillin.

During the antibiotic production phase, it is desired to maintain carbon-limited growth of *P. chrysogenum* in order to avoid carbon catabolite repression of penicillin synthesis. As a consequence, cell growth can be continuously estimated by means of a carbon balance; again, the supply of the carbon source can be controlled to meet the demands of the organism and with the added constraint of a desired specific growth rate. In this manner it is possible to maintain cells

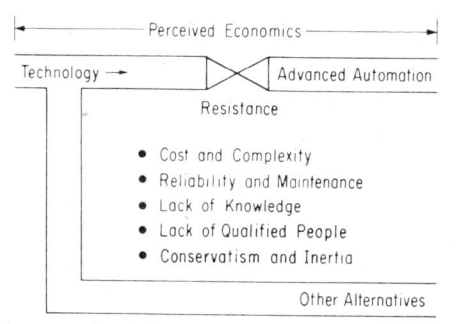

Fig. 6. Resistance to developments and application of computer control [4].

TABLE V
Research Needs in Computer Application to Fermentation Technology

1. Sensor Development

    Physical
    Biological

2. Elucidation of Biological Model

    Stoichiometry
    Bioenergetics
    Dynamics

3. Understanding of System Dynamics

4. Optimization Techniques

    (Suitable for biological systems)

5. Computer - Investigator - Fermentor

    Interaction

at a constant specific growth rate in a fed-batch fermentation. Results from such an exercise are shown in Figure 4. From these results it may be seen that it is possible to control growth at a variety of specific growth rates throughout the production phase. Thus, by means of computer control one can achieve a reproducible and prespecified fermentation pattern. This in itself does not provide optimization of penicillin production but it provides an important tool with which one can control the fermentation process and begin to evaluate the effect of operational parameters on penicillin production. An example of this is shown in Figure 5, which shows the specific rate of penicillin production as a function of time in the fermentations illustrated in Figure 4. One can see from these results that the growth rate chosen during the production phase has a distinct effect on the specific rate of penicillin production.

This example represents a beginning of the application of computers to complex fermentations and illustrates the types of problems one can address with the use of an on-line computer-coupled fermentation system. Many more examples of achievements with fermentation processes will be discussed in the following papers in this Symposium.

FUTURE DEVELOPMENTS

It is always with hesitation and trepidation that one attempts to predict future developments in any given field. Therefore, rather than attempt to predict where we will be at the time of the next conference on Computer Applications in Fermentation Technology, I would like to return to the question of where are the bottlenecks that will restrict our progress and what are the research needs that are required for us to move ahead. The bottlenecks to advancement in this area are illustrated in Figure 6, which is taken from a recent paper on computer applications in the chemical industry by Evans [4]. The driving force for improvement and achievement is the potential economics. The rate at which one gets there, however, becomes limited by the resistance. In this case, the resistance may take many forms. Identification of the major resistance would become a major point of discussion and I will leave the final choice of this resistance for you to identify in the context of your own situations. I personally feel that our lack of knowledge of the biological system and lack of suitable sensor technology combined with a dearth of qualified people at the present time to investigate these fields represent the major resistances. As evidenced by the number of people at this conference, I do not believe that conservatism and inertia are major restrictions.

In examining the research needs for the future I perceive five major areas as illustrated in Table V. These areas are not mutually independent and developments in one are likely to help others. Advancement of the frontier in fermentation technology will require advancement of these areas and I look forward with excitement to the next conference on Computer Application to Fermentation Technology.

References

[1] *Proceedings of The First European Conference on Computer Process Control in Fermentation* (Station de Genie Microbiologique, Dijon, 1973).
[2] D. Wilson, *In Search of Penicillin,* (Knopf, New York, 1976).
[3] H. Nyquist, *Bell Syst. Tech. J., 11,* 126 (1932).
[4] L. B. Evans, *Science, 195,* 146 (1977).
[5] F. V. Soltero and M. J. Johnson, *Appl. Microbiol., 1,* 52 (1953).
[6] P. Hosler and M. J. Johnson, *Ind. Eng. Chem., 45,* 871 (1953).
[7] S. Yamashita, H. Hoshi, and T. Inagaki, in *Fermentation Advances,* D. Perlman, Ed. (Academic, New York, 1969).
[8] S. Bacher and A. Kaufman, *Ind. Eng. Chem., 62,* 53 (1970).
[9] P. Grayson, *Proc. Biochem., 3,* 43 (1969).
[10] D. D. Dobry and J. L. Jost, *Ann. Rep. Ferment. Process, 1,* 95 (1977).
[11] W. A. Weigand, *Ann. Rep. Ferment. Process, 2,* 43 (1978).
[12] C. L. Cooney, H. Y. Young, and D. I. C. Wang, *Biotechnol. Bioeng., 19,* 55 (1977).
[13] C. L. Cooney, in *Microbial Growth on C_1-Compounds* (Society of Fermentation Technology, Japan, 1975).
[14] A. L. Demain, *Lloydia, 37,* 147 (1974).
[15] R. W. Squires, *Dev. Ind. Microbiol., 13,* 128 (1972).
[16] C. H. Pan, L. Hepler, and R. Elander, *Dev. Ind. Microbiol., 13,* 103 (1972).
[17] D. G. Mou, Ph.D. thesis, MIT, 1979.

On-Line Gas Analysis for Material Balances and Control

HENRY Y. WANG,[*] CHARLES L. COONEY, and DANIEL I. C. WANG
Massachusetts Institute of Technology, Cambridge, Massachusetts 02139

INTRODUCTION

Computer applications to process control are widely accepted in many chemical and petrochemical industries [1]. Presently, the fermentation industry is actively pursuing the use of computers to improve their processes. Unfortunately, the biological and physical complexity of industrial fermentations has inhibited work in this area as evidenced by the lack of information published on processes of industrial interest. However, if real merits are to be derived through applications of computer control, we must address commercially relevant processes with objective functions that are real and need to be optimized. Since computers provide capabilities for extremely rapid data acquisition, reduction, and storage, they are able to react faster and to make impartial decisions based on the information acquired and to optimize the specific fermentation process accordingly.

In considering the application of computers for the control of fermentation processes, it is important to ensure that the computer, the monitoring system, and the objectives are compatible with the level of scientific knowledge of the fermentation. After a detailed analysis of the types of fermentation processes that are currently suitable for the application of computer monitoring and control, it was concluded that certain prerequisites must be fulfilled. Specifically, the material and/or the energy balances of the system must be quantified with a certain degree of exactness [24]. The process selected as being most amenable to computer monitoring and control is the bakers' yeast production from molasses.

This paper will elaborate on an approach to develop a computer-aided control system and also on a possible extension of this approach to other commercial fermentation systems.

[*] Present address: Schering Corporation, Union, New Jersey 07083.

BIOLOGICAL MODEL BASED ON MATERIAL BALANCES

Although the utilization of on-line computer application and indirect measurement of cell growth have been proposed in the literature since the early 1970s [2,3], there are some weaknesses that can be cited. The most critical being that the mathematical models employed require the assumption that either the cell yield on oxygen or on carbon be constant. This approach is unrealistic since often these are the parameters that are to be optimized, especially in biomass production processes. In our approach, which was described in previous papers [4,5], the theoretical analysis, as well as the experimental executions, did not require assumptions for cell yield coefficients. We based our analyses upon using fundamental engineering principles that are material or component balancing. The only major assumption is that the stoichiometric equations for cell growth and product formation can be stated with a certain degree of exactness.

The stoichiometric relationship has been shown to be valid for yeast propagation [5,6]. Sucrose in molasses in the presence of oxygen along with other nutrients is converted to form yeast biomass, carbon dioxide, water, and other side products such as ethanol:

$$\text{(carbon-energy source)} + O_2 + NH_3 \rightarrow \text{(cells)} + H_2O + CO_2 + \text{(side product)(EtOH)} \quad (1)$$

Herbert [7] recently presented a comprehensive analysis on the stoichiometric aspects of cell growth. He indicated that the chemical composition of the microorganism can be affected by drastic changes in growth rate and the nature of the growth-limiting nutrients. In the case of bakers' yeast production, which

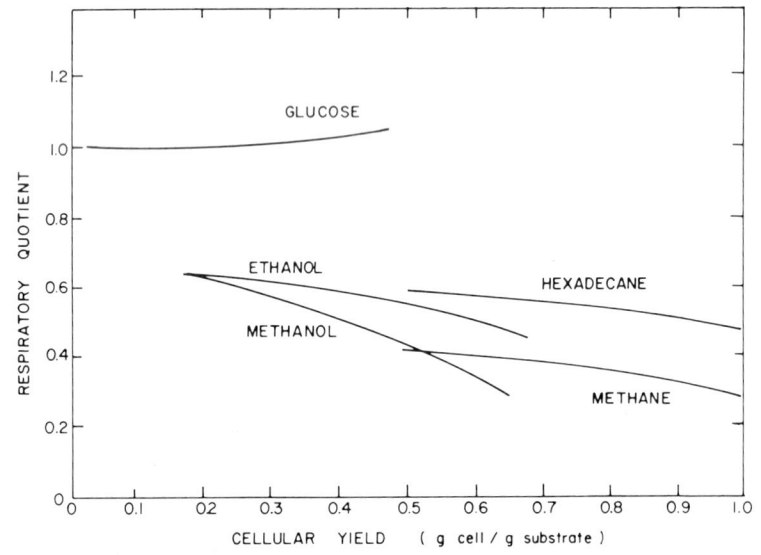

Fig. 1. RQ vs. cellular yield coefficient (theoretical analyses) for growth on different carbon-energy substrates. Cell composition is assumed constant ($C_6H_{10}NO_3$).

TABLE I

Comparison of Existing Sensors for Gases

Gas Measurements	Advantages	Disadvantages
Infrared Analyzer CO_2	Very specific	Drift Slow response time
Thermoconductivity CO_2	Less expensive	Interference by other gases
Mass Spectrometry CO_2 and O_2	Versatile, can be for other uses Fast response time	Expensive
Paramagnetic Analyzer (O_2)	Very specific	Drift, interference by by water vapor Slow response time
Electrolyte Fuel Cell (O_2)	Not interfered by water vapor	Interference by combustibles

is usually performed under reasonably standardized conditions with the object of obtaining a uniform product, we found that the cell composition is usually constant throughout its growth period. In any event, quantitative changes of cell composition can easily be incorporated into this biological model and stored into the computer if it is to be used for monitoring the process.

Our theoretical analyses [4,5] also indicated that eq. (1) can be reduced to four independent equations written as elemental balances on carbon, hydrogen, oxygen, and nitrogen. In order to solve for all these variables, three more independent equations or known values of these variables are required. Throughout our research we have employed the oxygen uptake rate (OUR), carbon dioxide evolution rate (CER), and ammonia addition as the measurable parameters.

TABLE II

Comparison of the Theoretical[a] RQ Value with the Experimental Value of RQ

Substrate Used	Cellular Yield	Respiratory Quotient Theoretical	Respiratory Quotient Experimental	Reference
Glucose	0.5	1.04	1.00	(von Meyenburg, 1969)
Ethanol	0.5	0.55	0.60	(Mor, et al, 1973)
Methanol	0.35	0.55	0.50	(Levine and Cooney, 1973)
n-Paraffin	1.00	0.45	0.50	(Sonoda, et al, 1973)
Methanol	1.00	0.30	0.194	(Harwood and Pirt, 1972)

[a] The theoretical RQ is based on the experimental cellular yield coefficient.

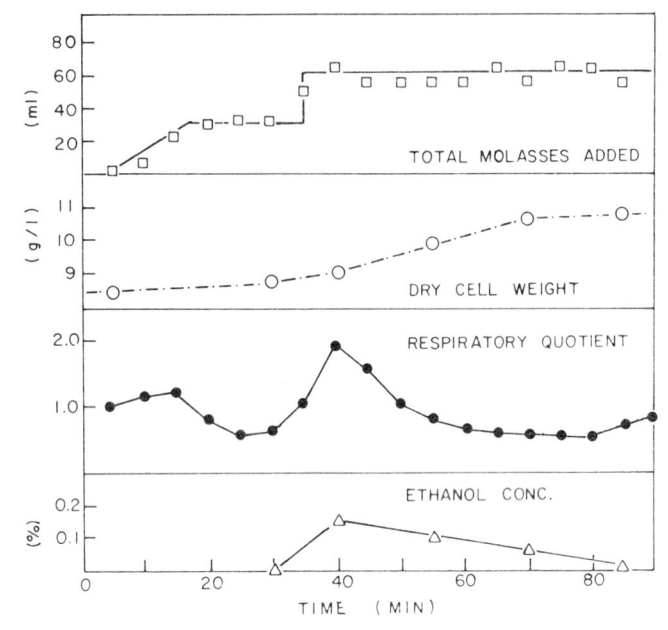

Fig. 2. Pulse experiment of molasses addition to show the response of respiratory quotient value. Molasses pulse was added at 30 min.

Other combinations, including measurement of fermentation heat, can also be used to monitor cell growth.

GAS EXCHANGE ANALYSES OF FERMENTATION PROCESS

Direct analysis of the inlet and outlet oxygen and carbon dioxide concentrations of an aerated fermentor in combination with the knowledge of the air flow rate will give the values of OUR and CER during fermentation. In general, the composition of the inlet air is quite constant. Using these variables for monitoring and control is not new. As early as 1956, Shu [8] published an article on controlling oxygen uptake rate in a fermentation by means of measuring oxygen concentration in the exit gas stream with a paramagnetic analyzer. Today, it is common practice in the pharmaceutical industry to monitor oxygen and carbon dioxide concentrations in the outlet air stream of the fermentors to assess fermentation status. Recent advances in sensor development also facilitate their uses in the fermentation industries. Advantages and disadvantages of some of this equipment are shown in Table I. In general, continuous measurement of exit CO_2 is a relatively simple problem as the results are partially unaffected by small changes in atmospheric pressure and relative humidity. On the other hand, the measurement of exit O_2 is definitely affected by changes in atmospheric pressure and relative humidity. Since the calculation of oxygen uptake rate requires knowing the difference between two fairly high concentrations of oxygen, accurate O_2 measurement becomes more difficult. This is especially

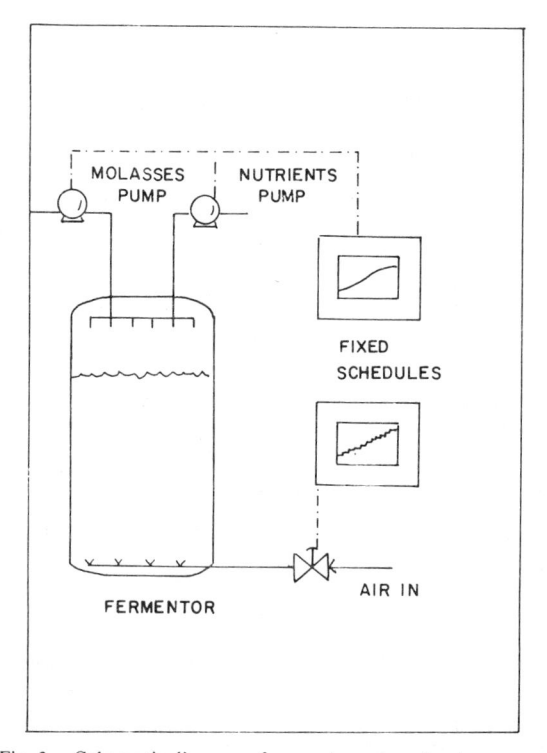

Fig. 3. Schematic diagram of current yeast production process.

true when using paramagnetic gas analyzers for measuring oxygen concentrations.

The respiratory quotient (RQ) is defined as the ratio of the CER to the OUR. Fiechter and von Meyenburg [9] described a continuous monitoring procedure of RQ during a yeast fermentation using gas analyzers. Although various investigators have used the respiratory quotient in the literature, experimental variations have made the interpretation of the RQ value obscure and undefined. Cell physiologists have long used RQ to gather information regarding nutrient metabolism. An approximate RQ value can always be found by knowing the molecular formula of the carbon-energy substrate used for growth. A more precise value for RQ can only be stated when the complete stoichiometry for growth and product formation are known. Our theoretical analysis indicates that the exact value of the respiratory quotient depends on the specific cellular yield coefficients as well as the composition of the carbon source (substrate) and the cells. If these components are constant, then RQ is related to the cellular yield coefficients:

$$RQ = f(Y_m) \qquad (2)$$

where Y_m is the molar cell yield coefficients.

A plot of respiratory quotient vs. cellular yield for growth on different car-

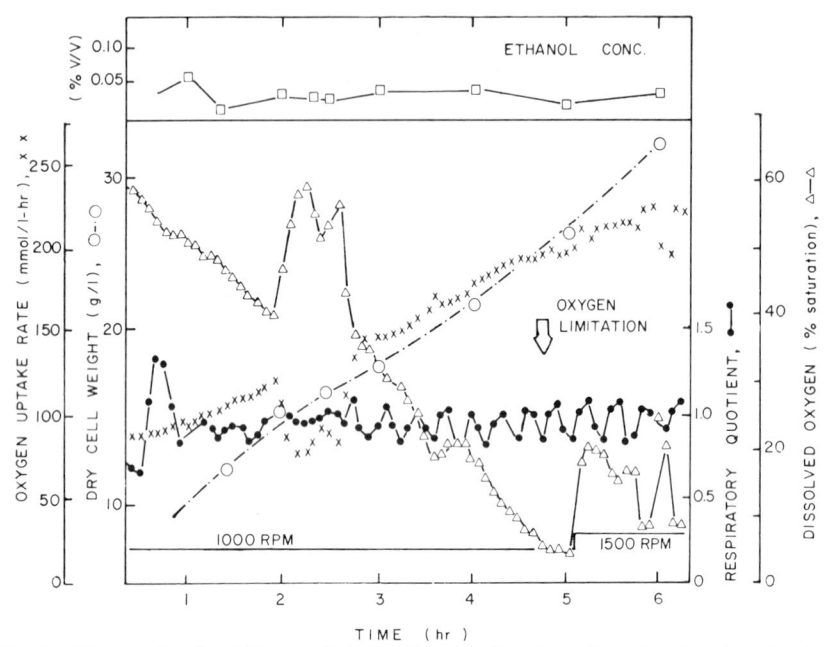

Fig. 4. Time profile of an RQ-controlled yeast fermentation. Control was based on the RQ value alone (RQ setpoint = 1.0).

bon-energy substrates is shown in Figure 1. The validity of the above analysis is shown in Table II, where the calculated RQ values for different carbon-energy substrates at different cell yields are compared with experimental values from the literature. The theoretical values of RQ are calculated from the experimentally observed values for cellular yield coefficients.

So far, product formation during fermentation has been ignored in the above calculations. The influence of product formation on the RQ value can be significant. If the product is extremely reduced as compared with the substrate, RQ with product formation will be higher than the RQ value in the absence of product formation (RQ_0). On the other hand, if the product is more oxidized than the substrate, the overall RQ will be smaller. The degree of variation will depend on the amount of product formed per unit of substrate utilized.

Growth of *Saccharomyces cerevisiae* on sucrose is generally accompanied by the formation of ethanol. Since the synthesis of ethanol through the glycolytic pathway only involves the production of CO_2, it could be expected that the RQ value would be an excellent indicator of ethanol formation [5] and can be expressed as

$$EPR = (RQ - RQ_0) \, OUR \tag{3}$$

where EPR is the ethanol production rate (mM/liter hr) and OUR is the oxygen uptake rate (mM/liter hr).

A similar relationship has also been derived by Aiba et al. [15], as is shown

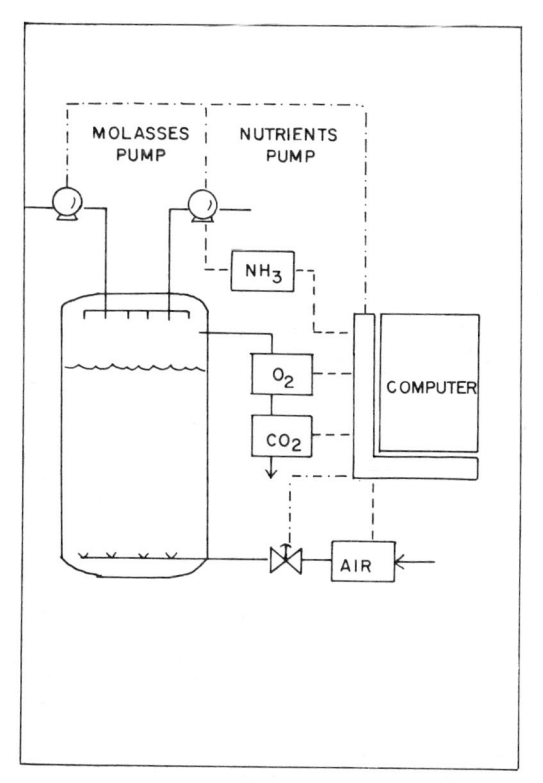

Fig. 5. Schematic diagram of the feed forward–feed back computer control scheme.

in Figure 1, where RQ_0 was found to have a value of 1.04 when the cellular yield is 0.5 g cell/g glucose. The sensitivity of using RQ as an indicator of ethanol production can be demonstrated from a pulse experiment in a yeast fermentation as shown in Figure 2. A pulse of molasses was added after the residual sugar concentration in the fermentor became low. Immediately a sharp increase of the RQ was noted. This transient in RQ correlates well with a simultaneous increase in the production of ethanol. The RQ eventually returns to its former value as ethanol reutilization proceeds.

CONTROL OF MOLASSES ADDITION IN YEAST FERMENTATION

Fed-batch or Zulauf yeast [16] propagation has been widely practiced in the bakers' yeast industry (Fig. 3). In this process, the molasses is fed incrementally according to a predetermined feed schedule during the entire fermentation. Thus the sugar concentration in the fermentor is kept very low and the expression of the Crabtree effect is prevented. In addition, the growth rate of the yeast becomes under external control depending on the sugar feed schedule.

Many sensors have been suggested in the literature for controlling yeast growth. The current problem is to choose the specific state variables that can

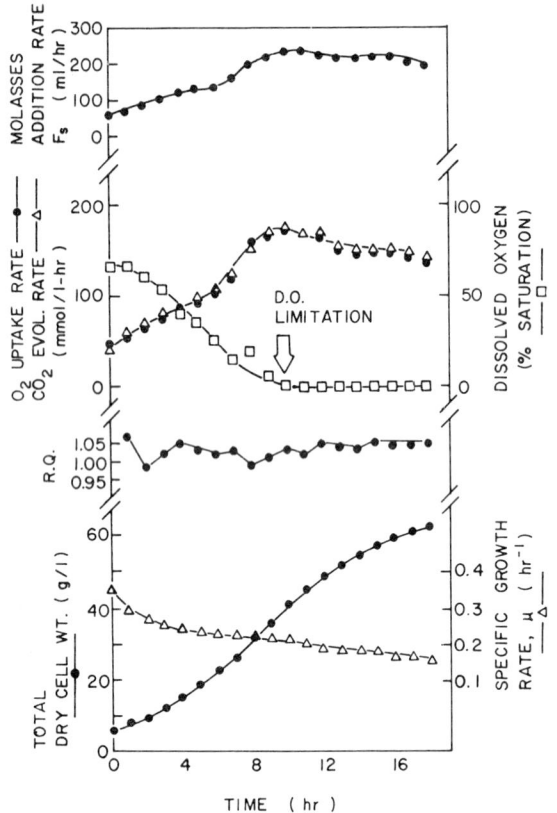

Fig. 6. Time history of a feed forward–feed back-controlled yeast fermentation.

be monitored easily, as sufficiently sensitive for control purposes. As early as 1959, Dietrich [20] suggested using RQ as a control parameter for yeast production. Aiba et al. [15] in a recent article controlled the bakers' yeast fermentation by means of limiting the RQ value within a range of 1.0–1.1. Unfortunately, their experiments were performed at a "pseudo" steady state and their final cell concentration never exceeded 10 g/liter. In industrial yeast production, 40 g/liter or greater must be produced at the end of the fermentation.

Initially, we also used the RQ value alone to control the yeast fermentation. The molasses addition rate was increased until the respiratory quotient was above 1.0. At the time, the molasses addition rate was decreased to maintain the RQ value at around 1.0. As shown in Figure 4, one experiment had been performed by using the RQ as the control parameter. In this case, the RQ value shown in the figure oscillated around 1.0. The ethanol formation was reduced to the minimum as shown in Figure 4 with a final ethanol concentration below 0.5 g/liter. However, the use of RQ control *per se* was not entirely satisfactory. As suggested by the OUR data, the growth of yeast was not uniform. This variation

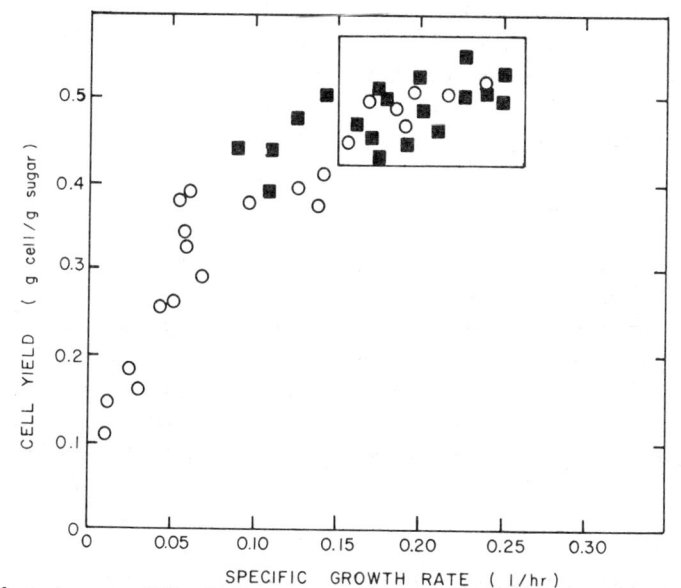

Fig. 7. Instantaneous cellular yields vs. the instantaneous specific growth rates for two yeast fermentations. (■) Run CBA is a computer-controlled fermentation, while (○) run SBG is a non-controlled fermentation.

in the OUR data was primarily due to over control and erratic behavior of the molasses pump. The oscillations generated in the molasses flow rate due to the fluctuations of the RQ value might decrease the overall cellular yield even though the ethanol formation was kept to a minimum.

Although we have shown so far that the respiratory quotient is theoretically and experimentally an excellent indicator of ethanol formation, the RQ value alone cannot be relied on as a good control parameter. Therefore, a better control strategy must be developed to upgrade the current yeast production process.

The development of this alternate control strategy will only be briefly presented, but a more detailed description of the entire control scheme can be found elsewhere [23]. A schematic diagram, shown in Figure 5, is used to show the operational features of this feed forward–feed back computer control system.

Indirect measurement of cell concentration and volume are continuously made by means of material balancing. The molasses addition rate (F_S) can then be estimated continuously by the following equation:

$$F_S = \frac{\mu(X \cdot V)}{Y_{X/S} \cdot S_f} \quad (4)$$

where $X \cdot V$ is the total yeast biomass in the fermentor, $Y_{X/S}$ is the cell yield coefficient (g cell/g sugar), and S_f is the sugar concentration in molasses (g/liter). Throughout the course of the fermentation, the molasses addition rate is constantly readjusted by means of a feedback loop:

$$F_{ss} = F_s[1 - K_c \, (\text{CER} - \text{RQ}_0 \cdot \text{OUR})] \quad (5)$$

F_{ss} is the corrected molasses flow rate and K_c is the controller gain of the feedback control loop. Theoretically and experimentally, the difference between the CO_2 evolution rate and the OUR is more sensitive, as well as more accurate than using RQ value as the control parameter [23]. Also, the control action should change as the yeast biomass concentration increases.

In Figure 6, the time history of a typical computer-controlled yeast fermentation is shown. The initial preset specific growth rate (μ) and the cellular yield coefficient ($Y_{X/S}$) were both adjusted to 0.25 hr^{-1} and 0.5 g cell/g sugar, respectively. A final cell concentration above 60 g/liter was achieved when oxygen limitation is not encountered.

These results lead us to speculate that the critical specific growth rate at which ethanol production starts to occur is dependent not only on the dissolved oxygen concentration in the fermentor [23], but also on the amount of cell biomass in the fermentor. The variation of specific growth rates and cellular yields during a yeast fermentation can be reduced if the molasses addition is achieved more smoothly and more precisely. This was accomplished as shown in Figure 7 by using the computer to control the molasses addition rate (see run CBA) as discussed above. Also shown in Figure 7, as run SBG when the incremental addition of molasses was not properly controlled. In this latter instance, both the specific growth rates and the cellular yield coefficients vary significantly. This indicates that frequent and sudden changes in molasses addition will have adverse effects on overall cellular yields. On the other hand, for run CBA, where automatic and proper control of molasses addition were maintained, the instantaneous specific growth rates are always in the region of 0.2 hr^{-1} and thus achieving a uniformly higher cellular yield coefficient of about 0.5 g cell/g sugar. With this computer control system, the final cellular yield and the volumetric productivity of the fed-batch process are constantly high regardless of changes in molasses quality, inoculum size, inoculum condition, and oxygen limitation [23]. The final cell yield in the controlled yeast fermentation is about 0.5 g cell/g sugar with overall volumetric productivities between 3 and 5 g/liter-hr depending on the initial biomass concentration. The control scheme not only minimizes ethanol formation, but also increases the uniformity of the fermentation process [23].

As a final concluding comment, it can be stated that computer technology in fermentation systems has been advancing quite rapidly in recent years. In the computer-aided fermentation system, one should strive not only to optimize the conversion yield and volumetric productivity, but also keep in mind the potential of achieving reproducibilities and process uniformity. These latter achievements have been shown to be possible for the case of yeast fermentation. Our savings can also come from the uniformity of the product and minimization and variation of operating parameters downstream related to the recovery operations.

References

[1] L. B. Evans, *Science, 195,* 1146 (1977).
[2] A. E. Humphrey, *Proceedings of the Labex Symposium* (Earls Court, London, 1971), p. 1–15.

[3] D. W. Zabriskie, "Real-time estimation of aerobic batch fermentation biomass concentration by component balancing and culture fluorescence," Ph.D. thesis, University of Pennsylvania, Philadelphia, 1976.
[4] C. L. Cooney, H. Y. Wang, and D. I. C. Wang, *Biotechnol. Bioeng., 19,* 55 (1977).
[5] H. Y. Wang, C. L. Cooney, and D. I. C. Wang, *Biotechnol. Bioeng., 19,* 69 (1977).
[6] J. S. Harrison, *Proc. Biochem., 2,* 41 (1967).
[7] D. Herbert, *Continuous Culture 6: Applications and New Fields.* A. C. R. Dean, D. C. Ellwood, C. G. T. Evans, and J. Melling (Eds.) (Ellis Horwood Co., London, 1976).
[8] P. Shu, *Ind. Eng. Chem., 48,* 2204 (1956).
[9] Fiechter and H. K. von Meyenburg, *Biotechnol. Bioeng., 10,* 535 (1968).
[10] H. K. von Meyenburg, *Arch. Mikrobiol., 66,* 289 (1969).
[11] J. R. Mor, A. Zimmerli, and A. Fiechter, *Anal. Biochem., 52,* 614 (1973).
[12] D. W. Levine and C. L. Cooney, *Appl. Microbiol., 26,* 982, (1973).
[13] N. Sonoda, J. Someya, N. Futai, N. Tagaya, and T. Muskemi, *J. Ferm. Technol., 51,* 479 (1973) [in Japanese].
[14] J. H. Harwood and S. J. Pirt, *J. Appl. Bacteriol, 35,* 597 (1972).
[15] S. Aiba, S. Nazai, and Y. Nishizawa, *Biotechnol. Bioeng., 18,* 1001 (1976).
[16] G. Reed and H. J. Peppler, *Yeast Technology* (Avi, Westport, CT, 1973).
[17] H. N. Sher, *Chem. Ind., 16,* 425 (1960).
[18] J. Hospodka, *Biotechnol. Bioeng., 8,* 18 (1966).
[19] T. Miskiewicz, J. Wilesniak, and J. Ziobrowski, *Biotechnol. Bioeng., 17,* 1829 (1975).
[20] K. R. Dietrich, *Ablanfuerwertung und Abwasserreinguag Heidelberg* (Springer-Verlag, Berlin, 1959), pp. 280–283.
[21] K. Rungaldier and E. Braun, U.S. patent No. 3,002,894 (1961).
[22] G. J. Fuld and C. G. Dunn, *Ind. Eng. Chem, 49,* 1215 (1957).
[23] H. Y. Wang, C. L. Cooney, and D. I. C. Wang, *Biotechnol. Bioeng.,* in press (1978).
[24] H. Y. Wang, Ph.D. thesis, MIT, 1977.

Analysis and Control of Mixed Cultures

R. T. HATCH, C. WILDER, and T. W. CADMAN
Department of Chemical Engineering, University of Maryland, College Park, Maryland 20742

INTRODUCTION

Mixed-culture processes dominate the environment and their dynamics play a key role in ecosystem biology. Although positive interactions among species can result in stable ecosystems that can be studied under controlled environments [1-9], these systems comprise a narrow segment of the very large number of potential mixed-culture systems that can be developed. Complex interactions among the various species, excessive time lags in the analysis of the populations, and natural selection for more competitive strains during longer-term continuous cultures make the study of mixed cultures under controlled conditions quite difficult. As a result of these complicating factors, the investigation of mixed cultures has been hindered by the lack of appropriate analytical instrumentation and control strategies [10].

In order to study the vast majority of mixed-culture systems, certain minimum instrumentation is necessary, capable of the following: rapid analysis of microbial populations; monitoring of environmental conditions; control of environmental conditions; and control of microbial populations. In addition to the instrumentation requirements, control strategies need to be developed for rapid and stable control of the mixed-culture system. Further advancement in the understanding of microbial kinetics is also necessary before the optimum control strategies can be developed. This is particularly true for the effects of transients and strain changes on microbial kinetics.

The greatest need in instrumentation has been the rapid analysis of microbial cultures. A number of investigators [8,11-13] have taken advantage of the differences in cell sizes to distinguish between cultures using particle size analyses. More recently, laser flow microphotometry has been developed for very rapid analysis of cell populations [14,15]. This instrumentation provides the potential of discrimination between species, rapid counting of individual cells, cell sorting, and analysis of the internal composition of the cell [16,17]. This investigation was undertaken to determine the utility of laser flow microphotometry for the analysis of yeast and bacterial populations and to examine the necessary control strategies for continuous, mixed microbial cultures.

MATERIALS AND METHODS

A model FC200R/FC4800A (Ortho Instruments) Cytofluorograf with a 50-mW, 488-nm argon laser and a 0.8-mW, 630-nm helium neon laser was interfaced to a Digital Equipment Corporation PDP 11/34 minicomputer by use of a high speed analog/digital (A/D) converter for the storage and analysis of the data from the photodiodes and photomultiplier tube. The photocells used to determine the flow of sample through the metering tube were also interfaced using the A/D converter to determine the sample flow rate. Cell broth samples were introduced into the sample chamber of the Cytofluorograf under computer control by use of a Gilson peristaltic pump. Electrodes were mounted in the sample chamber to gauge the amount of sample introduced.

Antisera for *Streptococcus mutans* ATC 6715-14 were supplied by NIH, Institute of Dental Research, National Caries Program. A mixture of *S. mutans* and *Streptococcus sanguis* ATC 10558 was centrifuged and resuspended in phosphate buffered saline. The antisera were diluted 512-to-1 in phosphate buffered saline and added in a 1:1 ratio with the resuspended bacteria. After 20 min, the sample was centrifuged and washed to remove the unbound antisera. The bacterial suspension was then introduced to the sample compartment of the Cytofluorograf and analyzed using the argon and helium lasers.

PROCESS CONTROL SYSTEM

For analysis and control of continuous, mixed cultures, a process control system was assembled as illustrated in Figure 1. The control system utilizes the PDP 11/34 with two moving head disks to permit high speed, raw data storage from the Cytofluorograf. Data inputs rates up to 10 kH are possible. However, rates of 0.2–1 kH provide the greatest accuracy. Simultaneously, closed loop control of pH, dilution rate, and liquid level, as well as supervisory control of temperature, were implemented. Ammonium salt and glucose feed rates were also under computer control using metering pumps to manipulate the respective nutrients.

MIXED-CULTURE PROCESS

The fermentation system as shown in Figure 2 consists of a 10-liter New Brunswick Scientific Magnaferm fermentor modified for continuous culture. The main nutrient feed is pumped from two 55 gal stainless-steel tanks through a pneumatic control valve, and multiple filters are used to remove all particles down to 0.2 μm. The flow rate is monitored by a Foxboro differential pressure cell. The liquid level is also monitored with a Foxboro differential pressure cell and regulated with a pneumatic control valve. The acid and base are metered using Radiometer solenoid valves, while glucose and ammonium salt are added by metering pumps and filtered through 0.2-μm Nucleopore membranes. The pH is monitored with an Ingold pH electrode and a Radiometer pH meter. The dissolved oxygen is monitored with a galvanic oxygen electrode supplied by New

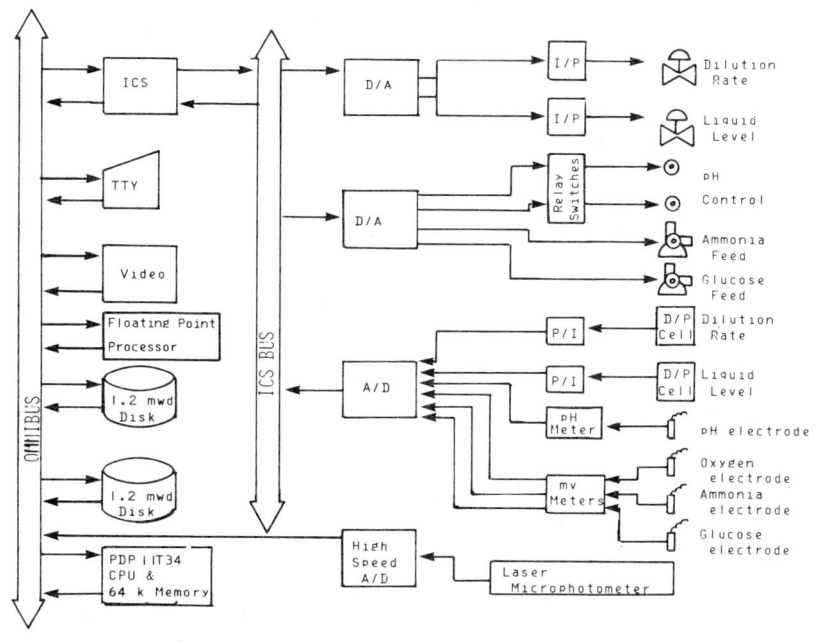

Fig. 1. Process control system.

Brunswick Scientific. An ammonia electrode supplied by Orion, Inc. and a glucose electrode supplied by Yellow Springs, Inc. are used to monitor the respective broth nutrient components off-line.

LASER FLOW MICROPHOTOMETER

As shown in Figure 3, the laser flow microphotometer consists of coherent light sources that irradiate a quartz flow channel (250-μm bore) at a right angle to flow. As a particle or cell passes through the flow channel, several signals are generated by the laser beams: 1) a 8-μsec timing signal; 2) light extinction signal of the 630-nm beam (30 μsec); 3) light scatter signal of the 630-nm beam (30 μsec); 4) light fluorescence from the particle in the red portion of the spectrum; 5) light fluorescence from the particle in the green portion of the spectrum. The cylindrical flow channel consists of a sheath water flow occupying the outer annular space of the cross section and a sample flow which comprises the central core. Both the sample flow and sheath flow can be regulated separately to restrict the cross section of the sample stream (thus lining up the particles in single file) and the velocity of the sample stream (thus eliminating turbulence and determining the dwell time of the particle in the beam, both of which affect the sensitivity). The light extinction and light scatter signals are monitored by photodiodes and the light fluorescence is monitored with photomultiplier tubes.

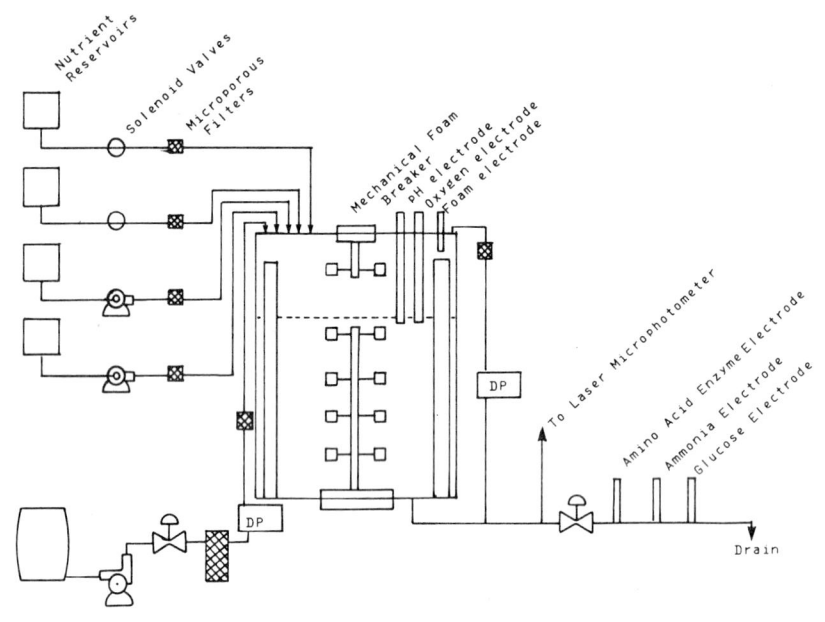

Fig. 2. Schematic of mixed-culture process.

RESULTS AND DISCUSSION

Laser Flow Microphotometer Analysis

The ability of the laser flow microphotometer to distinguish between a small bacterial species (*Corynebacterium glutamicum*) and a yeast (*Candida utilis*) was tested using the 630-nm laser. The light absorbance–frequency spectrum for this mixed population is shown in Figure 4. Since there is sufficient difference in size of the microbial cells, the different populations can be resolved quite easily. The light scatter–frequency spectrum is quite similar to Figure 4 and is therefore not included.

By storing the individual raw data points in the computer system, it is possible to divide the amount of light scatter by the amount of light absorbance for each particle and generate a ratio–frequency spectrum. As shown in Figure 5, this ratio is approximately 1.2 for *C. utilis* and approximately 1.0 for *Cor. glutamicum*. An important new parameter is therefore available to distinguish between different populations. Most often nutrient media cannot be prefiltered to remove the particulates. Under these conditions many inert particles may be present such as mineral precipitates, corrosion particles, precipitated proteins, etc., which are in the size range of microorganisms. In order to test the resolution of this instrument, a mixture of 1.01-μm and 3.9 + μm latex particles was added to *C. utilis* and analyzed using the 630-nm laser. The absorbance–frequency spectrum as shown in Figure 6 indicates a great deal of overlap in absorbance between the yeast and latex. These data alone are insufficient to discriminate

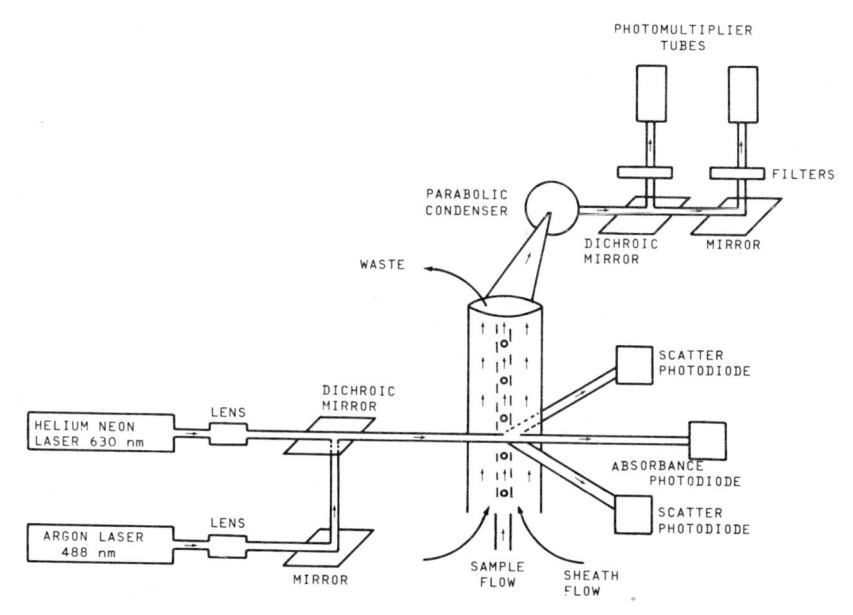

Fig. 3. Schematic of laser flow microphotometer.

between *C. utilis* and a comparably sized inert particle such as latex. After dividing the light scatter by the light absorbance, the ratio–frequency spectrum was constructed as shown in Figure 7. The yeast again had an average ratio of approximately 1.2. The 1.01-μm latex particles had an average ratio of approximately 0.4 and the 3.9-μm latex particles had an average ratio of approximately 0.7. The use of the ratio of light scatter to light absorbance makes it quite possible to determine in this case the nature of the individual particle and distinguish between inert particles and microbial cells. The living cells appear to have ratios ≥ 1.0 and the inert particles appear to have ratios ≤ 1.0. This may be due to the relative transparencies of latex particles, which are opaque, and microbial cells, which are translucent. There also appears to be some correlation between the ratio of scattered to absorbed light and particle size, with smaller sizes correlating with lower ratios.

The use of the 488-nm argon laser permits the excitation of fluorescence either by internal structures of cells (DNA, RNA, protein, etc. [18,19]) or surface molecules (Con-A). This makes possible the resolution of different species of the same genus through the use of fluorescently labeled antibodies. In order to illustrate the resolution of this assay *S. sanguis* and *S. mutans* were mixed together in approximately equal cell densities. The antiserum was added to the mixture and allowed to react. After removing the unbound antibodies, the mixture was analyzed using the 488-nm laser. A relatively tight distribution of cell fluorescence is shown in Figure 8. The fluorescent cell density was found to be 48,956 cells/ml. A total particle count of the mixture was also obtained

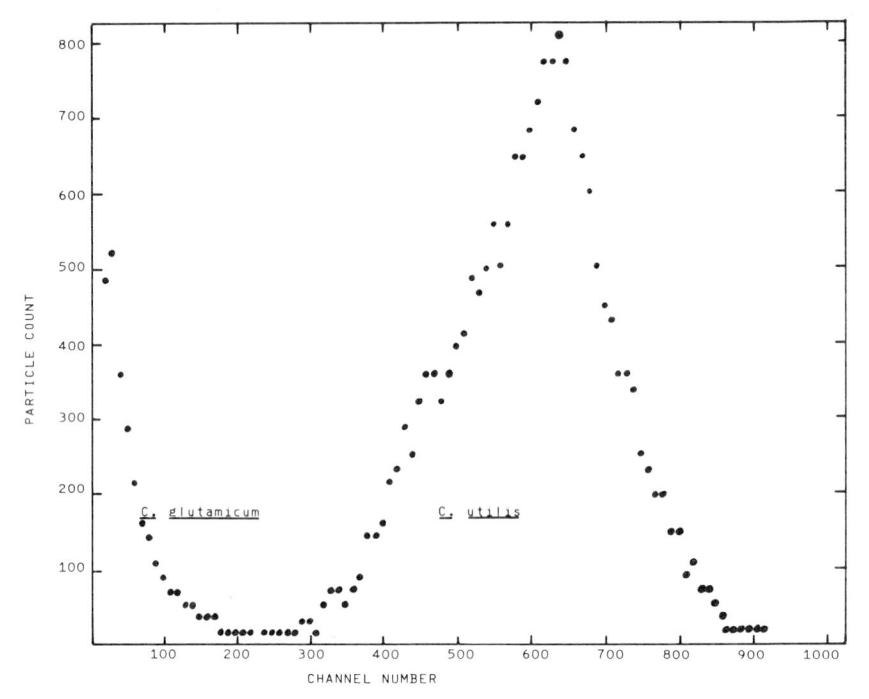

Fig. 4. Light absorbance–frequency spectrum for *Cor. glutamicum* and *C. utilis*.

using the 630-nm laser. There appeared to be a broader distribution of light absorbances than light fluorescence. The bacteria tend to form clumps or agglomerates which would give proportionally larger absorbances. The fluorescence, however, is directly related to the surface area available for antibody attachment. It would therefore be expected that the fluorescence distribution would be much tighter despite comparable clumping. The total cell density was found to be 83,035 cells/ml or about twice the fluorescent cell count.

The sensitivity of this instrumentation for the analysis of fluorescently labeled antibodies on micron-sized bacteria is clearly shown. Although the fluorescent antibody assay was developed for the fluorescent microscope, the laser flow microphotometry permits the use of only 5% of the normally required antisera concentration. As the test was run, several centrifugation steps and a 20-min reaction time were used. For real time control of a mixed culture, this procedure is unacceptable. It is possible, however, to optimize the reaction time and replace the centrifugation steps with either dialysis or filtration for a much quicker, automated procedure.

Mixed-Culture Control

A continuous, mixed culture of two microbial populations competing for the same limiting nutrient can only be maintained in stable, steady-state growth

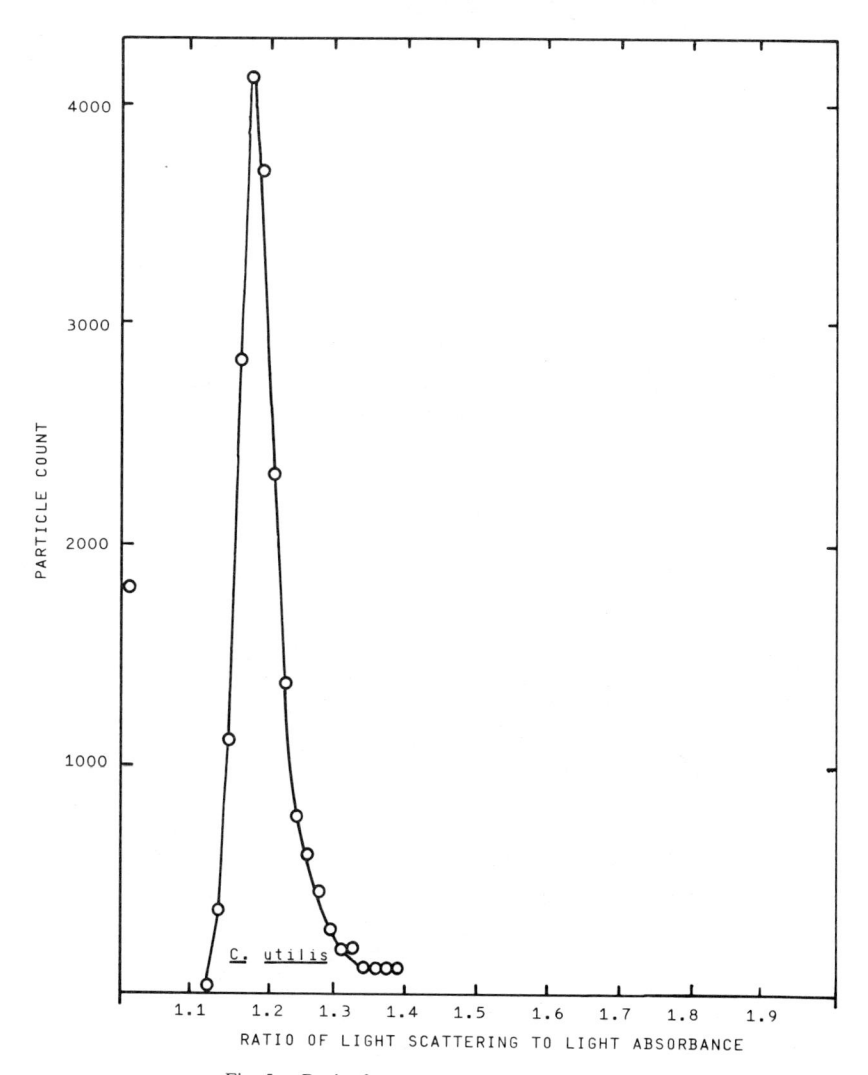

Fig. 5. Ratio–frequency spectrum for *C. utilis*.

if the cultures have the identical specific growth rate at the same limiting nutrient concentration. As is shown in Figure 9, the specific growth rates for *C. utilis* and *Cor. glutamicum* cross at 0.555 hr^{-1} for a limiting glucose concentration of 0.6 g/liter. In order to maintain the cultures at steady state, the dilution rate must be carefully regulated at this cross point. This cannot be done without control instrumentation. When it is desired to adjust the relative cell densities of the two cultures, or compensate for a change in the microbial kinetics (due to mutation or a change in environment), it would be necessary to implement a more sophisticated control strategy to effect the desired results.

Three feedback control strategies were theoretically investigated as indicated

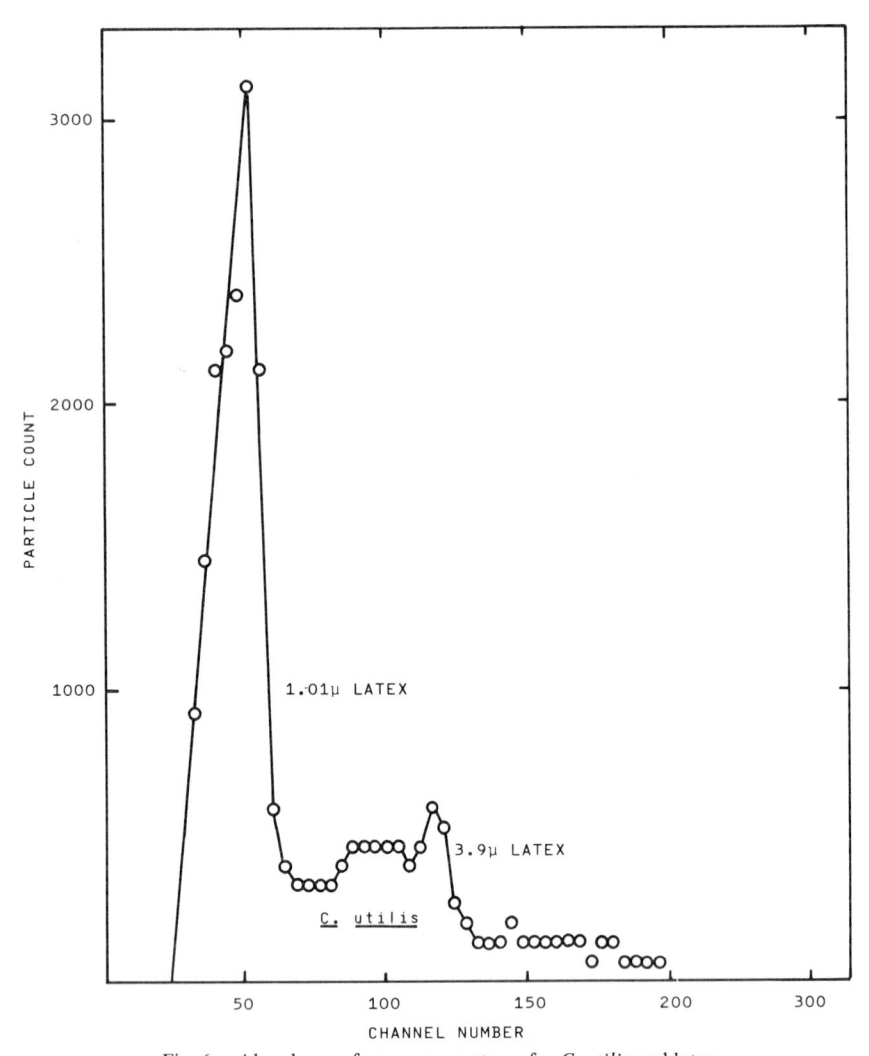

Fig. 6. Absorbance-frequency spectrum for *C. utilis* and latex.

in Table I [20] for the manipulation of steady-state cell densities, cell density ratios, and growth rates of the continuous mixed culture of *C. utilis* and *Cor. glutamicum* in competition for the limiting nutrient, glucose. Each strategy was based upon two manipulative variables, substrate feed rate (F_s) and dilution rate (D). Monod growth kinetics was assumed for each strategy. The first strategy, termed the classical control strategy, utilized the cell densities of culture 1 (X_1) and culture 2 (X_2) to control the two manipulative variables. The greatest stability occurred when F_s was set proportional to the deviation of the observed cell density from the setpoint cell density of the faster growing culture (X_1). The dilution rate was then set proportional to the deviation of the observed cell density

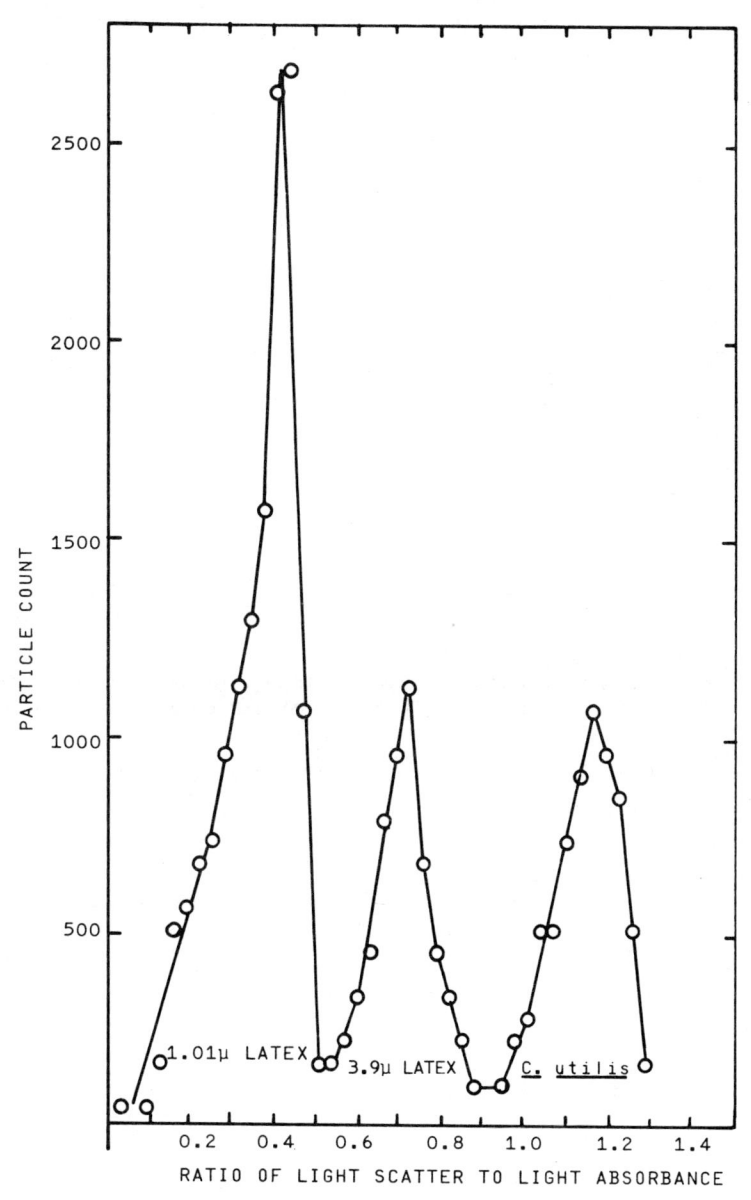

Fig. 7. Ratio-frequency spectrum for *C. utilis* and latex.

from the setpoint cell density of the slower culture (X_2). By using the resulting control equations it was possible to change both the cell density ratio of the two cultures as well as the total cell density to new specified values. In order to implement this control strategy, it is necessary to measure the independent cell densities of the two cultures. Although this classical control could be used to

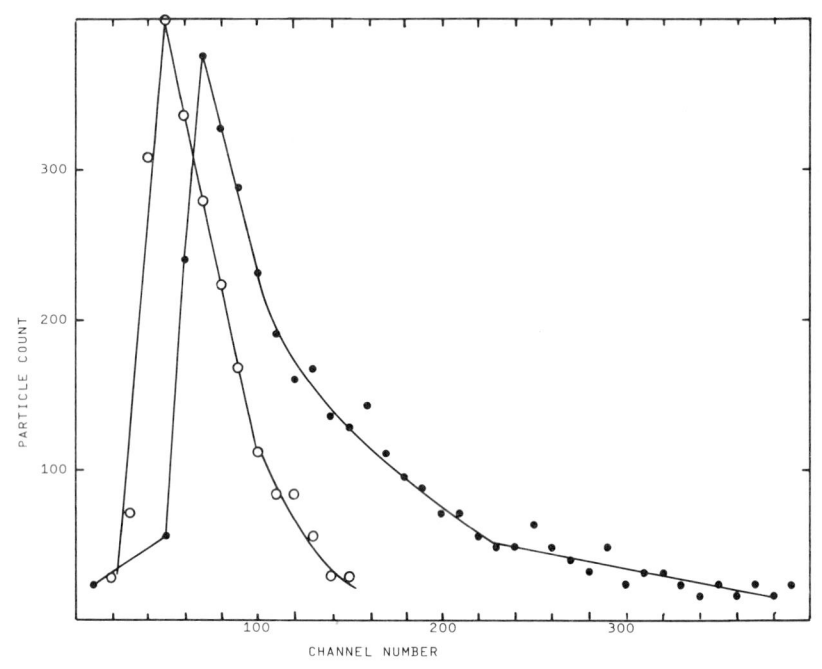

Fig. 8. Absorbance and fluorescence-frequency spectrum for a mixture of *S. sanguis* and *S. mutans*. (●) 630 nm laser, absorbance, 83,035/ml. (○) 488 nm laser, fluorescence, 48,956/ml.

change steady-state conditions, excessively long response times to change steady states, up to 50 hr, were predicted.

A more effective control strategy would utilize the total cell density to set the dilution rate since the dilution rate affects both species equally. The substrate concentration, in turn, affects the relative growth rates of the cultures. This suggests the use of the difference in cell densities to control the substrate feed rate. This "improved" control strategy results in a much faster response time for the change in cell density ratio and total cell density from one steady state to another. Once the basic Monod growth kinetic model was changed to account for a transient growth rate response, however, somewhat longer response times resulted along with damped oscillatory behavior of cell density with time. It was then necessary to adjust the control constants in order to maintain suitable control.

Further refinement in the control strategy resulted when the "separated" control algorithm was examined. By making use of the computer to calculate a cell mass balance across the fermentor, the difference in cell densities can be used to set the desired substrate concentration. The required F_s can then be calculated by incorporating the limiting nutrient consumption rates. The dilution rate is determined the same as in the improved control strategy. The separated control strategy predicted the fastest response of the three strategies for a change in total cell density with a minimum of oscillatory behavior for the cases incorporating transient response times.

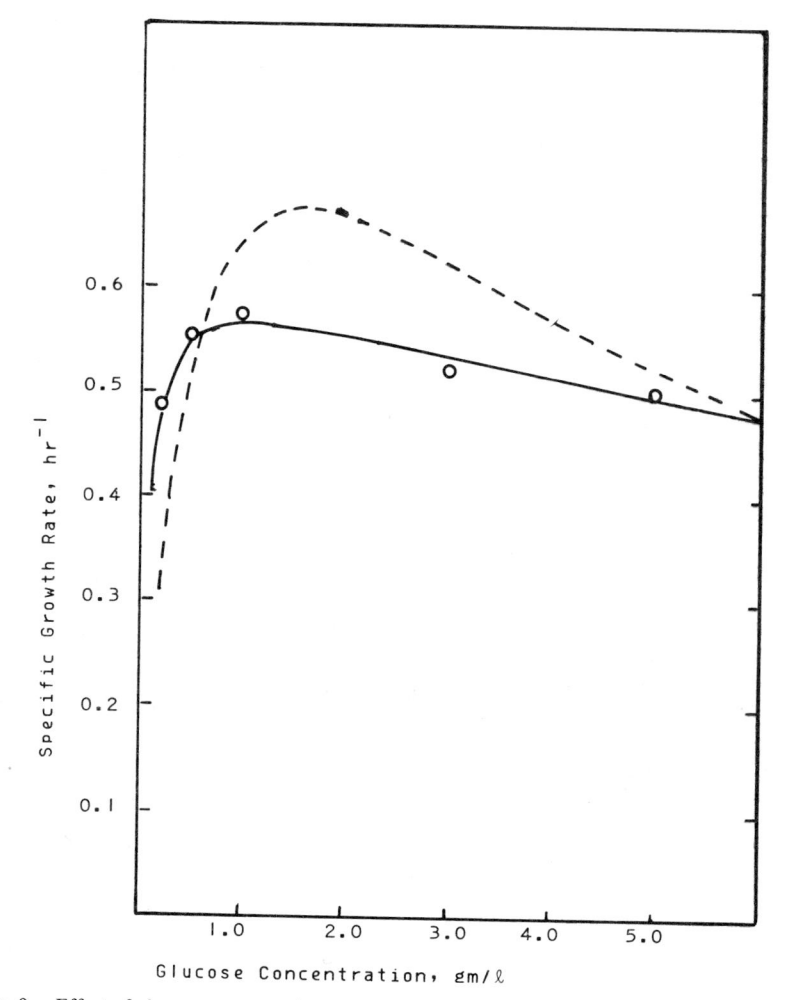

Fig. 9. Effect of glucose concentration on specific growth rates. (——) *C. utilis;* (- - -) *Cor. glutamicum.*

CONCLUSIONS

Advanced analytical instrumentation such as the laser flow microphotometer can be interfaced to available minicomputers for rapid analysis of microbial populations. As illustrated by the systems examined to date, the use of light scattering and absorbance can be very useful in distinguishing between microbial cells and inert particles as well as between microbial cells of comparable size. Fluorescent antibody assays can be used with this instrumentation to aid in the analysis of bacterial species of the same genus. The real time analysis of the different microbial populations can be used to implement relatively simple feedback control strategies for the on-line control of continuous, mixed cultures and the alteration of the individual population cell densities.

TABLE I
Feedback Control Strategies

Strategy	Control Equations	Response Time for Change	Comments
Classical	$F_s = -k_{C1}(X_1 - X_{1sp}) + F_{ss}$ $D = k_{C2}(X_2 - X_{2sp}) + D_{ss}$	50 hr for cell density ratio 40 hr for total cell density	
Improved Control	$F_s = -k_{C1}((X_1 - X_2) - (X_{1sp} - X_{2sp})) + F_{ss}$ $D = k_{C2}((X_1 + X_2) - (X_{1sp} - X_{2sp})) + D_{ss}$	25 hr for cell density ratio 15 hr for total cell density	Oscillatory behavior for non zero time constants in growth kinetic model
Separated Control	$\Delta S = -k_{C1}((X_1 - X_2) - (X_{1sp} - X_{2sp}))$ $D = k_{C2}((X_1 + X_2) - (X_{1sp} - X_{2sp})) + D_{ss}$ $F_s = \left[\dfrac{D_{ss}}{Y_1} + M_1\right](X_1 - X_{1sp}) +$ $\left[\dfrac{D_{ss}}{Y_2} + M_2\right](X_2 - X_{2sp}) +$ $\Delta S \left[\dfrac{C_1}{Y_1}X_1 + \dfrac{C_2}{Y_2}X_2 + D\right] + F_{ss}$ $+ S_{ss}(D - D_{ss})$	4 hr for total cell cell density	Minimum oscillatory behavior $C_1 = \left(\dfrac{d\mu_1}{dS}\right)_{S=S_{ss}}$ $C_2 = \left(\dfrac{d\mu_2}{dS}\right)_{S=S_{ss}}$

Nomenclature

C_n linearized growth rate change with respect to substrate concentration (hr^{-1})
D dilution rate (hr^{-1}) (liter/hr dilution stream/liter fermentor volume)
D_{ss} dilution rate at steady state
F_s substrate feed rate (g/hr liter)
F_{ss} substrate feed rate at steady state
k_{C1} control constant for substrate feed rate [g/liter hr/(g cell/liter)]
k_{C2} control constant for dilution rate [hr^{-1}/(g cell/liter)]
M maintenance constant for substrate consumption (g/g cell hr)
S limiting nutrient concentration (g/liter)
X_1 cell density of species 1 (g/liter)
X_2 cell density of species 2 (g/liter)
$X_{n\,sp}$ desired cell density of species n ($n = 1$ or 2)
Y_n cell mass yield on substrate for species n (g cell/g substrate)
μ specific growth rate (hr^{-1})

The authors gratefully acknowledge the support of this research by the National Science Foundation under grant No. ENG 76-18480. The authors would also like to recognize the support of the Minta Martin Foundation and AMOCO Chemical Company for the acquisition of computer control hardware.

References

[1] K. Hotta and S. Takao, *J. Ferment. Technol. (Jpn.), 51*, 2 (1973).
[2] U. Pawlowsky and J. A. Howell, *Biotechnol. Bioeng., 15*, 889 (1973).
[3] U. Pawlowsky and J. A. Howell, *Biotechnol. Bioeng., 15*, 897 (1973).
[4] U. Pawlowsky and J. A. Howell, *Biotechnol. Bioeng., 15*, 905 (1973).
[5] T. G. Wilkinson, H. H. Topiwala, and G. Hamer, *Biotechnol. Bioeng., 16*, 41 (1974).

[6] R. Sudo, K. Kobayashi, and Aiba, *Biotechnol. Bioeng., 17,* 167 (1975).
[7] P. van den Ende, *Science, 181,* 562 (1973).
[8] C. C. Chao and P. J. Reilly, *Biotechnol. Bioeng., 14,* 75 (1972).
[9] I. H. Lee, A. G. Fredrickson, and H. M. Tsuchiya, *Biotechnol. Bioeng., 18,* 513 (1976).
[10] H. R. Bungay and M. L. Bungay, *Adv. Appl. Microbiol., 10,* 269 (1968).
[11] I. H. Lee, A. G. Fredrickson, and H. M. Tsychiya, *Biotechnol. Bioeng., 18,* 513 (1976).
[12] J. F. Drake and H. M. Tsuchiya, *Appl. Microbiol., 26,* 9 (1973).
[13] A. Shindala, H. R. Bungay, N. R. Kreig, and K. Culbert, *J. Bacteriol., 89,* 693 (1965).
[14] R. T. Hatch and T. W. Cadman, "Computer control of continuous mixed culture processes," 172 National Meeting of the ACS, San Francisco, 1976.
[15] R. T. Hatch and T. W. Cadman, *Conference on Mechanisms and Kinetics of Uptake and Utilization of Substrates for Single-Cell Protein Production* (Nauka, Moscow, USSR, 1977).
[16] J. Bailey, D. McQuitty, and J. Fazel-Madjlessi, "Measurements of composition distribution in *Bacillus subtilis* populations using flow microfluorometry," #Q26, ASM Annual Meeting, Washington, D.C., 1977.
[17] J. Bailey, J. Fazel-Madjlessi, D. McQuitty, L. Y. Lee, J. C. Allred, and J. A. Oro, *Science, 198,* 1175 (1977).
[18] K. J. Hutter and H. E. Eipel, *FEMS Microbiol. Lett., 3,* 35–38 (1978).
[19] K. J. Hutter, Th. Görtz, H. Oldiges, and H. E. Eipel, *Chemosphere, 1,* 51–58 (1978).
[20] C. T. Wilder, R. T. Hatch, and T. W. Cadman, *Biotechnol. Bioeng.,* accepted for publication.

IMPAC/FORTRAN Schemes for Fermentation Data

D. D. DOBRY and J. L. JOST

The Upjohn Company, Kalamazoo, Michigan 49001

INTRODUCTION

For the past three years, the Upjohn Company has used a Foxboro FOX 2/30 process-control computer for data logging and process control in our fermentation pilot plant. This computer is equipped with an operating system that apparently allows simultaneous background–foreground program execution, so that user-written programs can be run in any one of several priority modes without disrupting process control. Most of the on-line data analysis performed with this system is accomplished by the system software that came with the machine. However, there have been a number of cases in which user-written programs have played an important role.

The general objectives of on-line data analysis in our system have been to provide data for storage and later recall for calculations or as graphs and to provide calculated quantities for use in process control. Programs to meet these objectives can be written in assembler language, FORTRAN IV (with process-control extensions), or in IMPAC, a process-control language written by the Foxboro Company for use with their computer systems. Programs in assembler language are written for us by a computer support group called Systems-Control and Automation, but programming in FORTRAN IV and IMPAC is mainly done by personnel associated with the fermentation pilot plant. IMPAC programming is done with an interrogative compiler so that the programmer needs to know the physical and mathematical details of what is to be done but not much about programming *per se*. FORTRAN IV programs can be simply written with this system, but must be developed through a multistep process (using separate system programs to edit, compile, assemble, link, and load the program) during which the programmer must keep track of at least five drum files.

IMPAC is intended to be the principal process-control language with this system, and IMPAC programs are assigned the highest priority of any user-accessible software. The time-sharing system of the computer devotes the first part of each second to processing IMPAC programming and the remainder of each second to processing other programming. Because of this, IMPAC programming is the most secure kind of programming to use for process control. Using IMPAC programming has the further advantage of not requiring drum storage space for the completed program. The IMPAC compiler is arranged so that it is virtually impossible to make a mistake of the sort that will shut down the computer. Finally, IMPAC programs allow for easy interaction by the operator through special displays on the operator's CRT. For example, operating parameters such as setpoints, tuning constants (in the case of control programs), or mathematical constants (in the case of computational programs) can be easily entered or changed.

On the other hand, using IMPAC severely limits the scope of what can be programmed. There are 21 kinds of IMPAC blocks, eight specifically control blocks, five to handle special problems in interfacing, and the remainder to perform mathematical operations. The mathematical blocks can combine various inputs with addition, subtraction, multiplication, or division, can delay a signal by a dead time, and can scale a number by addition to or multiplication by various constants (which can be simply changed either through the operator's CRT or by a supervisory program). However, the system contains a limited maximum number of IMPAC blocks, so that any operation that must be performed on data on a tank-by-tank basis, or that involves complicated calculations, might be better done with FORTRAN or assembler programs.

Two examples will be given in this presentation of the use of IMPAC programming for on-line data handling. In the first case, the temperature setpoint of an antibiotic fermentation was automatically changed to produce a relatively constant level of dissolved oxygen. Doing this required that the input from the dissolved oxygen probe be corrected for changes in temperature.

In the second example, a combination of IMPAC and FORTRAN programming was used to calculate oxygen uptake rates from data taken with a paramagnetic oxygen analyzer. In this case the data were corrected for drifts in atmospheric pressure that may have occurred while the data were being taken.

REGULATION OF DISSOLVED OXYGEN BY TEMPERATURE MANIPULATION

The dissolved oxygen level in a fermentation broth is a function of the rates at which oxygen is being supplied from the sparged gas stream and used up by the metabolism of the fermenting organism. Therefore, it can be expected that a temperature shift in the direction that increases the metabolic rate will decrease the dissolved oxygen level, and a temperature shift in the direction that decreases the metabolic rate will increase the dissolved oxygen level. This can be used to keep the dissolved oxygen in a fermentation from dropping too far during the period of most rapid growth and metabolism.

Equipment

Dissolved oxygen was measured with a membrane-covered polarographic probe as described by Borkowski and Johnson [1]. The current developed by the probe was shunted through a 240-Ω resistor and the voltage developed in this was sensed by a Foxboro SPEC 200 model 2 AI-T2V input card calibrated from 0 to 10 mV. The 0–10-V signal generated by the input card was connected to a Foxboro INTERSPEC analog-to-digital converter and read through a Foxboro FOX 2/30 computer system.

Temperature was measured with a Foxboro model PR-1N platinum resistance temperature device. The resistance of the device was used to generate a current signal in the 4–20-mA range (with a Foxboro model E-94 transmitter) which was converted to a 0–10-V signal by a Foxboro SPEC 200 model 2AI-I2V input card. The 0–10-V signal was treated as was that for dissolved oxygen.

Temperature was controlled by the FOX 2/30 computer. The output signal for control was converted to an analog signal (0 to 10 V) by the Foxboro Interspec interface and fed to a Foxboro model 2AO-V2I output card, which generated a 4–20-mA current signal representing the control valve position. This signal was then converted to a 3 to 15 psig pneumatic signal which (by a Foxboro model 69 TA-1 converter) powered the control valve.

The fermentor is a 250-liter stirred stainless-steel tank. Figure 1 shows a schematic view of the fermentor and associated instrumentation.

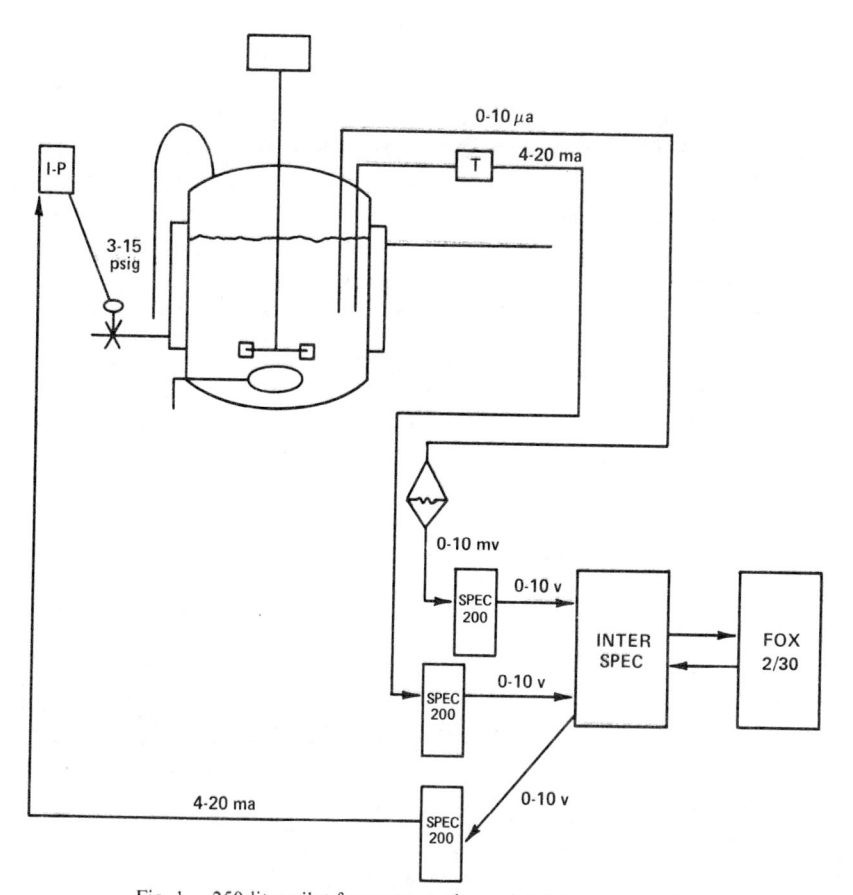

Fig. 1. 250-liter pilot fermentor and associated instrumentation.

Experimental

The response of the dissolved oxygen probe to temperature was determined in the following manner. The probe was installed in a 250-liter fermentor. The fermentor was filled to the operating volume with water, agitated, and sparged with air. The fermentor was kept open to the atmosphere. The temperature in the fermentor was initially 20.5°C and was raised in steps of 2°C every 5 hr (by program control) until it reached about 43°C. Both temperature and oxygen probe output (as the voltage across the resistor) were recorded at 5-min intervals. The voltage developed across the resistor was found to be linearly dependent on the temperature ($R = 0.9996$).

The factor necessary to correct readings taken at various temperatures back to what the readings would have been had they been taken at 20.5°C was computed for the 13 experimental points taken at stable values of temperature and probe output. One of the features of the IMPAC language is its ability to develop four-term polynomials to linearize block inputs: A table of temperatures and correction factors was used to develop such a polynomial so that a block could be generated that would convert temperatures to correction factors.

Figure 2 shows a block diagram of the IMPAC programming. The temperature signal was used twice, once to compute a correction factor for the oxygen probe output and once in the temperature control block itself. Blocks L0002 and L0003 were scaled (and their constants were chosen) so that the temperature setpoint supplied to T0014 could be no higher than 32°C and no lower than 26°C.

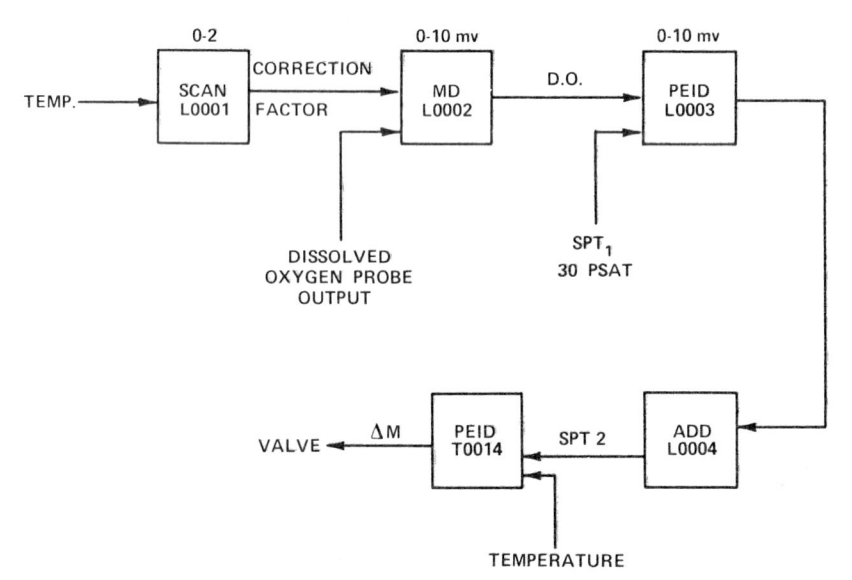

Fig. 2. Block diagram: IMPAC programming to control dissolved oxygen by temperature setpoint manipulation.

Results and Discussion

Figure 3 shows plots of dissolved oxygen and temperature for an antibiotic fermentation run with this scheme and the dissolved oxygen plot of a control. In the experimental run, dissolved oxygen was to have been controlled to a setpoint of 30% air saturation. The dissolved oxygen was in fact not held very constant, but the control scheme did prevent the most extreme fluctuations seen with the control including the severe oxygen deficiency between 24 and 48 hr.

This control scheme was implemented without adding any extra equipment to the tank and without doing any programming in either FORTRAN or assembler language. The IMPAC programming used to condition the data and affect the control could have been written in about 1 hr.

The same general techniques could be used to control dissolved oxygen or other measured or calculated variables (respiratory quotient, for example) by manipulating any of a number of controllable variables (e.g., air flow, back-pressure, stirring rate, feed additions, or pH).

The data-handling scheme was successful in that the temperature of the fermentation was automatically varied to keep the dissolved oxygen values reasonably constant.

COMPUTATION OF OXYGEN UPTAKE RATES

The oxygen uptake rate is a direct function of the metabolism of a living culture. Initially, this rate may be proportional to the culture's growth rate. For a fermentation that exhibits non-growth-associated product formation, the

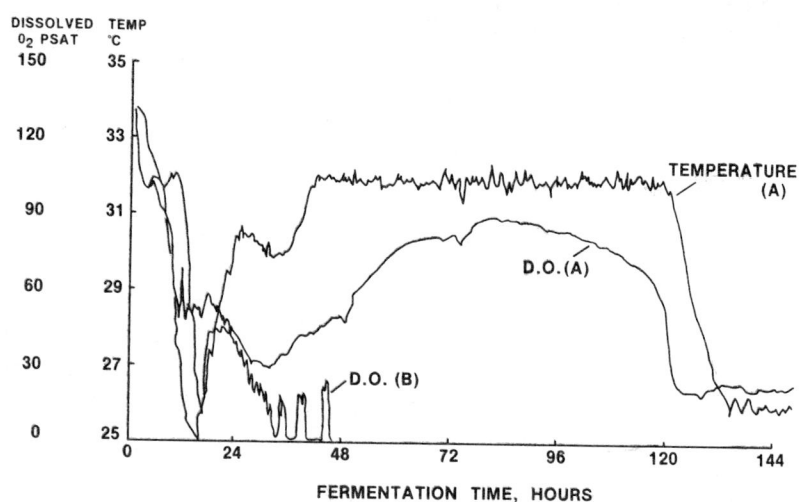

Fig. 3. Dissolved oxygen and temperature from an antibiotic fermentation. Temperature line and dissolved oxygen (A) are from a fermentation in which the dissolved oxygen was manipulated by adjustments to the temperature setpoint. Dissolved oxygen line (B) is from the first 48 hr of a control fermentation.

oxygen uptake rate may be correlated to product formation in the latter stages of the fermentation process. For seed growth, the oxygen uptake rate may be a convenient on-line measurement of the physiological state and a useful indication of when to best harvest the tank. Unusual oxygen uptake patterns may be an on-line indication of contamination. For these reasons an on-line measure of oxygen uptake can be very useful in both the fermentation pilot plant and production plant.

Equipment

The oxygen concentration in the fermentor off gas was measured by a Mine Safety Appliances (MSA) model 802 paramagnetic oxygen analyzer. The voltage developed by this analyzer was sensed by a Foxboro SPEC 200 model 2AI-T2V input card calibrated from 0 to 10 mV. The 0–10-V signal generated by the input card was interfaced to the FOX 2/30 computer system as previously described for the dissolved oxygen input.

The oxygen analyzer was connected to the effluent gas lines from the pilot-plant fermentors through a manifold. At any one time, only a single gas stream was measured for oxygen content. The switching between the various effluent gas streams and the reference air stream was controlled by a system of three-way Automatic Switch Company (ASCO) solenoid valves. When not being analyzed, the effluent gas streams were vented to the atmosphere through flow resistors. The interface between the ASCO solenoid valves and the FOX 2/30 computer system was the digital output module (DOM). This module consists, in part, of latching relays which, when closed, activate the individual solenoid valves through a 110-V circuit. The latching relays may be activated by IMPAC software or through user-written FORTRAN programs.

All the gas streams were dried with silica gel before entering the analyzer. The manifold was constructed so as to minimize the hold-up volume of gas between the analyzer and the vents. This volume was such that it took about 5 min of purging with the gas to be analyzed before accurate readings could be obtained.

This system collected five values for oxygen uptake rate for each of six fermentors every 2 hr. The values were taken 1 min apart.

Experimental

The effluent gas lines from the pilot-plant fermentors enter the control room at a manifold where any six can be connected. These six streams are sampled consecutively and a reference air stream is analyzed between samples. The reference air stream is sampled this often to correct for atmospheric pressure changes.

The fermentor off gas monitoring software consists of IMPAC blocks and supervisory programs written in FORTRAN. The IMPAC blocks are of two types, computational and monitoring. The computational blocks receive the input from the oxygen analyzer through the SPEC 200–INTERSPEC system, and compute on-line the oxygen uptake (mmol oxygen/liter air flow/hr) in engineering units. To compute this oxygen uptake, it is necessary to measure the oxygen content of the air before and after measuring the oxygen content of the effluent gas stream. Since it takes about 10 min to purge the lines and then take a series of measurements, the calculated oxygen uptake is available at a time after the effluent gas stream is measured. Because of the timing of the measurements, it is convenient to delay the effluent gas measurements 20 min.

Figure 4 displays the flow diagrams of the IMPAC computational blocks. There are only three types of blocks used: (1) SCAN blocks, which display the measurement and may pass it to another block; (2) dead time (DT) blocks, which delay the measurement before passing it on; (3) add–subtract (ADSB) blocks, which combine two measurements by addition or subtraction.

There are two paths of information that flow through the system to block W0020: ABEF and AC. When reference air is being analyzed, measurement M_2 represents the oxygen content of the air at present and M_1 that of 10 min in the past. M_3 represents the oxygen content of air 20 min in the past. Thus the output of block W0020 is the average of the two reference air measurements. This measurement, M_4, is then subtracted from the sample oxygen measurement (measured 10 min previously) in block W0019. The output of W0019 is the oxygen uptake and is passed through another dead time block, W0021, for the convenience of operation. The SCAN block, W0014, then contains the oxygen uptake value of a fermentor that was valid 20 min previously. However, the measurement in W0014 is valid only for 10 min. For 10 min both before and after this period, the displayed value in W0014 is meaningless. The supervisory programs to be described store the valid measurements contained in W0014.

A set of supervisory programs, written in FORTRAN, coordinate the switching of the solenoid valves of the manifold, error checking, and the display of the data

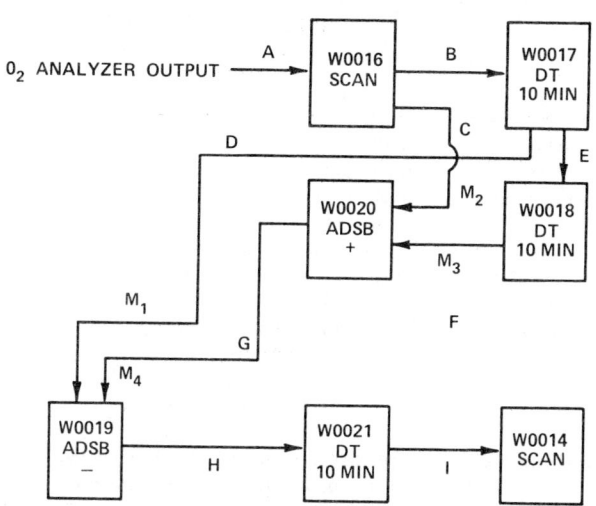

Fig. 4. Block diagram: IMPAC programming to calculate oxygen uptake values.

in other SCAN blocks. Bookkeeping is done by the MANAGER program. This program calculates which tank is to be monitored, coordinates the switching of the appropriate solenoid valves, schedules the DATA program to run, and then reschedules the MANAGER to run. The DATA program collects measurements from the final computational block, W0014, at the appropriate time. The program then stores the valid oxygen uptake measurements in other SCAN blocks. The MANAGER program stores the bookkeeping data in a core COMMON area for fast access. This information includes the total number of tanks being analyzed, the tank presently being analyzed, a mask of which tanks are being analyzed, and a parameter indicating whether or not the reference air stream is to be analyzed next.

Various other supervisory programs were written to support the basic system. The functions they perform include initialization of core COMMON, error checking, and modification of the COMMON area. Another option of the system is to allow the execution of a process control program while the appropriate and current oxygen uptake data are being displayed. Table I lists the various programs and their functions.

Results and Discussion

Typical oxygen uptake data from a fermentation are presented in Figure 5. For the initial three days, little growth occurred and the oxygen uptake rate was zero. At about 90 hr, significant growth occurred to alter the oxygen level of the effluent gas. The oxygen uptake rate and the culture's metabolic rate increased rapidly until a peak was reached at about 97 hr. Then the rate decreased, presumably as a nutrient was taken up.

The system described provided adequate measurement of the oxygen uptake

TABLE I
Various Programs and Their Functions

Program name	Function
INITIAL	initializes core COMMON
SET SCANS	initializes SCAN blocks
MANAGER	schedules DATA to run; controls status of solenoid valves
DATA	stores valid data in SCAN blocks
CHECKER	checks core COMMON for invalid parameters
REPORT	dumps contents of core COMMON
FILLER	modifies core COMMON
EDITOR	enters control programs to be run when oxygen data are available on-line in a drum file
SCHEDULE	schedules control programs to be executed at the appropriate time when data are available

OXYGEN UPTAKE
$\frac{mmoles}{1.\ HR.}$

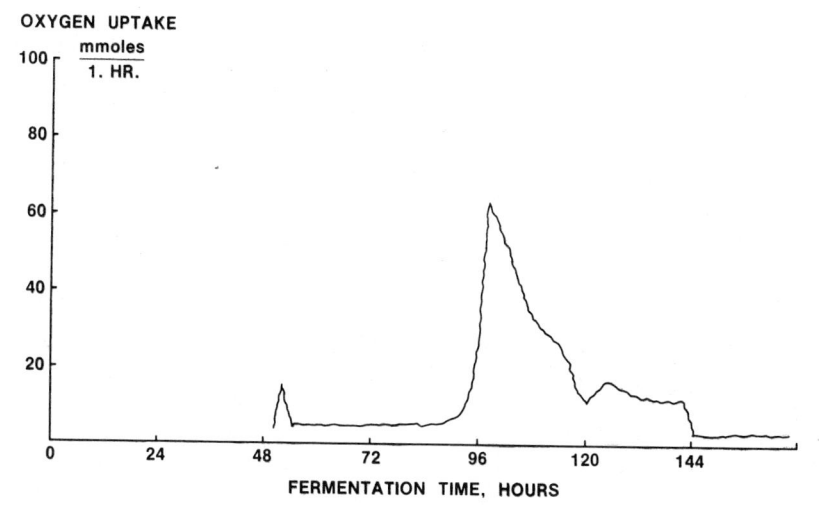

Fig. 5. Oxygen uptake vs. time in an antibiotic fermentation.

rate for up to six fermentors. The data were obtained on-line. However, the analytical–mechanical apparatus used required that the oxygen uptake rate could only be calculated 10 min after the oxygen content of the fermentor had been measured. Although this system is obsolete as a method of measuring oxygen uptake, it is described here as an example of how on-line data can be corrected for drfting sensor outputs.

Reference

[1] J. Borkowski and M. J. Johnson, *Biotechnol. Bioeng.*, 9, 635 (1967).

Applications of Mass–Energy Balances in On-Line Data Analysis

L. E. ERICKSON

Department of Chemical Engineering, Kansas State University, Manhattan, Kansas 66506

INTRODUCTION

On-line data analysis is important in biochemical engineering operations. Culture transfers to larger fermentors in batch processes, improvements in product yield, early identification of problems such as contamination, and process control are some of the reasons why on-line data analysis is needed. In this work applications of material and energy balances and associated regularities in on-line data analysis are investigated. Energy requirements for growth, maintenance, and product formation are examined and energetic yield parameters are employed. Important regularities [1–4] that have been identified and quantified by Minkevich and Eroshin and co-workers [5,6] are used in this work. These regularities are particularly important in on-line data analysis where only limited data are available.

THEORY

Consider the balance equation of microbial growth

$$CH_mO_l + aNH_3 + bO_2 = y_c CH_pO_nN_q + zCH_rO_sN_t + cH_2O + dCO_2 \tag{1}$$

where CH_mO_l denotes the elemental composition of the organic substrate, $CH_pO_nN_q$ denotes the elemental composition of the biomass, and $CH_rO_sN_t$ gives the average elemental composition of the extracellular products. The subscripts denote numbers of atoms of hydrogen, oxygen, and nitrogen per carbon atom. The coefficients y_c, z, and d give the fraction of organic substrate carbon converted to biomass, products, and CO_2, respectively.

The concept of reductance degree, γ, or number of equivalents (equiv) of available electrons/gram atom carbon [1–6] is used in this work with $C = 4$, $H = 1$, $O = -2$, and $N = -3$. The reductance degree of the organic substrate, biomass, and products may be found, respectively, as follows:

$$\gamma_s = 4 + m - 2l \tag{2}$$

$$\gamma_b = 4 + p - 2n - 3q \tag{3}$$

$$\gamma_p = 4 + r - 2s - 3t \tag{4}$$

Note that CO_2, H_2O, and NH_3 have reductance degrees of zero.

Equation (1) may be used as the basis of a total mass balance, a carbon balance, an available electron balance, a nitrogen balance, a hydrogen balance, an oxygen balance, and an energy balance; however, only five independent balances may be written. The carbon balance is

$$y_c + z + d = 1.0 \qquad (5)$$

The available electron balance is

$$\gamma_s - 4b = y_c\gamma_b + z\gamma_p \qquad (6)$$

When the organic substrate does not contain nitrogen, the nitrogen balance is

$$a = y_c q + zt \qquad (7)$$

A total mass balance and an energy balance may be written to complete the set of five independent balances.

Equations (5)–(7) are exact as written. The regularities identified by Minkevich and Eroshin [1–6] may be used to estimate the weight fraction carbon in dried biomass, σ_b, and the reductance degree of biomass, γ_b. If average values rather than measured values are used, then the balances will no longer be satisfied exactly. The third regularity, which states that the heat of reaction per electron transferred to oxygen is approximately constant, may be used to obtain the energy balance

$$Q_o\gamma_s - 4Q_o b = y_c Q_o \gamma_b + z Q_o \gamma_p \qquad (8)$$

where Q_o is the heat evolved/equiv available electrons transferred to oxygen. This equation is not independent of eq. (6) when Q_o is assumed to be constant; however, when both heat evolution and oxygen consumption are measured, the value of Q_o may be calculated and compared to the average value of 26.95 kcal/g-equiv O_2 consumed. If the measurements are accurate, good agreement should be obtained.

The carbon balance, available electron balance, and nitrogen balance may be used in a variety of ways in on-line data analysis. These balances may be used to check the accuracy or consistency of experimental results. Variables that are not measured may be estimated using these balances. These balances also lead to relationships among yield parameters and maintenance coefficients.

Jefferis and Humphrey [7] first proposed using the yield-maintenance model of Marr et al. [8] and Pirt [9] in on-line data analysis. Recently, Zabriskie and Humphrey [10] have used the yield-maintenance model for real-time estimation of biomass concentration. Erickson et al. [3,4] have illustrated how regularities and balances may be used in estimating the yield and maintenance parameters.

Energy for growth, maintenance, and product formation is usually provided by the organic substrate. The available electron balance and the energy balance enable the fractional allocation of energy supplied by the organic substrate to be examined. Equations (6) and (8) may be written in the form

$$4b/\gamma_s + y_c(\gamma_b/\gamma_s) + z(\gamma_p/\gamma_s) = 1.0 \tag{9}$$

or

$$\epsilon + \eta + \xi_p = 1.0 \tag{10}$$

where $\epsilon = 4b/\gamma_s$ is the fraction of available electrons in the organic substrate that is transferred to oxygen, $\eta = y_c\gamma_b/\gamma_s$ is the fraction of available electrons or energy that is incorporated into biomass, and $\xi_p = z\gamma_p/\gamma_s$ is the fraction of available electrons or energy that is incorporated into extracellular products.

Pirt [9] has used the model

$$1/Y_S = 1/Y_S^{max} + m_S/\mu \tag{11}$$

to consider growth and maintenance. This model may also be written in the form [3]

$$1/\eta = 1/\eta_{max} + m_e/\mu \tag{12}$$

where η_{max} is the "true" biomass energetic yield associated with the growth process based on the portion of substrate energy associated with growth. Substrate energy is consumed for maintenance and growth, and η_{max} gives the biomass energetic yield when maintenance energy is considered separately from the energy utilized for growth.

When a, y_c, and z are zero in eq. (1), the resulting equation describes the maintenance process. Similarly, eq. (1) with $z = 0$ may be used to describe the growth process. Corresponding carbon balances and available electron balances may also be written for maintenance and for growth. These balances may be used to relate the yield parameters as has been shown elsewhere [3]. The "true" biomass energetic yield associated with the growth process is related to "true" mass yields as follows:

$$\eta_{max} = (\sigma_b\gamma_b/\sigma_s\gamma_s)Y_S^{max} \tag{13}$$

$$\eta_{max} = (\gamma_b/\gamma_s)y_c^{max} \tag{14}$$

$$1/y_o^{max} = (\gamma_b/4)(1/\eta_{max} - 1) \tag{15}$$

$$1/Y_O^{max} = (2\sigma_b\gamma_b/3)(1/\eta_{max} - 1) \tag{16}$$

$$1/y_d^{max} = \gamma_b/\gamma_s\eta_{max} - 1 \tag{17}$$

$$1/Y_D^{max} = (44\sigma_b/12)(\gamma_b/\gamma_s\eta_{max} - 1) \tag{18}$$

All of the above are "true" biomass yields; Y_S^{max} is g dry biomass/g organic substrate, y_c^{max} is g biomass carbon/g substrate carbon, y_o^{max} is g atoms biomass carbon/g mol O_2, Y_O^{max} is g dry biomass/g O_2, y_d^{max} is g atoms biomass carbon/g mol CO_2, and Y_D^{max} is g dry biomass/g CO_2.

Maintenance parameters are also related. For example, the corresponding maintenance parameters for eqs. (13)–(18) are

$$m_S = (\sigma_b\gamma_b/\sigma_s\gamma_s)m_e \tag{19}$$

$$m_c = (\gamma_b/\gamma_s)m_e \tag{20}$$

$$m_o = (\gamma_b/4)m_e \tag{21}$$

$$m_O = (2\sigma_b\gamma_b/3)m_e \tag{22}$$

$$m_d = (\gamma_b/\gamma_s)m_e \tag{23}$$

$$m_D = (44\sigma_b\gamma_b/12\gamma_s)m_e \tag{24}$$

where m_e is g equiv available electrons transferred to oxygen/g-equiv available electrons in biomass (hr), m_S is g organic substrate/g dry biomass (hr), m_c is g organic substrate carbon/g biomass carbon (hr), m_o is g mol O_2/g atom biomass carbon (hr), m_O is g O_2/g dry biomass (hr), m_d is g mol CO_2/g atom biomass carbon (hr), and m_D is g CO_2/g dry biomass (hr).

The process of microbial product formation may also be examined with respect to energetic yield or available electron yield. If y_c is assumed to be zero in eqs. (1), (5), and (6), a "true" product energetic yield coefficient, ξ_p^{max}, may be calculated based on that portion of the total substrate energy associated with product formation. The balance equations may be used to relate mass yields and energetic or available electron yields. For example, the corresponding relationships are

$$\xi_p^{max} = (\sigma_p\gamma_p/\sigma_s\gamma_s)Y_{P/S}^{max} \tag{25}$$

$$\xi_p^{max} = (\gamma_p/\gamma_s)y_{p/s}^{max} \tag{26}$$

$$1/y_{p/o}^{max} = (\gamma_p/4)(1/\xi_p^{max} - 1) \tag{27}$$

$$1/Y_{P/O}^{max} = (2\sigma_p\gamma_p/3)(1/\xi_p^{max} - 1) \tag{28}$$

$$1/y_{p/d}^{max} = \gamma_p/\gamma_s\xi_p^{max} - 1 \tag{29}$$

$$1/Y_{P/D}^{max} = (44\sigma_p/12)(\gamma_p/\gamma_s\xi_p^{max} - 1) \tag{30}$$

where $Y_{P/S}^{max}$ is g product/g organic substrate, $y_{p/s}^{max}$ is g product carbon/g organic substrate carbon, $y_{p/o}^{max}$ is g atoms product carbon/g mol O_2, $Y_{P/O}^{max}$ is g product/g O_2, $y_{p/d}^{max}$ is g atoms product carbon/g mol CO_2, and $Y_{P/D}^{max}$ is g product carbon/g CO_2.

It is important to realize that this approach divides the organic substrate that is used into growth associated, maintenance, and product associated fractions. Similarly, there is oxygen consumption and CO_2 production associated with each of the three processes of growth, maintenance, and product formation. The identification of the fractional allocation of organic substrate, oxygen, and CO_2 production may be determined if η_{max}, m_e, and ξ_p^{max} are known and if η and ξ_p are known. Values of the biomass energetic yield, η, and the product energetic yield, ξ_p, can be determined from measured variables (biomass, product, and organic substrate concentrations, oxygen consumption, CO_2 production). Estimation of the parameters η_{max}, m_e, and ξ_p^{max} from experimental data is discussed elsewhere [3,4,9,11]. When there is no product formation, linear regression may be used to estimate η_{max} and m_e; however, when growth maintenance and product formation must be considered, simultaneously, it is not as easy to obtain accurate estimates of η_{max}, m_e, and ξ_p^{max}. An example of the estimates that can be obtained with limited data is presented elsewhere [4].

One of the reasons for using yield and maintenance parameters such as η_{max}, m_e, and ξ_p^{max} is that these parameters can be related to the energetic yields associated with various biochemical pathways. Although some progress has been made in this regard [12-14], further work is needed. Another reason for using these parameters is that they are frequently relatively constant. Although these parameters have been found to be relatively constant in much of the work in the literature [3,4,9], more work is needed. In determining η_{max}, m_e, and ξ_p^{max}, it is important to use consistent experimental data. The carbon and available electron balances may be used to check the consistency of the experimental data and identify sets of data that are not consistent [3,4]. These balances may also be used to determine whether or not extracellular products are present in appreciable amounts.

ON-LINE DATA ANALYSIS

The balances, regularities, and yield and maintenance parameters introduced above may be used in on-line data analysis. Measurement errors are always present in fermentation processes. In addition, some variables are difficult or impossible to measure and values for these variables may be determined indirectly.

The carbon and the available electron balances may be used to check the consistency of experimental data if sufficient measurements are made. When extracellular products are not present, three variables need to be measured, and when products are present, four variables need to be measured in order to check for consistency. Organic substrate consumption, biomass production, oxygen consumption, carbon dioxide production, heat evolution, and product formation are the variables that may be easily used in eqs. (5), (6), and (8).

Values of the weight fraction carbon in biomass, σ_b, and the reductance degree of biomass, γ_b, are needed when eqs. (5) and (6) are used. The average values from the work of Minkevich and co-workers [6] are σ_b = 0.462 g carbon in biomass/g dry biomass and γ_b, = 4.291 g equiv available electrons/quantity biomass containing 1 g atom carbon. If heat evolution data are used, the regularity, Q_o = 26.95 kcal evolved/g equiv available electrons transferred to oxygen, may be used. The coefficient of variation is 4% for Q_o and γ_b and 5% for σ_b [6]. When repeated work is being carried out with a particular culture, values of these regularity parameters may be determined experimentally and used in place of the average values found by Minkevich and co-workers.

Examples where the consistency of experimental data has been examined have been presented elsewhere [1-4,15]. These examples show that the values of the regularities σ_b, γ_b, and Q_o are sufficiently accurate to be very useful in examining the consistency of experimental data.

Zabriskie and Humphrey [10] have used the biomass yield and maintenance parameters together with oxygen measurements to estimate biomass production.

When values of η_{max} and m_e are known, and no extracellular products are produced, measurement of one variable (oxygen consumption or carbon dioxide production, for example) is sufficient for prediction of the other variables (organic substrate consumption, biomass production, heat evolution). If initial conditions are known, real-time integration may be used to arrive at biomass and organic substrate concentrations. If kinetic parameters are also known, real-time integration may be independently carried out using kinetic information and compared with the results obtained using yield and maintenance parameters and experimental measurements. When the yield results give larger biomass productivities than are possible based on kinetic information, this may be because of product formation or measurement errors. If product formation is present, additional variables need to be measured to accurately predict process performance.

When two variables are measured (O_2 consumption and CO_2 production, for example) and extracellular products are expected to be present in only very small amounts, known values of η_{max} and m_e may be used with the oxygen consumption data to predict values for organic substrate consumption, biomass production, CO_2 production, and heat evolution. The predicted CO_2 production can be compared to the measured value. This information may be used to determine whether extracellular products may be present in significant amounts. For carbohydrates, the respiratory quotient is approximately 1 when extracellular products are not produced, but values significantly different from 1 frequently occur when extracellular products are produced. Real-time integration using known initial conditions and utilization of kinetic information as a check are also desirable in this case.

In large fermentors, oxygen consumption, CO_2 production, and heat evolution are easily measured. The ratio of heat evolution to oxygen consumption can be compared to the value of 26.95 kcal evolved/g equiv oxygen consumed. Good agreement provides a check on the accuracy of these two measurements. However, once this check has been made, utilization of the available information is very similar to the case described above.

When products are present in the above cases with O_2 and CO_2 measurements, additional measurements are desirable. The energy allocation to product formation needs to be known in this case. Direct measurement of the quantity of products formed is frequently the best way to estimate the energy allocated to product formation. In some cases, the respiratory quotient may give an estimate that may be sufficiently accurate to be useful. For example, in the case of ethanol production from glucose the oxygen measurement can be used with η_{max} and m_e to predict biomass production and CO_2 evolution associated with biomass production and maintenance. The difference between the measured CO_2 production and the estimated CO_2 production due to growth and maintenance is the estimated CO_2 production due to ethanol (EtOH) formation. This value may be used to predict the ethanol formed. A carbon balance or available electron balance then gives the substrate consumption associated with the oxygen consumed, biomass formed, CO_2 evolved, and EtOH produced. While this approach

APPLICATIONS OF MASS-ENERGY BALANCES

is not expected to be very accurate, in some cases it may be sufficiently close to be useful.

Frequently, oxygen consumption and CO_2 evolution data are available continuously, and other measurements are also available, but a significant measurement time lag exists. Two types of on-line data analysis may be used in this case. Using the full set of measurements, the carbon and available electron balances should be used to check for data consistency and for unmeasured extracellular products. Real-time integration in batch fermentation should be carried out using this more complete set of data. These integration results should be used as initial conditions for on-line data analysis with the continuously measured variables to provide current estimates of all the variables of interest.

In fermentations in which the organic substrate does not contain nitrogen, the nitrogen balance, eq. (7), may also be used in on-line data analysis. Minkevich and co-workers [6] have examined the nitrogen content in biomass and found the weight fraction to be 0.089 g nitrogen/g dry biomass with a coefficient of variation of 20%. Because of the large variance, knowledge of the weight fraction nitrogen in the culture being investigated is desirable. Since ammonia is frequently supplied as needed for pH control, the ionic balance and dissolved nitrogen should be considered in using nitrogen balances. Recently, Minkevich [16] has considered the ionic balances associated with the growth of microorganisms.

Nitrogen balances may be used to examine the consistency of experimental results. For example, the measured or estimated biomass production and the measured nitrogen addition may be used to estimate the weight fraction nitrogen in the produced biomass. This result may be compared to values obtained from prior nitrogen analyses of dry biomass or the average value of Minkevich and co-workers [6].

When checks are made for data consistency using these balances and when real-time integration is used with kinetic information as well as yield information, there will be times when the results are not consistent and decisions will need to be made regarding which set of information is not accurate. In some cases biochemical engineers who are familiar with the fermentation will be able to identify the reason for the inconsistent results and select the more accurate results as a basis for further on-line data analysis. However, in other cases, it may be desirable to continue on line data analysis using both sets of results until it is clear which set is more accurate.

RESULTS AND DISCUSSION

In order to illustrate the application of mass and energy balance regularities in on-line data analysis, experimental data from the laboratory of Fiechter were used [17,18]. *Candida tropicalis* was cultured on *n*-hexadecane in batch and continuous culture. The continuous-culture data [17] were analyzed to obtain an estimate of the true energetic growth yield η_{max} and the maintenance parameter m_e. These values were then used together with the batch-culture oxygen consumption rate data and initial biomass concentration of Hug and Fiechter [18] to estimate the biomass concentration and CO_2 evolution rate. The results are shown in Figure 1. The agreement between the predicted values (solid lines) and the experimentally measured values is relatively good.

In the process of analyzing this data, carbon balances and available electron balances were used to examine the consistency of the experimental data. The consistency of the experimental data was relatively good, and extracellular products were not present in significant amounts. Values of $\eta_{max} = 0.43$ and $m_e = 0.01$ hr^{-1} were used in Figure 1. The values of $\sigma_b = 0.493$ and $\gamma_b = 4.385$, which have been reported for *C. tropicalis* growing on *n*-alkanes [5,19], were used in this work.

In Figure 2 predicted and measured values of cell concentration and *n*-hexadecane concentration are presented for the batch-culture data of Hug and Fiechter [18]. The carbon dioxide evolution data were used together with initial conditions and the above parameter values to obtain the curves shown in Figure 2. Agreement of predicted and measured values is relatively good.

The results indicate that good estimates of cell concentration and organic substrate concentration can be obtained from gas analysis data in batch fermentation when the data are consistent and when extracellular products are not formed.

CONCLUSIONS

Material and energy balances and biomass regularities may be used in on-line data analysis. An available electron balance is presented that relates oxygen

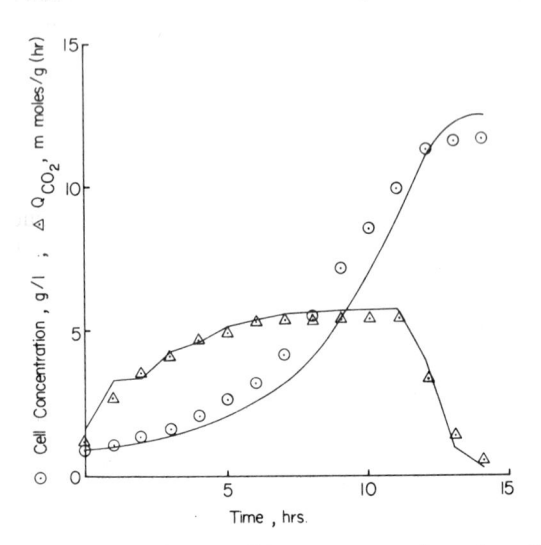

Fig. 1. Comparison of experimental values of biomass concentration and specific CO_2 evolution rate with values estimated using oxygen consumption measurements and $\eta_{max} = 0.43$ and $m_e = 0.01$ hr^{-1}.

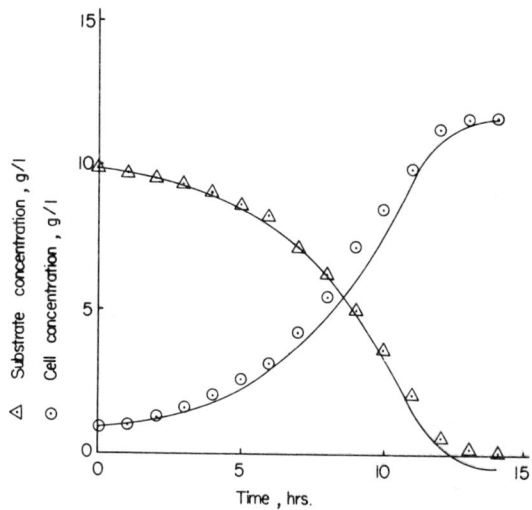

Fig. 2. Comparison of experimental values of biomass concentration and n-hexadecane concentration with values estimated using CO_2 evolution measurements and $\eta_{max} = 0.43$ and $m_e = 0.01$ hr^{-1}.

consumption, organic substrate consumption, biomass production, and product formation. This balance can be used to estimate heat evolution and energetic yields of biomass and products.

Organic substrate is used for growth, maintenance, and product formation. "True" biomass energetic yields and "true" product energetic yields, which are based on the fraction of the organic substrate associated with each of these processes, are related to "true" mass yields using mass and energy balances. The application of these yields and balances in on-line data analysis is discussed. It is shown that variables of interest may be estimated by using indirect measurements and that the consistency of the results may often be checked. The carbon balance and the available electron balance can be used on-line to check the consistency of measurements.

Nomenclature

a	mol NH_3/quantity organic substrate containing 1 g atom C (g mol/g atom C)
b	mol O_2/quantity organic substrate containing 1 g atom C (g mol/g atom C)
c	mol H_2O/quantity organic substrate containing 1 g atom C (g mol/g atom C)
d	mol CO_2/quantity organic substrate containing 1 g atom C (g mol/g atom C)
l	atomic ratio of oxygen to carbon in organic substrate (dimensionless)
m	atomic ratio of hydrogen to carbon in organic substrate (dimensionless)
m_c	rate of organic substrate consumption for maintenance (g atom C/g atom C in biomass hr)
m_D	rate of CO_2 evolution for maintenance (g/g biomass hr)
m_d	rate of CO_2 evolution for maintenance (g mol CO_2/g atom C in biomass hr)
m_e	rate of organic substrate consumption for maintenance (g equiv available electrons/g equiv available electrons in biomass hr or kcal/kcal biomass hr)
m_O	rate of O_2 uptake for maintenance (g/g biomass hr)

m_o	rate of O_2 uptake for maintenance (g mol O_2/g atom C in biomass hr)
m_S	rate of organic substrate consumption for maintenance (g/g dry biomass hr)
n	atomic ratio of oxygen to carbon in biomass (dimensionless)
P	atomic ratio of hydrogen to carbon in biomass (dimensionless)
Q_o	heat evolution in fermentation/equiv O_2 uptake (kcal/g-equiv)
q	atomic ratio of nitrogen to carbon in biomass (dimensionless)
r	atomic ratio of hydrogen to carbon in products (dimensionless)
s	atomic ratio of oxygen to carbon in products (dimensionless)
t	atomic ratio of nitrogen to carbon in products (dimensionless)
Y_D^{max}	biomass "true" yield based on CO_2 evolution (g dry biomass/g CO_2 evolved)
Y_O^{max}	biomass "true" yield based on O_2 consumption (g dry biomass/g O_2 consumed)
$Y_{P/D}^{max}$	product "true" yield based on CO_2 (g product/g CO_2 evolved)
$Y_{P/O}^{max}$	product "true" yield based on O_2 consumption (g product/g O_2 consumed)
$Y_{P/S}^{max}$	product "true" yield based on organic substrate consumption (g product/g substrate)
Y_S	biomass yield on organic substrate (g dry biomass/g substrate)
Y_S^{max}	biomass "true" yield based on organic substrate (g dry biomass/g substrate)
y_c	biomass carbon yield (fraction of organic substrate C in biomass) (dimensionless)
y_c^{max}	biomass "true" carbon yield (fraction of growth-associated organic substrate C in biomass) (dimensionless)
y_d^{max}	biomass "true" yield based on CO_2 evolution (g atom biomass C/g mol CO_2)
y_o^{max}	biomass "true" yield based on O_2 consumption (g atom biomass C/g mol O_2)
$y_{p/d}^{max}$	product "true" yield based on CO_2 evolution (g atom product C/g mol CO_2)
$y_{p/o}^{max}$	product "true" yield based on O_2 consumption (g atom product C/g mol O_2)
$y_{p/s}^{max}$	product "true" yield based on organic substrate (fraction of product associated organic substrate C in product) (dimensionless)
z	fraction of organic substrate carbon in products (dimensionless)
γ_b	reductance degree of biomass (equiv available electrons/g atom C)
γ_p	reductance degree of products (equiv available electrons/g atom C)
γ_s	reductance degree of organic substrate (equiv available electrons/g atom C)
ϵ	fraction of energy in organic substrate that is evolved as heat (dimensionless)
η	fraction of energy in organic substrate that is converted to biomass or biomass energetic yield (dimensionless)
η_{max}	"true" biomass energetic yield (dimensionless)
μ	specific growth rate (hr^{-1})
ξ_p	fraction of energy in organic substrate that is converted to products (dimensionless)
ξ_p^{max}	"true" product energetic yield (dimensionless)
σ_b	weight fraction carbon in biomass (dimensionless)
σ_s	weight fraction carbon in organic substrate (dimensionless)
σ_p	weight fraction carbon in products (dimensionless)

This work was supported in part by the National Science Foundation (Grant No. ENG 77-16999). The help of Y. H. Lee is appreciated.

References

[1] L. E. Erickson, I. G. Minkevich, and V. K. Eroshin, *Biotechnol. Bioeng., 20,* 1601 (1978).
[2] L. E. Erickson, S. E. Selga, and U. E. Viesturs, *Biotechnol. Bioeng., 20,* 1629 (1978).
[3] L. E. Erickson, I. G. Minkevich, and V. K. Eroshin, *Biotechnol. Bioeng., 21,* 575 (1979).
[4] L. E. Erickson, *Biotechnol. Bioeng., 21,* 725 (1979).
[5] I. G. Minkevich and V. K. Eroshin, *Folia Microbiol., 18,* 376 (1973).
[6] I. G. Minkevich, V. K. Eroshin, T. A. Alekseeva, and A. P. Tereshchenko, *Mikrobiol. Promst. Ref. Sb., 2*(144), 1 (1977) [in Russian].

[7] R. Jefferis and A. E. Humphrey, "Indirect measurement of cell biomass by oxygen material balancing," National Meeting of the American Chemical Society, Chicago, IL, 1972.
[8] A. G. Marr, E. H. Nilson, and D. J. Clark, *Ann. N.Y. Acad. Sci., 102,* Art. 3, 536 (1962).
[9] S. J. Pirt, *Proc. R. Soc. London Ser. B, 163,* 224 (1965).
[10] D. W. Zabriskie and A. E. Humphrey, *AIChE J., 24,* 138 (1978).
[11] S. J. Pirt, *Principles of Microbe and Cell Cultivation* (Halsted, Wiley, New York, 1975).
[12] I. G. Minkevich and V. K. Eroshin, *Stud. Biophys., 49,* 43 (1975).
[13] I. G. Minkevich and V. K. Eroshin, *Stud. Biophys., 59,* 67 (1976).
[14] I. G. Minkevich and V. K. Eroshin, *Usp. Sovrem. Biolo., 82,* 103 (1976) [in Russian].
[15] Y. H. Lee, *Proceedings of the Eighth Biochemical Engineering Symposium,* C. E. Dunlap, Ed. (University of Missouri, Columbia, MO 1978).
[16] I. G. Minkevich, *Biotechnology and Bioengineering Symposium* (Zinatne, Riga, 1978), Vol. 1, p. 98 [in Russian].
[17] H. W. Blanch and A. Einsele, *Biotechnol. Bioeng., 15,* 861 (1973).
[18] H. Hug, and A. Fiechter, *Arch. Mikrobiol., 88,* 77 (1973).
[19] V. K. Eroshin and A. I. Belyanin, *Prikl. Biokhem. Mikrobiol., 4,* 704 (1968) [in Russian].

Instrumentation for Fermentation Monitoring and Control

ÅKE UNDÉN

AB Fermenta, Strängnäs, Sweden

The variables that characterize a cultivation process can largely be divided into those that define the physical state of the process, and those that measure the composition of the fermentor contents, and the composition and magnitude of the flows into and out of the vessel (Table I) [1]. It is normally desired to control the physical state variables at a constant value, or in the case of feed rates, perhaps according to a predetermined program. Such control is often advantageous in avoiding measurement errors caused by the interactions among the physical state variables: e.g., the galvanic probe, which is normally used for dissolved oxygen measurement, has a high temperature coefficient. Most of the instruments used to measure physical variables are readily available for bench- and plant-scale equipment. For example, liquid consumption is typically measured with a pressure transducer or a load cell under the storage vessel, or in larger scale, with an in-line flowmeter. Similarly the broth weight or volume is derived from weighing the fermentor or measuring differential pressure over its height. A simple alternative, which we have used [2], is to provide the storage vessel with a conductometric probe with good linearity.

Most development efforts are devoted to variables that are related to culture growth and productivity. These variables are listed in the second column of Table I. For instance, when controlling pH, the acid or base addition rate can be a measure of growth or product formation.

Gas analyses have the particular merit of being outside the zone that requires sterilization. Being quite responsive at normal gas flow rates, they provide on-line data of the culture state with only a few minutes delay. The outlet gas flow rate is usually not directly measured. It is calculated from the inlet flow rate and corrected for changes in gas composition [3]. It is equally important, however, to correct the measurement values for pressure changes. Suppose, for example, that the ambient pressure is 770 mm Hg at calibration time, and it falls to 750 mm Hg later in the day. An oxygen concentration of 20.9% at the calibration time will appear to be 20.4% after the change in ambient pressure, which is a reduction of 2.5%. This will cause an apparent increase in the oxygen uptake rate and an apparent decrease in the carbon dioxide production rate. This can lead to a large error in the respiratory quotient (RQ), which is not acceptable if the RQ is to be used for control purposes [4]. The problem is easily solved with

TABLE I

Variables that Should be Measured or Controlled in a Well-Instrumented Fermentation Unit

Physical state	Gas and liquid analysis
Temperature	pH
Pressure	Redox potential
Agitation speed	Dissolved oxygen
Power input	Dissolved carbon dioxide
Air flow rate	Effluent gas oxygen content
Liquid feed rates:	Effluent gas carbon dioxide content
nutrients	Cumulative amount of liquid addition:
precursors	acid
inducers	base
Broth weight	antifoam
Broth volume	Concentrations of:
Viscosity	cells
	medium constituents
	metabolites
	cellular components
	products

an on-line pressure transducer provided the means for pressure correction is under program control [2]. Occasional drift correction is of course also required. An alternative method, although more complicated, would be to have automatic calibration of the analyzers at regular time intervals, such as described by Swartz and Cooney [5].

The total biomass concentration, as estimated from the broth turbidity [2,6,7], is often used as an indicator of cultivation progress. A more important measurement could well be the active biomass [7], and this could be derived from the gas balance [8], the measurement of some cellular constituent [9], or heat evolution [10] data. The difference between total and active biomass might well be an important control variable relating closely to the well-being of the culture.

Material balancing and indirect measurements have proven to be efficient replacements for true measurements of some metabolites and products. However, the introduction of new and practical direct-measuring sensors is likely to bring a new dimension to fermentation monitoring. This may come by replacing a cumbersome and possibly more error-prone indirect method, or by bringing formerly inaccessible information to the operator's continuous attention. The specificity of enzymes as utilized in enzyme electrodes [11] or enzyme thermistors [12] promises to provide solutions here although it may be necessary to place such devices outside the fermentor. At present, these devices mainly serve off-line for discrete sample analysis, or to supplement continuous-flow-type analyzers because of calibration requirements and limited stability. When one chooses to measure variables outside the fermentation vessel, it is necessary to ensure that nutrient limitations do not occur during the assay that could distort the measurement. Problems of this type have been encountered, for example, in the development of microcalorimeters for use with fermentation equipment [13]. Indirect measurement techniques may be the best solution when limitations such as these are unavoidable [9].

These comments have summarized the current status in the monitoring of fermentation processes and have suggested some of the problems that remain. The papers that follow describe some new and promising approaches in this technology.

References

[1] D. Y. Ryu and A. E. Humphrey, *J. Appl. Chem. Biotechnol., 23,* 283 (1973).
[2] Å. Undén, Ph.D. thesis, Royal Institute of Technology, Stockholm, 1977.
[3] A. Fiechter and K. von Meyenburg, *Biotechnol. Bioeng., 10,* 535 (1968).
[4] L. K. Nyiri, G. M. Toth, C. S. Krishnaswami, and D. V. Parmenter, *Workshop: Computer Applications in Fermentation Technology 1976* (Verlag Chemie, Weinheim, 1977), p. 37.
[5] J. R. Swartz and C. L. Cooney, *Process Biochem., 13*(2), 3 (1978).
[6] R. P. Jefferis, H. Winter, and H. Vogelmann, *Workshop: Computer Applications in Fermentation Technology 1976* (Verlag Chemie, Weinheim, 1977), p. 141.
[7] A. G. Lane, in *Proceedings of the 1st European Conference on Computer Process Control in Fermentation* (INRA, Dijon, 1973).
[8] A. E. Humphrey, in *Proceedings of the 1st European Conference on Computer Process Control in Fermentation* (INRA, Dijon, 1973).
[9] D. W. Zabriskie and A. E. Humphrey, *Appl. Environ. Microbiol, 35,* 337 (1978).
[10] D.-G. Mou and C. L. Cooney, *Biotechnol. Bioeng., 18,* 1371 (1976).
[11] H. Nilsson, K. Mosbach, S.-O. Enfors, and N. Molin, *Biotechnol. Bioeng., 20,* 527 (1978).
[12] K. Mosbach, B. Danielsson, A. Borgerud, and M. Scott, *Biochim. Biophys. Acta, 403,* 256 (1975).
[13] R. Eriksson, personal communication, 1976.

Affinity Sensors for Individual Metabolites

JEROME S. SCHULTZ and GREGORY SIMS
Department of Chemical Engineering, University of Michigan, Ann Arbor, Michigan 48109

INTRODUCTION

The principles used in developing affinity sensors are similar to those used in radioimmune assay (RIA) procedures; both are based on the displacement of a known labeled solute from a specific binding site by a similar unlabeled solute of unknown concentration. To develop a typical RIA assay procedure, a specific antibody is prepared to a particular chemical species (antigen), e.g., insulin, by appropriate immunological procedures [1]. After purification, the antibody is immobilized by attachment to latex microspheres. To determine the unknown amount of insulin in a sample, an aliquot is added to a tube containing the immobilized antibody along with a known quantity of radioactively labeled insulin. Due to exchange between the labeled and unlabeled insulin, some radioactivity is displaced into the solution. The tube is centrifuged and the radioactivity in the supernatant is measured. The amount of released radioactivity can be related to the unknown amount of insulin in the sample by appropriate calibration curves.

The essential elements of the procedure are a) a specific binding agent, b) competition between the labeled and unlabeled species for the binding site, and c) subsequent separation of the solution from the binding agent.

The same concepts can be used to fabricate a specific sensor for metabolites but based on optical rather than radioisotope detector techniques. A conceptual approach for an affinity sensor for glucose is illustrated in Figure 1. In this case, conconavalin A was selected as the specific binding agent. This protein, one example of a lectin isolated from jack beans, has binding sites which are highly selective for sugars and carbohydrates [2]. The binding constants between Con-A and some typical sugars are shown in Table I. Although each molecule of Con-A has two binding sites, they appear to act independently [3] and according to simple biomolecular kinetics for simple sugars [4].

In the conceptual fiber optic sensor shown in Figure 1, Con-A is immobilized on the surface of fiber bundles carrying the excitation light beam. The coating is thin enough to allow light to penetrate into the transducer chamber. Also placed in this chamber is a small amount of a ligand that competes with glucose for Con-A binding sites. Since the response signal is produced by the displacement of this compound, it must be distinguishable from glucose in some manner.

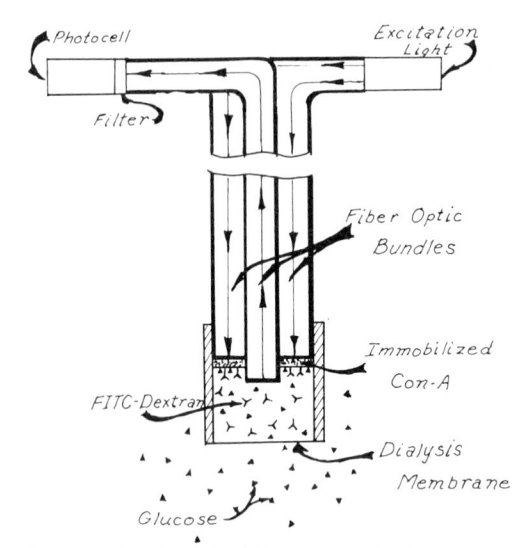

Fig. 1. Schematic diagram of a fiber-optic affinity sensor for the measurement of glucose (not to scale). Increasing the concentration of glucose in the external solution, increases the concentration of glucose in the sensor element. Fluorescent dextran is displaced from the immobilized Con-A and a greater emission intensity is measured.

For this case, a commercially available fluorescein-labeled dextran (FITC dextran) was chosen. Dextrans are linear polysaccharides of glucose which reversibly bind to Con-A owing to presence of the free moieties located at the terminus of the polymer branches [5].

The sensor element communicates with external solutions by means of a dialysis membrane. The membrane porosity is selected to allow free diffusion of low-molecular-weight species and to prevent the leakage of the high-molecular-weight competitive ligand, in this case glucose and FITC-dextran, respectively.

The total amount of Con-A and FITC-dextran in the element is constant, and the internal concentration of glucose will vary according to its concentration in the external solution.

Thus the following reversible reactors take place within the sensor element:

$$\text{glucose} + \text{Con-A} \rightleftharpoons \text{Con-A-glucose} \qquad (1)$$

$$\text{FITC-dextran} + \text{Con-A} \rightleftharpoons \text{Con-A-FITC-dextran} \qquad (2)$$

When the sensor element is exposed to higher concentrations of glucose reaction (1) is displaced to the right and reaction (2) is displaced to the left, resulting in an increase in free FITC-dextran. This increase provides a larger amount of fluorescence in view of the emission optical fibers and thus an increase in the output of the photodetectors. Conversely, a reduction in external glucose concentration results in a decrease in signal.

TABLE I
Affinity Constants of Sugars with Con-A

Sugar	K^a (mol/liter)
Methyl-d-D-mannoside	5.0×10^{-5}
Isomaltose	1.8×10^{-4}
Maltose	3.5×10^{-4}
Fructose	7.3×10^{-4}
Glucose	1.7×10^{-3}
Sucrose	1.9×10^{-3}

[a] $K = \text{(sugar)(Con-A)}/\text{(Con-A-sugar)}$.

The approach can be used for a wide variety of organic species provided suitably specific affinity proteins and competing detector ligands can be found or prepared. This does not appear to be a major shortcoming since many selective binding agents have already been documented for affinity chromatography [6] and procedures for preparing new antibodies are fairly standard [1].

The feasibility of this approach for making affinity sensors was evaluated by using the Con-A glucose system as a model.

MATERIALS AND METHODS

Conconavalin-A–Sepharose (Pharmacia) is obtained as a gel, and the Con-A content is given as 10 mg/ml gel. Fluorescein-labeled dextran (Pharmacia FITC-dextran) was obtained with a mean molecular weight of 70 000.

Binding studies: Approximately 1 ml Con-A suspension (containing 0.45 ml gel) plus 5 ml of approximately $3 \times 10^{-7} M$ FITC-dextran solution are placed in a centrifuge tube. Aliquots of concentrated glucose solutions are added to individual tubes to obtain a range of glucose concentration. The volume of all tubes is brought up to 10 ml with $0.2 M$ phosphate buffer at pH 7.3. The tubes are mixed and then centrifuged after 15 min. Aliquots of the supernatant are removed the fluorescence is measured without dilution (Perkin Elmer fluorescence spectrophotometer), and then returned to their respective tubes.

Approximately 0.1 ml 50 mg/ml methyl-mannoside is added to each tube, mixed, and allowed to set for 1 hr. After centrifugation, the emission intensity of the supernatant is obtained again for each tube. The purpose of the methyl-mannoside addition is to displace all the FITC-dextran from the Con-A and serve as a check on possible irreversible binding of the dextran.

RESULTS AND DISCUSSION

A typical calibration curve relating emission intensity to FITC-dextran concentration is shown in Figure 2. This calibration curve shows several characteristics of fluorescent species; first, the sensitivity is quite high (concentrations are in the micromolar range); second, the response becomes nonlinear at higher concentrations due to emission self-absorption by FITC-dextran. Thus, for this

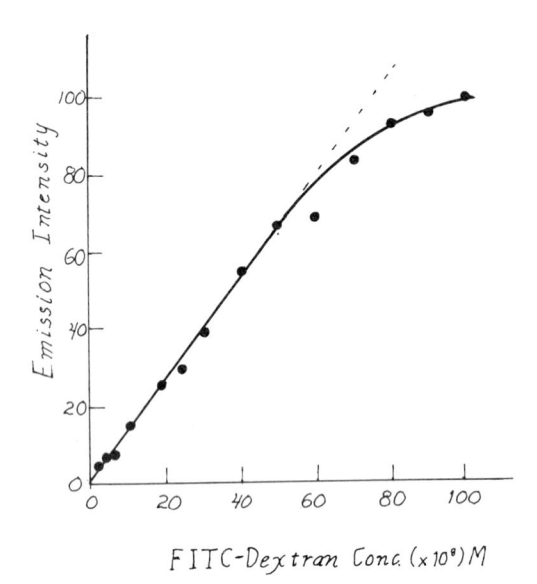

Fig. 2. Calibration curve relating emission intensity to FITC-dextran concentration.

application, it is desirable to design the sensor so that the free FITC-dextran concentration is below $6 \times 10^{-9} M$, to ensure a linear response.

Simple experiments to validate the concept of affinity sensors were undertaken using commercially available Con-A bound to Sepharose beads. Various amounts of glucose were added to test tubes containing a fixed amount of Con-A and FITC-dextran. The amount of fluorescent dextran released from the immobilized Con-A on glucose addition was determined by removing aliquots of the supernatant after centrifuging and placing the sample in a fluorescent spectrophotometer. The results of this type of experiment are shown in Figure 3.

Note that the curve does not start at the origin. As indicated by eq. (2), even in the absence of glucose some free FITC-dextran exists in solution in equilibrium with the Con-A bound FITC-dextran. As the concentration of glucose increases, the amount of free dextran increases also. This is expected due to the competition between reactions (1) and (2). The saturation expected at high glucose concentrations where all of the dextran is displaced from Con-A is also apparent in Figure 3. As a result of this saturation behavior, the sensitivity of the method is better at lower glucose concentration.

To check for possible nonspecific binding of dextran to the immobilized Con-A, methyl-mannoside was added to each tube as a final step. Con-A has a much greater affinity for methyl-mannoside than for either glucose or dextran (Table I). As can be seen, all of the fluorescent dextran is recovered by this treatment and thus nonspecific binding does not appear to be important.

Also appearing in Figure 3 is a computed curve for the expected response if all the binding sites on Con-A and FITC-dextran acted independently. This curve is based on the assumption that the mechanism of the binding reactions are as

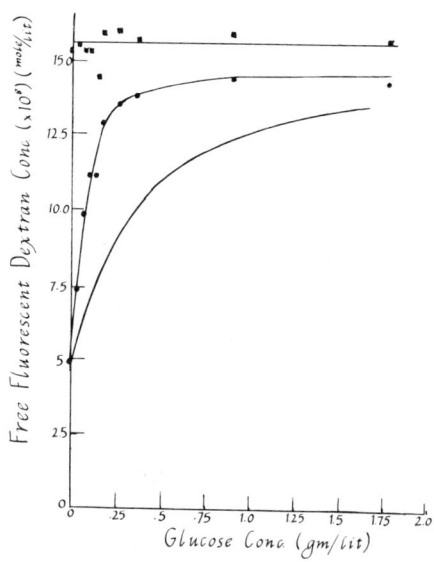

Fig. 3. Displacement of FITC-dextran from Con-A bound to sepharose by increasing concentrations of glucose. (●) Free FITC-dextran concentration after glucose additions. (———) Calculated response based on simple competition between glucose and dextran for Con-A sites. (■) Free FITC-dextran concentration after the methyl-mannoside addition.

shown in eqs. (1) and (2). Actually, Con-A has two binding sites and FITC-dextran may have up to 20 available binding sites [7]. Thus the binding equilibrium may be much more complicated than shown in eqs. (1) and (2). More extensive modeling of this system is underway. Fortunately for this application, the measured calibration curve is steeper at low concentrations of glucose than the simple theoretical model, resulting in better characteristics for sensitivity to glucose in this range.

The response and sensitivity of the sensor can be altered by manipulating the parameters that determine the characteristics of the system, namely, the specific binding agent and its concentration, the assay ligand and its concentration, and the geometry of the detector unit. In these model studies we evaluated the effects of varying the concentration of the specific binding agent, Con-A, and the concentration of the assay ligand, FITC-dextran.

Of course, when the FITC-dextran concentration is increased, the intensity of emission increases as well, but this is at the expense of some loss in relative sensitivity. In Figure 4 the intensity is expressed on a relative basis to emphasize the shape of the curves, and it is seen that at lower concentrations of FITC-dextran, the response characteristics are better from an analytical viewpoint. On the other hand, changing the Con-A concentration does not appreciably result in major alterations in the sensitivity of the method (Fig. 5).

These model experiments have illustrated the behavior expected of affinity sensors. Even though at this writing the complete sensor has not been fabricated, the potential applicability of this approach has been demonstrated. There are

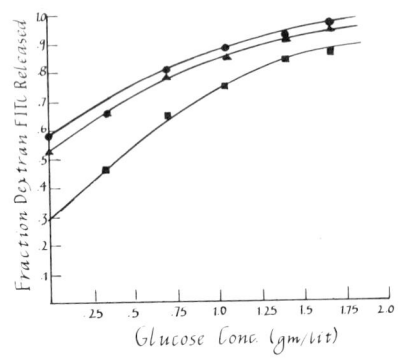

Fig. 4. Effect of FITC-dextran concentration on the sensitivity of the affinity sensor method for glucose measurement. Dextran-FITC concentration ($\times 10^8 M$): (●) 40; (▲) 30; (■) 15.

several advantages of affinity sensors over other devices in current use and development. First, the method is based primarily on equilibrium phenomena, whereas other sensors, such as those based on glucose oxidase, are very sensitive to artifacts that may affect enzyme activity. Also, because the response of the affinity sensor is based on a competitive equilibrium, the rate of glucose transport across the dialysis membrane is not critical to the final response. The response of enzyme electrodes are directly dependent on diffusion rates across the membrane; thus fouling of the membrane can have a major influence on the response of these types of electrodes.

The response time of affinity sensors depends on the transport rate of the metabolite across the membrane and the kinetics of the binding reactions. Data on the kinetics of binding on Con-A [7] show that the rate constant for the reverse reaction (the slow step) is on the order of 4 sec^{-1}, corresponding to a time con-

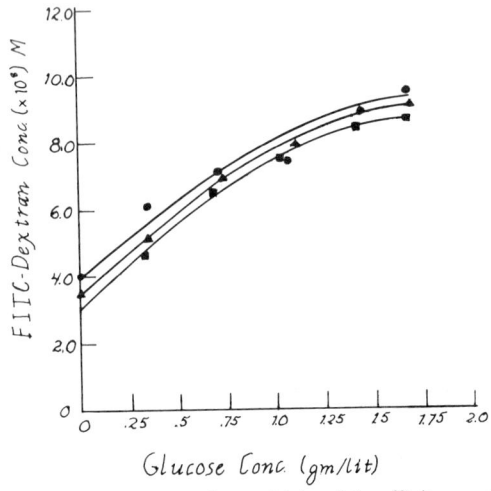

Fig. 5. Effect of Con-A concentration on the sensitivity of the affinity sensor method for glucose measurement. Con-A concentration ($\times 10^5 M$): (●) 1.3; (▲) 2.0; (■) 2.6.

stant on the order of less than 1 sec. The limiting factor in response rate of affinity sensors would appear to be the ratio of membrane transport to sensor volume given by the ratio PA/V, where P is membrane permeability in cm/sec; A is membrane area in cm^2, and V is the sensor volume in cm^3.

Affinity sensors have the potential to be developed further so as to be sensitive to a number of metabolites simultaneously. This can be implemented by including two or more binding agents in the same sensor with their respective appropriate assay ligands. The major consideration is that the assay ligands must be selected with sufficiently different optical characteristics so that they can quantified in the same solution.

A potential limitation to affinity sensors is the interference of other metabolites that, and might, react with the binding agent to displace the assay ligand. Table I shows the affinity constants for a number of sugars with Con-A. If these sugars are present in equivalent concentrations to glucose, they also will displace dextran and give erroneous results.

SUMMARY

Specific metabolite sensors may be constructed for species when a specific binding agent and optically detectable competing assay ligands are available.

The authors wish to thank William Armstrong, Michael Powers, W. Kim, and Arun Hejmadi for their participation in this project. This work was partially supported by the National Institutes of Health Research Grant No. GM15152 and the Diabetes Research and Training Center Grant No. AM20572.

References

[1] B. F. Erlanger, *Pharmacol. Rev., 25,* 271 (1973).
[2] L. L. So and I. J. Goldstein, *J. Immunol., 99,* 158 (1967).
[3] L. L. So and I. J. Goldstein, *Biochim. Biophys. Acta, 165,* 398 (1968).
[4] R. D. Gray and R. H. Glew, *J. Biol. Chem., 218,* 7547 (1973).
[5] E. H. Donnelly and I. J. Goldstein, *Biochem. J., 118,* 679 (1970).
[6] S. Colowick and N. Kaplan, in *Affinity Techniques, Methods in Enzymology, XXXIV,* W. B. Jakoby and M. Wilchek, Eds. (Academic, New York, 1974).
[7] I. J. Goldstein, personal communication.

Cultivation of Microorganisms with a DO-Stat and a Silicone Tubing Sensor

TAKESHI KOBAYASHI, TAKUO YANO, HIRONORI MORI, and SHOICHI SHIMIZU

Department of Food Science and Technology, Faculty of Agriculture, Nagoya University, Nagoya 464, Japan

INTRODUCTION

In the cultivation of methanol-utilizing microorganisms, a high concentration of methanol in the culture medium not only inhibits growth of micoorganisms, but also prolongs the lag phase after inoculation. On the other hand, a low concentration of methanol in the culture medium is favorable for growth of microorganisms and diminishes the amount of methanol evaporated from the medium with aeration. In this case, however, the methanol in the culture medium is consumed rapidly and the final concentration of microorganisms is low.

For this reason, it is desirable that methanol be fed continuously or intermittently to a batch culture. For fed-batch culture on methanol, a feedback control system has been reported based on the measurement of the methanol concentration in the exhaust gas from the fermentor with a flame ionization detector [1]. In this system, effects of operational conditions such as air-flow rate and impeller speed on methanol concentration in the exhaust gas from the fermentor cannot be ignored.

In a fed-batch culture, the dissolved oxygen tension (DO) in the culture medium may become a growth-limiting factor as the microorganisms grow. Thus, the development of a sensor to measure the methanol concentration in the culture medium and an apparatus to maintain constant DO levels (DO-stat) throughout the cultivation is needed in fed-batch culture. DO measurements can also be used as a control indicator for feeding methanol, since the DO abruptly increases when the methanol in the medium is completely depleted. A similar situation applies to the cultivation of ethanol-assimilating microorganisms.

In the present work, the DO-stat system was examined for its application in systems designed to produce high cell densities, and as a carbon source feeding indicator [2]. A silicone tubing sensor [3] was then developed for measuring the carbon substrate concentration in the culture medium. The characteristics of this sensor were examined. Finally, it was demonstrated that the combined use of these two systems was effective in maintaining the methanol concentration at which the microorganism showed the maximum specific growth rate while obtaining a high density of biomass.

MATERIALS AND METHODS

Microorganisms

The organisms used in this study were *Protaminobacter ruber* (a methanol-assimilating bacterium), which was isolated from the activated sludge of a petrochemical company in Yokkaichi [4], and *Candida brassicae* (an ethanol-assimilating yeast), which was donated by Dr. M. Kagami and Dr. Y. Amano [5].

Cultivation

The compositions of the basal media are shown in Table I. For the cultivation of *P. ruber*, the pH was automatically controlled at 6.5 with 33% ammonium hydroxide. The dissolved oxygen tension was automatically controlled at 2 ∼

TABLE I
Compositions of Mediums for *C. brassicae* and *P. ruber*

C. brassicae		*P. ruber*	
Component	Concentration	Component	Concentration
KH_2PO_4	12 (g/l)	KH_2PO_4	8 (g/l)
K_2HPO_4	1	Na_2HPO_4	4.8
Na_2HPO_4	0.4	$(NH_4)_2SO_4$	2
$(NH_4)_2SO_4$	2	$MgSO_4 \cdot 7H_2O$	1
NH_4Cl	0.5	$CaCl_2 \cdot 2H_2O$	50 (mg/l)
Yeast extract (Oriental)	6	$FeSO_4 \cdot 7H_2O$	10
		$CoSO_4 \cdot 7H_2O$	10
$MgSO_4 \cdot 7H_2O$	2.4	$MnSO_4 \cdot nH_2O$	5
$FeSO_4 \cdot 7H_2O$	40 (mg/l)	$Na_2MoO_4 \cdot 2H_2O$	4
$CaCl_2 \cdot 2H_2O$	40	$ZnSO_4 \cdot 7H_2O$	2
$MnSO_4 \cdot nH_2O$	10	$CuCl_2 \cdot 2H_2O$	2
$AlCl_3 \cdot 6H_2O$	10	H_3BO_3	2
$CoCl_2 \cdot 6H_2O$	4	$AlCl_3 \cdot 6H_2O$	2
$ZnSO_4 \cdot 7H_2O$	2	methanol	variable
$Na_2MoO_4 \cdot 2H_2O$	2	pH	6.5
$CuCl_2 \cdot 2H_2O$	1		
H_3BO_3	0.5		
ethanol	variable		
pH	5.5		

3 ppm with the DO-stat to be discussed below. The temperature of the culture was controlled at 30°C.

The seed culture was prepared as follows: A loop of the microorganism was used to inoculate 100 ml of the basal medium (2 vol % methanol) in a shake flask (500 ml), which then was cultured on a reciprocal shaker at 30°C for two days. Centrifuged cells of the seed culture were used to inoculate a jar-fermentor (working volume 1 liter, Iwashiya Co. Ltd., type MB).

For the cultivation of *C. brassicae,* the pH was controlled at 5.5. The temperature was controlled at 36°C. The seed culture was shaken for one day. Other conditions were the same as in the case of *P. ruber.*

Analytical Method

Methanol and ethanol concentrations were determined using a gas chromatograph equipped with a flame ionization detector and a 2-m stainless-steel column containing Flusin T (PEG 1000, 20%, 60–80 mesh) and Chromosorb 101 (80–100 mesh), respectively.

The growth of the microorganisms was determined by measuring the optical density of the culture broth at 570 nm with a Shimadzu Spectronic 20 photometer. The cell concentration was also determined by drying and weighing the centrifuged cells. The ratio of optical density to dry cell weight (g/liter) was 2.05 for *P. ruber* and 1.59 for *C. brassicae.*

DO was measured with an oxygen sensor (Beckman, Fieldlab oxygen analyzer). All the chemicals used were of reagent grade. Water was deionized after distillation.

DO-Stat

The instruments were made to order by Sanko Electronics Co. Ltd., Nagoya, and were designated as DO-stats.

The DO-stat type A controls the DO level by switching between the low and the high rotation speeds of the impeller. The low- and high-speed settings were adjustable. In the early stage of cultivation, the lower rotation speed of the impeller was used. As the DO decreased as the microorganism grew, the rotation speed was switched from the low to the high speed when DO became lower than the preset level. When the DO became higher than the preset level, the rotation speed was switched over from the high speed to the low speed. In this way, the DO was maintained within a suitable range. When the DO level could not be maintained at the upper rotation speed, the low- and high-speed settings on the controller were increased manually.

The control of the carbon substrate addition was achieved by turning on a pump when the DO decreased below a preset level on the DO-stat. The amount of each addition could be varied by adjusting the pump flow rate or the duration of the pumping interval for each addition. The control apparatus was also provided with a sensitivity adjustment which delays the operation of the feed pump between addition periods for a predetermined interval, usually 5 sec. Incorrect operations due to electrical noise and attachment of air bubbles to the tip of the oxygen sensor can be avoided in this manner.

The DO-stat type B controls the DO level by choosing the appropriate opening of the gas-flow rate control valve and the rotation speed of the impeller. Opening and closing of the gas-flow rate control valve were automatically performed using a servomotor. When pure oxygen was mixed with air, the direction of the control valve adjustment was reversed to that used when air alone was used to aerate the fermentor. The impeller speed was controlled proportionally between preset lower and upper limits. By a combination of these two means, used simultaneously or separately, the DO was maintained at a constant level for a fairly long cultivation time without manual interventions.

Silicone Tubing Sensor

As shown in Figure 1, silicone tubing ($\phi = 1$ mm, thickness 0.25 mm, Fuji Kobunshi Co.) was fitted to two hypodermic needles. Air was introduced at one end of the tubing by a peristaltic pump. The gas at the other end, which contained methanol or ethanol, was passed through a stainless-steel tube ($\phi = 2$ mm) to the gas sampler (1.9 ml sample volume) attached to a gas chromatograph (Hitachi-Seisakusho Co., model 163). The methanol or ethanol concentration in

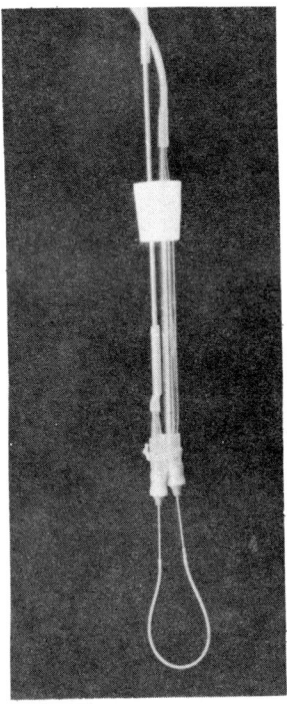

Fig. 1. Photograph of silicone tubing sensor. Silicone tubing ($\phi = 1$ mm, thickness 0.25 mm, length 10 cm) was connected with a hypodermic needle, and then stainless-steel tubing. (Reprinted from Ref. 3 with permission of the Publisher.)

the gas sampler was measured intermittently with the gas chromatograph. The peak height of the chromatogram was used as an indicator of the methanol or ethanol concentration in the culture medium. The analytical time of the gas chromatography was about 3 min. A gas chromatogram was run every 5 min. When the peak heights decreased, the speed of the feed pump of the carbon source was increased gradually, and vice versa.

RESULTS AND DISCUSSION

Figure 2 shows a schematic diagram for the different operating modes for both of the DO controllers. In illustration 1, the gas-flow rate is fixed and the DO level is controlled by switching between the lower and the upper rotation speeds (on-off control). In the procedures used in illustration 2, the rotation speed controller operates at first when DO becomes lower than the preset level. The gas-flow rate controller is not manipulated when DO is controllable between the preset lower and upper rotation speed limits. Otherwise, the gas-flow rate controller operates after the preset time, which was adjusted by the low-level DO timer, has expired, and the rotation speed of the impeller is gradually reduced to a standard rotation speed between the lower and upper speed limits. The rate of change of the rotation speed was adjusted by the return-time knob. In illustration 3, the rotation speed controller and the gas-flow-rate controller operate simultaneously. In illustration 4, rotation speed is fixed and DO level is controlled only by the gas-flow-rate controller. Methods 2–4 were implemented using DO-stat type B.

Typical examples for methods 1 and 2 are shown in Figures 3 and 4 for the cultivation of *P. ruber*, respectively. In both cases, the DO-stat worked well and DO level was controlled within a suitable range. In method 1, however, manual modifications of pure oxygen and air flow rates were frequent, especially in the

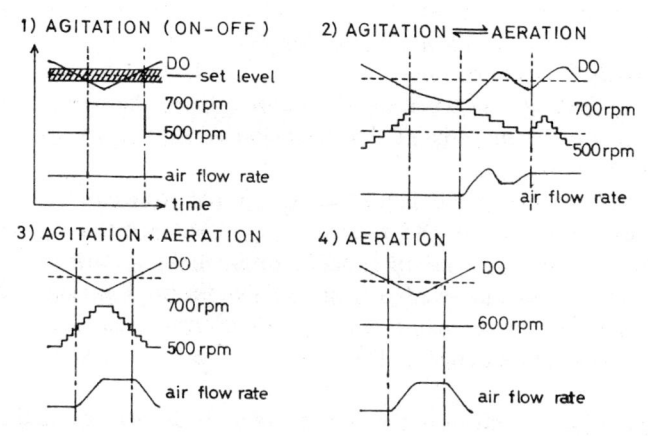

Fig. 2. Schematic diagram for the order of operations of rotation speed controller and gas-flow-rate controller.

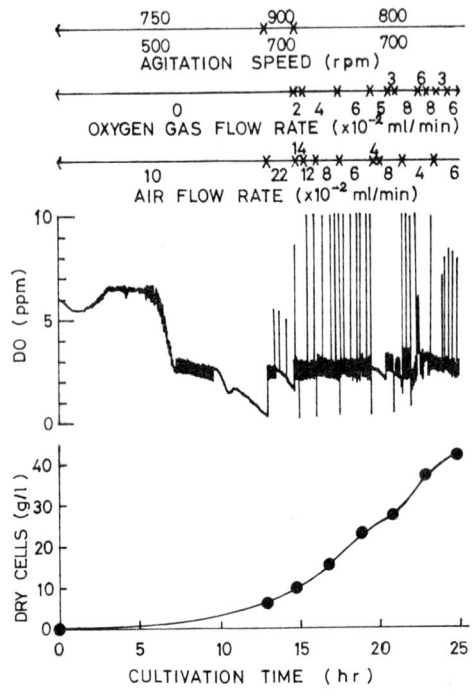

Fig. 3. Typical example of the cultivation of *P. ruber* with the DO-stat for method 1. Upper and lower figures in the agitation speed show upper and lower rotation speed of the impeller, respectively.

late stage of cultivation. In methods 2–4, manual modifications of the preset values were seldom required (examples for methods 3 and 4 are not shown).

As shown in Figures 3 and 4, whenever the carbon source in the culture medium was about to be completely used up and therefore became a growth-limiting factor, the DO increased abruptly and then decreased to the former level after feeding additional carbon source. In this way, the DO can be used as a control indicator for feeding the carbon source. Figure 5 shows the result of cultivation of *C. brassicae* on feeding ethanol. The final biomass reached 138 g dry cells/liter after 15 hr of cultivation.

In this cultivation method, lack of ethanol in the broth can be detected as a signal of abrupt increase of DO, but the ethanol concentration cannot be measured. Only the maximum ethanol concentration in the culture medium can be set by varying the amounts of ethanol pumped into the broth during each addition period. It is not possible to maintain the ethanol concentration at a constant level using these control procedures. Therefore a silicone tubing sensor was developed.

Figure 6 shows the effects of the flow rate of air to fermentor and the agitation speed on the peak heights of gas chromatograms for a solution of constant methanol concentration obtained by using 5 cm tubing. In the system of Reuss

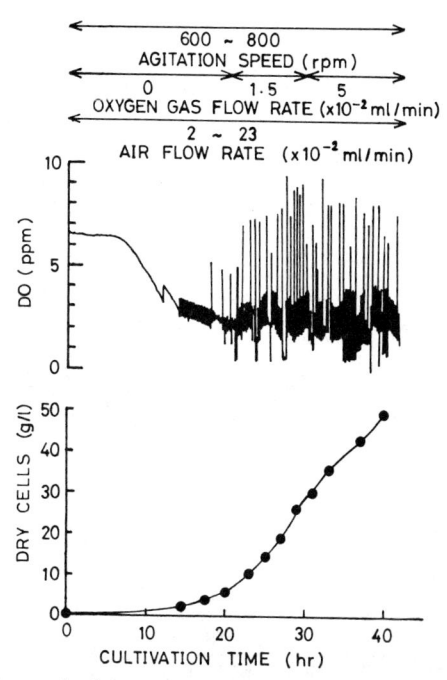

Fig. 4. Typical example of the cultivation of *P. ruber* with the DO-stat for method 2.

et al. [1] the methanol concentration in the exhaust gas from a fermentor was influenced by both of these factors. In the case of the silicone tubing method, however, both factors had little influence on the methanol concentration in the gas passed through the silicone tubing.

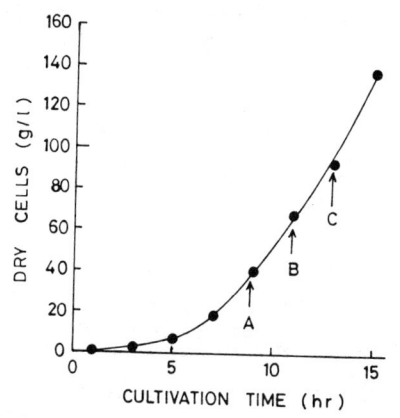

Fig. 5. Cultivation result of *C. brassicae* with the DO-stat on feeding ethanol. Additions of metal ions and yeast extract are shown at the arrows. (A) 3.5 mg $CuCl_2 \cdot 2H_2O$; 14 mg $CoCl_2 \cdot 6H_2O$; and 3 g yeast extract. (B) 7 mg $ZnSO_4 \cdot 7H_2O$; 7 mg $Na_2MoO_4 \cdot 2H_2O$; and 2 g KH_2PO_4. (C) 70 mg $FeSO_4 \cdot 7H_2O$; 17.5 mg $MnSO_4 \cdot nH_2O$; 14 mg $CoCl_2 \cdot 6H_2O$; 3.5 mg $CuCl_2 \cdot 2H_2O$; 1 g $MgSO_4 \cdot 7H_2O$; and 3 g yeast extract.

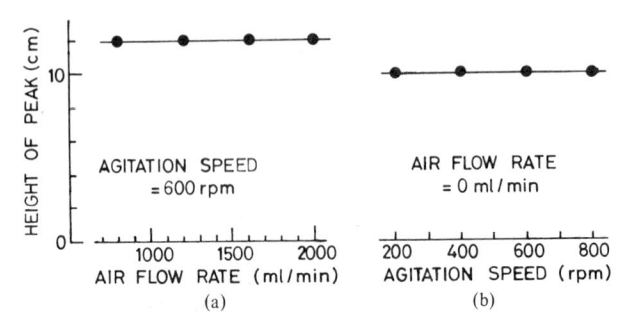

Fig. 6. Effects of air-flow rate to fermentor (a) and agitation speed (b) on peak height of chromatogram for methanol at 30°C and 5 cm tubing length. Volumetric air-flow rates inside silicone tubing for (a) and (b) were 1.6 and 5.0 ml/min, respectively. (Reprinted from Ref. 3 with permission from the Publisher.)

The relation between tubing length and peak height for methanol is shown in Figure 7. Tubing lengths over 50 cm seemed to be unnecessary. A higher peak may be obtained in a thinner tubing with the same tubing length. The dotted lines in Figure 7 were derived using the assumption that mass transfer resistance for methanol exists in only the silicon tubing wall [3].

The response time of the silicone tubing sensor was measured and is shown in Figure 8 for methanol. The lag time for the response was about 1.5 to 2 min, and 5 min were required for 63.2% ($=1-e^{-1}$) response. The unsteady state for a step response was analyzed as a cascade of delay and first-order systems [3],

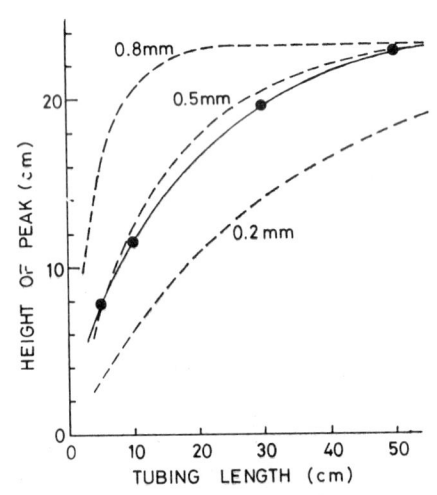

Fig. 7. Effect of tubing length of sensor on peak height of chromatogram for methanol at 30°C, outer radius of tubing 1 mm and volumetric air-flow rate inside tubing 3.1 ml/min. (---) These were derived for three different inner radii of tubing from the assumption that mass transfer resistance for methanol exists in only the silicone tubing wall. (Reprinted from Ref. 3 with permission of the Publisher.)

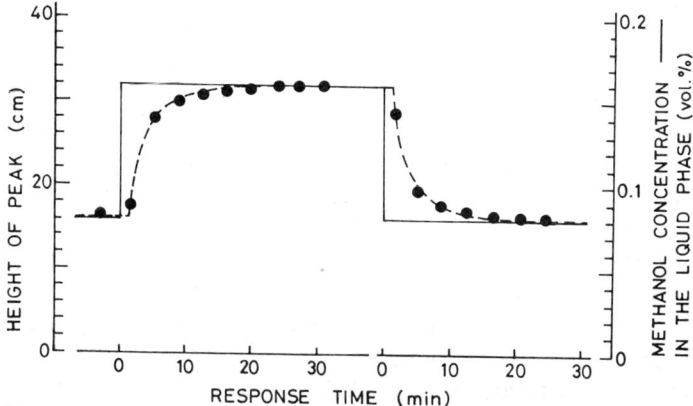

Fig. 8. Step response of the silicone tubing sensor at 30°C, tubing length 30 cm, and volumetric air-flow rate inside tubing 3.1 ml/min. (---) Evaluated from unsteady state analysis for a cascade of delay and first-order systems. (Reprinted from Ref. 3 with permission of the Publisher.)

and the dotted line in Figure 8 shows the result. In the case of ethanol, 33 min were required for a 63.2% response.

This silicone tubing sensor had splendid durability for autoclaving; after 15 autoclaving cycles the peak height was approximately the same as that observed initially.

Control of the methanol concentration in the culture medium with the silicone tubing sensor was performed for the cultivation of *P. ruber*. In Figure 9, the methanol concentration was controlled at 1.0 vol %. Methanol feeding was started after 15.8 hr, after which the methanol concentration was successfully controlled. This tubing sensor method may be suitable for cultivation of a microorganism showing inhibition kinetics for growth. The control of the ethanol concentration in the culture medium with the silicone tubing sensor was rather difficult because the response rate was slow.

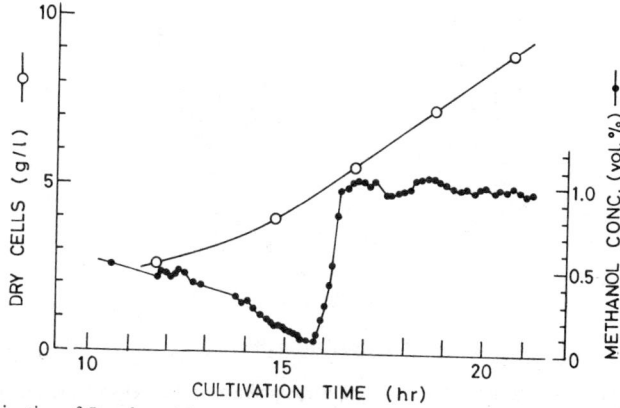

Fig. 9. Cultivation of *P. ruber* with a constant methanol concentration (1.0 vol %) controlled with the silicone tubing sensor at 30°C. (Reprinted from Ref. 3 with permission of the Publisher.)

The results obtained by the above experiments were combined. With the silicone tubing sensor, the methanol concentration in the culture medium was measured during the cultivation, and the methanol concentration could be controlled within a narrow range.

The silicone tubing sensor was used to control the feeding of methanol. An auxiliary feeding system using the DO-stat was added as a backup for when the response of the silicone tubing sensor became too slow with respect to the process dynamics, or when a methanol limitation was detected. The DO-stat was also used to control the addition of pure oxygen to air for aeration of the fermentor so that it was possible to cultivate high concentrations of microorganisms without toxicity due to high oxygen tension. The cultivation results with the combined use of these two systems are shown in Figure 10. These combined systems were excellent and the final biomass reached 78 g dry cells/liter after 40 hr cultivation. DO level was maintained at 2 ~ 3 ppm throughout the cultivation.

These control systems have some areas in need of improvement. One is the improvement of the response rate of the silicone tubing sensor, which depends upon the thickness of silicone tubing. A thinner silicone tubing is required for a more rapid response. If a sensor with a more rapid response rate is found, then the cultivation of an ethanol-assimilating microorganism would be possible with the combined use of these two systems. Another aspect requiring further development is the necessity of a fully automatic system. In the present experiments, gas collected in the gas sampler was fed manually to the gas chromatograph. The peak height was also read manually. The speed of the methanol feeding pump was also controlled manually. A fully automatic system, which consists of an automatic gas sampler, an automatic apparatus for the measurement of methanol peak height or area, and a controller for the speed of the methanol feeding pump, will save labor costs considerably.

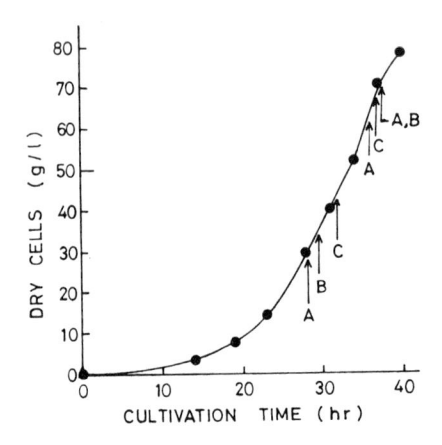

Fig. 10. Cultivation result of *P. ruber* with combined use of the DO-stat and the silicone tubing sensor on feeding methanol. Additions of metal ions are shown at the arrows. (A) 5 mg MnSO$_4$·nH$_2$O; (B) 10 mg FeSO$_4$·7H$_2$O and 2 mg ZnSO$_4$·7H$_2$O; (C) 400 mg MgSO$_4$·7H$_2$O.

References

[1] M. Reuss, J. Gnieser, H. G. Reng, and F. Wagner, *Eur. J. Appl. Microbiol., 1,* 295 (1975).
[2] T. Yano, T. Kobayashi, and S. Shimizu, *J. Ferment. Technol., 56,* 416 (1978).
[3] T. Yano, T. Kobayashi, and S. Shimizu, *J. Ferment. Technol., 56,* 421 (1978).
[4] S. Shimizu, T. Kobayashi, K. Sato, K. Ohmiya, M. Mori, T. Nishimura, and M. Sasaki, *J. Agri. Chem. Soc. Jpn., 52* (10), 477 (1978).
[5] Y. Amano, O. Yoshida, and M. Kagami, *J. Ferment. Technol., 53,* 264 (1975).

Redox Potential as a State Variable in Fermentation Systems

L. KJAERGAARD and B. B. JOERGENSEN

Department of Applied Biochemistry, The Technical University of Denmark, Block 223, DK-2800 Lyngby, Denmark

INTRODUCTION

Several investigations have revealed that the redox potential, E_h, yields more information about the oxidative status in aerobic or partially aerobic microbial cultures than the concentration of dissolved oxygen, DO [1-4]. In a typical batch fermentation the redox potential E_h, drops rather fast from about 225 mV to 50-100 mV during the short time interval immediately after the time when DO becomes zero. After this time, the E_h may drop further if there is a sufficient substrate concentration. Wimpenny [4] and Kjaergaard [2] have shown that a number of metabolic activities show a great dependency on the redox potential under these circumstances.

Since the redox potential is derived from the combined effects of the oxidative status of the culture and the culture DO, it would be difficult to maintain a constant redox potential by regulating the dissolved oxygen concentration alone. This is especially true at low DO levels (0-1%) where the sensitivity and accuracy of DO electrodes are unsatisfactory for control applications.

Therefore, in order to maintain the redox potential at values corresponding to DO values from 0-1% saturation, it can in some cases be necessary to use other ways of regulation.

In this paper it will be shown that, the redox potential can be partially regulated by letting the deviation from a preset value of E_h direct the addition of glucose.

MATERIALS AND METHODS

Microorganism

The facultative aerobic *Bacillus licheniformis* (NCIB 8061) was used and maintained on nutrient agar slants by monthly subculture.

Media

A carbon-limited synthetic medium was used in these experiments as described by Evans et al. [5]. The initial medium contained 1 or 2% (w/v) glucose and the feed medium contained 40% (w/v) glucose. The concentrations of salts in both media corresponded to that described by Evans et al. [5] for a 5% (w/v) glucose medium.

Cultural Conditions

The organisms were grown in a 3 liter Biotec FL 103 laboratory fermentor with a working volume of 2 liter. The culture was maintained automatically at pH 7.5 by addition of either $4N$ NaOH or $4N$ H_2SO_4. The temperature was maintained at 37°C, and the air-flow rate adjusted to 0.5 v/v/m using a flowmeter. The redox potential and the concentration of oxygen were measured as previously described [2].

The fermentor containing 2 liter substrate with 1 or 2% (w/v) glucose was inoculated with 20–50 ml of a growing culture of *B. licheniformis,* grown in a shake flask in a medium identical to that in the fermentor. After growth overnight (16 hr), the glucose in the initial medium was depleted; at this time (defined as $t = 0$) glucose was added in order to maintain the redox potential at a constant value.

Equipment for Regulating the Redox Potential

Because of the introductory status of these experiments, the equipment used for regulating the redox potential was extremely simple. From previous experiments [6] it was known that as soon as the glucose is exhausted from the medium, the oxygen consumption decreases and, consequently, the redox potential and the concentration of dissolved oxygen increase. Therefore, the millivoltmeter, on which the redox potential was measured, was provided with a controller that closed a relay, activating a pump, so that glucose could be added to the fermentor automatically.

The equipment that has been described previously [1] effects a simple on–off regulation of the glucose addition. Since the added glucose is used by the microorganisms before a new pulse is added, the growth is regulated too.

Analytical Procedures

The cell concentration, the acetic acid concentration, and the glucose concentration were measured as previously described [7].

The outlet air from the fermentor was sent through an apparatus for measuring the content of CO_2 (% CO_2) and O_2 (% O_2) (URAS 2T and MAGNOS 2T, Hartmann and Braun, West Germany). The pH, DO, E_h, % CO_2, % O_2 used, and the total amount of added glucose, were continuously recorded.

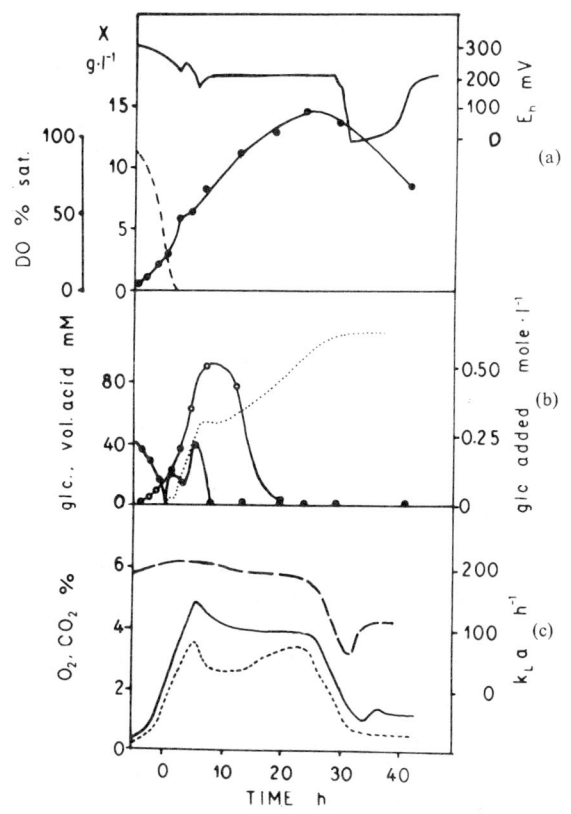

Fig. 1. Growth of *B. licheniformis* with the redox potential E_h maintained at 217 mV by addition of glucose. Time zero is the time when the E_h regulation by glucose addition is started. (a) (—●—) Concentration of cell mass, X; (——) E_h; (- - -) concentration of dissolved oxygen (DO). (b) (—●—) Concentration of glucose; (—○—) concentration of volatile acids; (·····) amount of glucose added. (c) (- -) $k_L a$; (——) percent oxygen used in the outlet air; (·····) percent carbon dioxide in the outlet air.

THEORY

As described above the state with a constant redox potential will not be obtained before the cells have passed through at least a part of the exponential growth phase. Since the addition of glucose is regulated by means of a parameter indicating the level of the DO in the medium, it is obvious that the growth cannot be exponential when the regulator is functioning. When the DO is constant or very low, then the oxygen supply, OS, is also constant as indicated in eq. (1):

$$OS = k_L a (DO^* - DO) \qquad (1)$$

where $k_L a$ (hr^{-1}) is the volumetric oxygen transfer coefficient and DO* is the saturation concentration of dissolved oxygen.

The oxygen consumption rate Q_{O_2} (mmol/liter hr) is normally considered

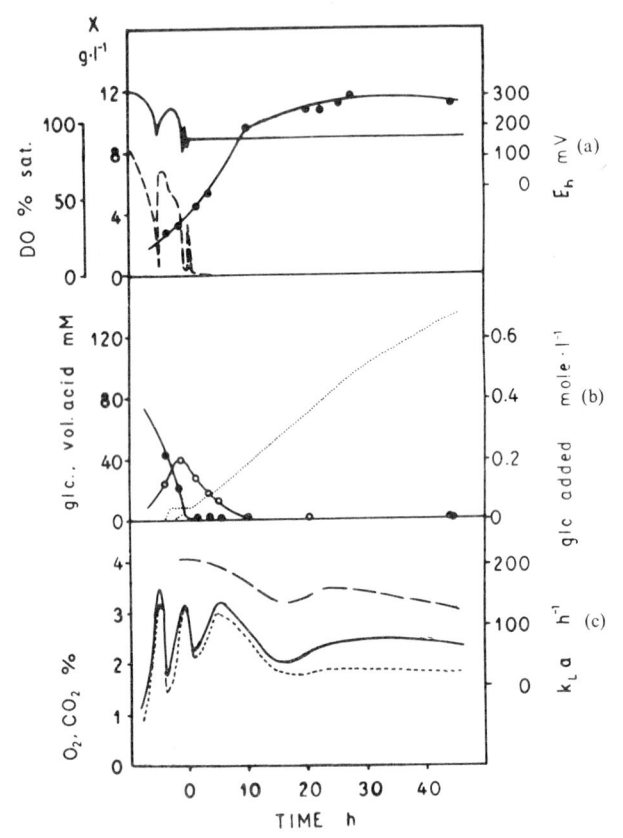

Fig. 2. Growth of *B. licheniformis* with the redox potential E_h maintained at 167 mV by addition of glucose. Time zero is the time when the E_h regulation by glucose addition is started. Symbols same as in Figure 1.

to depend on the cell concentration X (g/liter) and the specific growth rate μ (hr^{-1}):

$$Q_{O_2} = aX + b\frac{dX}{dt} = (a + b\mu)X \qquad (2)$$

where a is the maintenance coefficient with respect to oxygen and b is the reciprocal yield constant with respect to oxygen. Both are considered as constants.

Since $Q_{O_2} = $ OS is constant for DO less than about 1%, it is obvious that dX/dt and μ must diminish for increasing X. Consequently, the fraction of the oxygen used for maintenance metabolism will increase as growth increases.

The added glucose will be used either for synthesis of cell constituents or for energy yielding metabolism. Assuming a simple metabolism where the glucose not used for cell formation is completely oxidized to CO_2 with respiratory quotient (RQ) = 1.0, the consumption rate of glucose will be

$$Q_{glc} = \tfrac{1}{6} aX + (1/Y_{glc})\mu X \qquad (3)$$

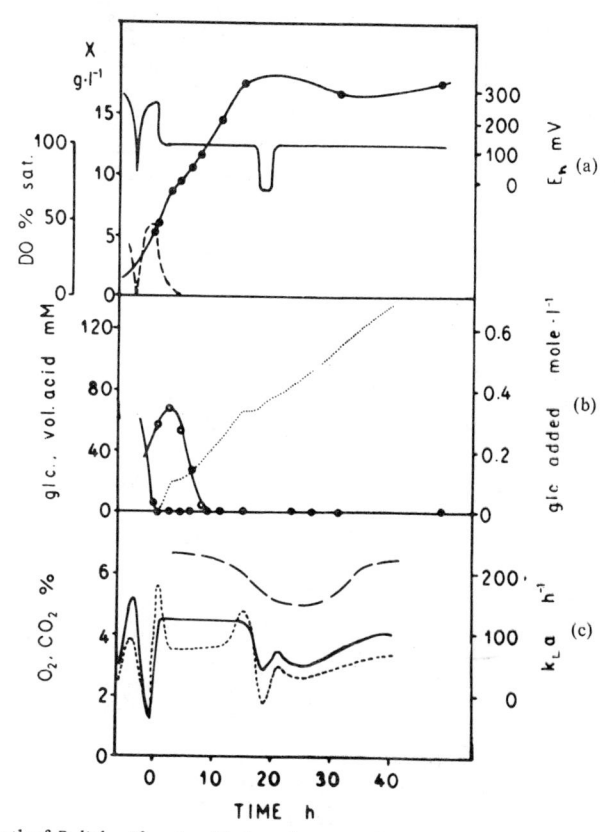

Fig. 3. Growth of *B. licheniformis* with the redox potential E_h maintained at 117 mV by addition of glucose. Time zero is the time when the E_h regulation by glucose addition is started. Symbols same as in Figure 1.

where Y_{glc} is the true yield constant with respect to glucose. The quantity $1/Y_{glc}$ is composed of two parts: the carbon that is incorporated directly into the mass of the cell and the carbon that is oxidized to provide energy for the growth processes. By letting g_x be the mass fraction of the cell biomass that is carbon, and g_s be the mass fraction of the substrate that is carbon, then the substrate that is converted directly to cellular carbon constituents is given by $(g_x/g_s)\mu x$. Since this quantity corresponds directly to the oxygen consumed for growth-related processes [second term of eq. (2)], then

$$\tfrac{1}{6} b\mu X = (1/Y_{glc} - g_x/g_s)\mu X$$

or

$$(1/Y_{glc})\mu X = \tfrac{1}{6} b\mu X + (g_x/g_s)\mu X \tag{4}$$

and therefore by inserting $Q_{O_2} = OS$ and eqs. (2) and (4) into eq. (3),

$$Q_{glc} = \tfrac{1}{6} OS + (g_x/g_s)\mu X \tag{5}$$

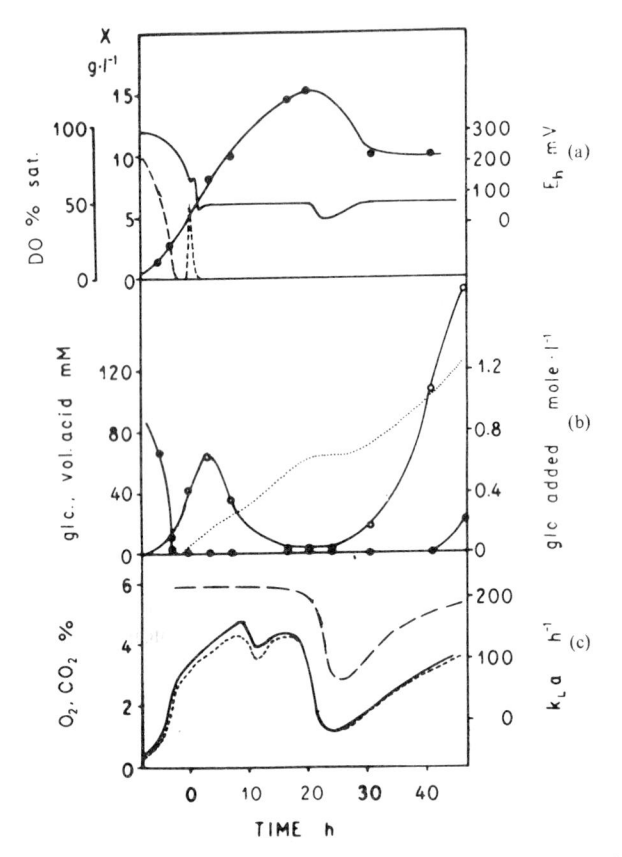

Fig. 4. Growth of *B. licheniformis* with the redox potential E_h maintained at 67 mV by addition of glucose. Time zero is the time when the E_h regulation by glucose addition is started. Symbols same as in Figure 1.

As $\mu X = dX/dt$ decreases for increasing X, Q_{glc} will diminish during growth with a constant OS.

From eq. (2) it can be seen that the maximum amount of cells obtainable will be $X_m = OS/a$, so that the lowest value of Q_{glc} will be

$$Q_{glc_{min}} = \frac{1}{6} OS \qquad (6)$$

From this it is obvious that in a state where the oxygen supply is kept constant, there will be a decreasing rate of glucose consumption but always with $Q_{glc} \geqq \frac{1}{6}OS$.

RESULTS AND DISCUSSION

In a series of experiments it has been demonstrated that the redox potential can be kept at the prechosen value for a period. In Figures 1–5 the progress of the bacterial growth, the glucose consumption, the redox potential, and the

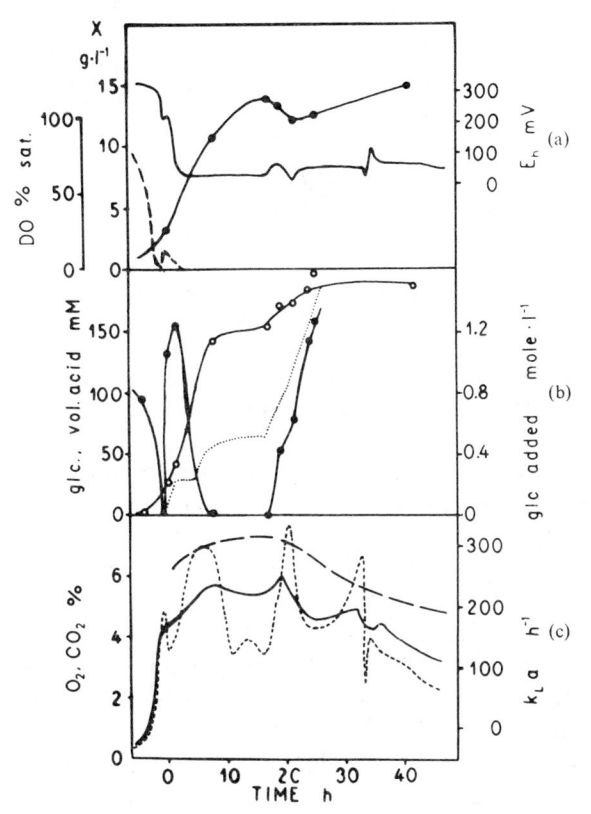

Fig. 5. Growth of *B. licheniformis* with the redox potential E_h maintained at 17 mV by addition of glucose. Time zero is the time when the E_h regulation by glucose addition is started. Symbols same as in Figure 1.

concentration of volatile acids are shown for experiments in which the redox potentials were chosen to be 217, 167, 117, 67, and 17 mV.

It is obvious that only for $E_h = 167$ mV is it possible to keep the redox potential constant for more than 24 hr. Furthermore, it is obvious that after the regulation fails there is (for $E_h = 217, 117,$ and 67 mV) a period during which the redox potential tends to be constant again, but with a somewhat lower glucose consumption rate.

Although the E_h was constant in the experiments with $E_h = 217$ and 167 mV, some pronounced fluctuations in the DO value were seen, as indicated in Figure 6. These fluctuations may be explained by the on/off nature of the regulator. If a proportional regulation was used instead, these fluctuations might have been less pronounced.

In Table I is shown how much DO fluctuates as well as the average amount of oxygen used during the period with constant E_h. Furthermore, the average OS values for the different experiments are calculated in relation to the maximum oxygen supply, which would be obtained if there were no fluctuation in

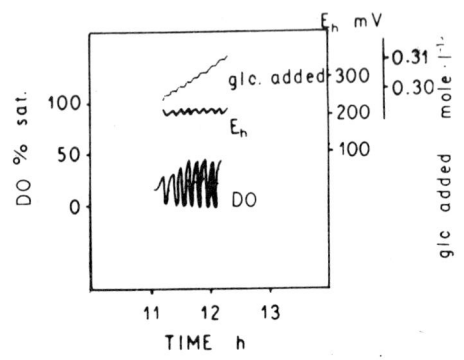

Fig. 6. Fluctuations in the concentration of dissolved oxygen, DO, in a culture of B. licheniformis, where the redox potential is regulated by glucose addition.

DO. It can be seen that for decreasing E_h values the consumption of oxygen increases, which was expected since OS/OS_m increases.

The reason for the drop in E_h after a certain time may presumably be found in the $k_L a$ value. In Figures 1–5 it can be seen that during the last part of the period with constant E_h, the % O_2 used in the outlet air decreases, which means that the O_2 consumption rate decreases. Since DO = 0 and the DO* is constant or increases slightly when OS decreases, the only reason for the decrease of OS can be a decreasing $k_L a$ value. The $k_L a$ value calculated from eq. (1) is shown in the c figures in Figures 1–5. Concomitantly with the decrease in Q_{O_2} there will be a decrease in the specific oxygen consumption rates q_{O_2} since the cell

TABLE I

Fluctuations of the Concentration of Dissolved Oxygen (DO-Fluc.), the *In Situ* Observed Oxygen Consumption Rate (Q_{O_2}), and the Average Oxygen Supply (OS) Relative to the Maximum Oxygen Supply (OS_m) for Experiments in which *B. licheniformis* Was Grown Batchwise under a Constant Redox Potential E_h

E_h mV	DO-fluc. % of sat.	Q_{O_2} $\frac{mmol}{l \cdot h}$	OS/OS_m
217	0–50	3.5	0.75
167	0–20	3.5	0.9
117	0–5	4.5	0.98
67	0–1	4.5	0.99
17	~ 0	5	1

TABLE II

Observed Specific Glucose
Consumption Rates q_{glc} for
B. licheniformis Grown
Batchwise at a Constant
Redox Potential E_h

E_h mV	q_{glc} $\frac{mmol}{g \cdot h}$
217	1.1
167	1.1
117	1.6
67	2.0

concentration at this time is still slightly increasing. When the q_{O_2} becomes as low as the constant a in eq. (2), the q_{O_2} cannot be lower until some of the cells have ceased to respire. Therefore, when this level is reached the concentration of dissolved oxygen tends to be even lower and the E_h, consequently, will decrease. It is inexplainable why this has not happened in the experiment with E_h = 167 mV.

The above-mentioned decreasing glucose consumption rate as growth proceeds cannot be seen easily from the figures. This is because the specific growth rate is very low—less than 0.05 hr^{-1}—and, consequently, the amount of glucose used for growth is low. Another complicating factor is the use of the volatile acid produced earlier. It is obvious that when the cells use acetic acid as well as glucose for growth and maintenance, the glucose consumption rate will not be a good measure for the metabolic activity.

In Table II the specific glucose consumption rates during the periods in which the concentration of volatile acids is constant (not obtained for E_h = 17 mV) are listed. From Table II it is significant that for decreasing redox potentials the q_{glc} increases. This may be due to the same effects as observed by Wimpenny [4] and Kjaergaard [2]; i.e., that the energy-yielding catabolism of glucose is strongly dependent on the E_h.

The same tendency is obvious from the curves for the concentrations of volatile acids where it is significant that at $E_h \geq 67$ mV the volatile acid, produced during the exponential growth phase and during the growth that takes place during the first period of regulation, is used up rapidly. At E_h = 17 mV the volatile acid is not used again, and even more is produced. It may also be noted that at E_h = 67 mV [Figure 4(b)] the cells produce more volatile acid during the second regulation period.

CONCLUSION

It has been demonstrated that it is possible to use the redox potential as a state variable in fermentation systems. It can give information about the oxidative status in microbial cultures especially during the so-called partially aerobic phase, where there is a great difference between cultures having E_h values from 167 mV down to 17 mV, especially in their capability of producing volatile acids as a byproduct. However, presumably due to changes in some physical–chemical properties, the volumetric oxygen transfer coefficient diminishes during the redox regulated growth and after a certain period with a constant redox potential E_h drops to a rather low value. Therefore, if it is desirable to establish a redox-stat it is necessary to have some device to maintain a fixed value of k_La. A redox-stat might be of great importance for investigations of the capability of microbial cultures to produce enzymes, antibiotics, etc. which often are produced by some kind of a fed-batch method.

Nomenclature

a	maintenance coefficient with respect to oxygen
b	reciprocal yield constant with respect to oxygen
DO	concentration of dissolved oxygen
DO*	saturation concentration of dissolved oxygen
E_h	redox potential
g_s	mass fraction of the cell biomass that is carbon
g_x	mass fraction of the substrate that is carbon
k_La	volumetric oxygen transfer coefficient
OS	oxygen supply
OS_m	maximum oxygen supply
Q_{glc}	glucose consumption rate
Q_{O_2}	oxygen consumption rate
q_{glc}	specific glucose consumption rate
q_{O_2}	specific oxygen consumption rate
RQ	respiratory coefficient
X	cell concentration
X_m	maximum cell concentration
Y_{glc}	true yield constant with respect to glucose
μ	specific growth rate

The authors are most gratefully indebted to Professor O. B. Joergensen for having offered excellent working conditions and to Mrs. Berit Lassen for her skilled technical assistance.

References

[1] L. Kjaergaard, in *Advances in Biochemical Engineering*, T. K. Ghose, A. Fiechter, and N. Blakebrough, Eds. (Springer-Verlag, Berlin, 1977), Vol. 7, p. 131.
[2] L. Kjaergaard, *Eur. J. Appl. Microbiol.*, 2, 215 (1976).
[3] E. A. Andreeva, *Mikrobiologiya*, 43, 780 (1974).
[4] J. W. T. Wimpenny, *Biotechnol. Bioeng.*, 11, 623 (1969).
[5] C. G. T. Evans, D. Herbert, and D. W. Tempest, in *Methods of Microbiology*, J. R. Norris and D. W. Ribbons, Eds. (Academic, London, 1970), Vol. 2, p. 277.
[6] L. Kjaergaard, Ph.D. thesis, The Technical University of Denmark, Lyngby, Denmark, 1976.
[7] R. Larsen and L. Kjaergaard, *Eur. J. Appl. Microbiol. Biotechnol.*, 5, 177 (1978).

Indirect Fermentation Measurements as a Basis for Control

JAMES R. SWARTZ and CHARLES L. COONEY
*Department of Nutrition and Food Science, Massachusetts Institute of Technology
Cambridge, Massachusetts 02139*

INTRODUCTION

Computer control has become a valuable tool in the chemical process industry [1]. However, as Dobry and Jost [2] point out, there has been little demonstration of indirect monitoring and on-line control for fermentation processes. The study described here uses material balances for indirect monitoring and demonstrates the application of computer control to continuous culture.

The process examined is the growth of *Hansenula polymorpha* DL-1 in continuous culture with methanol as the sole carbon and energy source. The system is relatively well defined and off-line measurements can readily be used to evaluate the results of computer monitoring and control. Also, since the system is unstable, the process can benefit from computer control. Figure 1 shows the oxidative pathway for methanol and the concentrations of methanol, formaldehyde, and formic acid at which growth is slowed. Culture instabilities can be caused solely by the accumulation of methanol [3,4]. In addition, formaldehyde

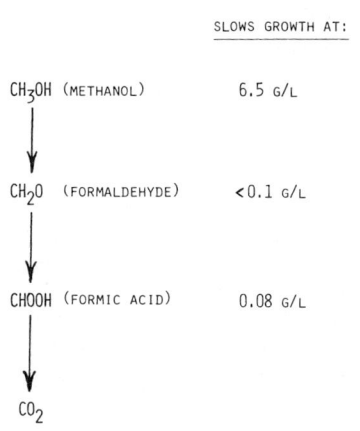

SLOWS GROWTH AT:

CH_3OH (METHANOL)   6.5 G/L

CH_2O (FORMALDEHYDE)   <0.1 G/L

CHOOH (FORMIC ACID)   0.08 G/L

CO_2

Fig. 1. Oxidation pathway for methanol catabolism in *H. polymorpha* DL-1. Also shown are the concentrations of methanol, formaldehyde, and formic acid that begin to slow the specific growth rate of the organism as determined in shake-flask experiments.

and formic acid can be excreted following periods of methanol accumulation [5,6], and these inhibitory products also contribute to culture instabilities.

Computer control is therefore needed to stabilize the process. The key to that stabilization is control of methanol accumulation. Methanol can accumulate in the fermentor any time the feed rate exceeds the maximum utilization rate by the organism. This occurs when methanol is fed too fast during culture start-up and when the metabolic rate of the organism is reduced by environmental influences. Thus, when methanol accumulates, the feed rate must be decreased to allow the organism to reduce the accumulation. Also, the control system should restore the culture to its desired steady state as soon as possible after the culture recovers.

For such control, the computer must reliably assess the residual methanol concentration, specific growth rate, and dilution rate. Thus, the stoichiometry of cell formation from methanol must be continuously monitored. In this paper, we describe the development of a system to indirectly monitor and to control a continuous methanol–single-cell protein (SCP) process.

MATERIALS AND METHODS

The organism used is *Hansenula polymorpha* DL-1 as isolated by Levine and Cooney [7]. It is grown at 40°C and pH 4.5 on a methanol–mineral salts medium containing 59 g/liter methanol; 10.3 g/liter KH_2PO_4; 2.0 g/liter $MgSO_4$; 0.55 g/liter $CaCl_2 \cdot 2H_2O$; 0.1 g/liter NaCl; trace salts; 1.2×10^{-4} g/liter biotin, and 1.2×10^{-2} g/liter thiamine HCl.

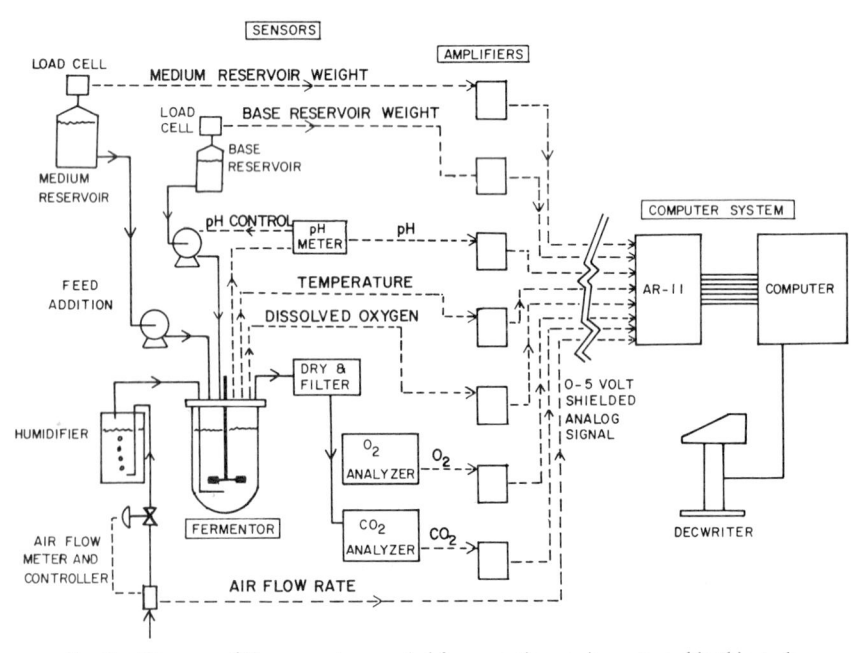

Fig. 2. Diagram of the computer-coupled fermentation equipment used in this study.

A diagram of the equipment used is shown in Figure 2. The fermentor has a 2-liter working volume and is computer coupled via the sensors and amplifiers shown. The computer is a Digital Equipment Corporation (Maynard, MA) PDP 11/10 with 24K of memory. The system is also designed for on-line measurement of heat of fermentation and for automatic recalibration of the Leeds and Northrup paramagnetic O_2 analyzer (model 7083, North Wales, PA) and the MSA Lira model 303 infrared CO_2 analyzer (Pittsburgh, PA). The recalibrations are normally performed every 12 hr as the computer switches a gas of known composition to the analyzers and monitors the analyzer outputs.

RESULTS AND DISCUSSION

Selecting an Indirect Monitoring Technique

Since we wished to use a fundamental, generally applicable approach, we began with the general stoichiometric equation

$$aCH_3OH + bO_2 + cNH_3 \rightarrow$$
$$dC_6H_{11}N_{0.84}O_{3.5} + eCO_2 + fH_2O + \text{heat} + gC_UH_wN_YO_V \quad (1)$$
$$\text{(cells)} \qquad\qquad\qquad\qquad\qquad\qquad\quad \text{(products)}$$

The coefficients are expressed as molar rates (mole/hr) and a general formula is presented for the extracellular products. Heat is expressed in kcal/hr. The elemental and energy balance equations can be written

$$\text{C:} \qquad a = 6d + e + gU \qquad (2)$$

$$\text{O:} \qquad a + 2b = 3.5d + 2e + f + gV \qquad (3)$$

$$\text{H:} \qquad 4a + 3c = 11d + 2f + gW \qquad (4)$$

$$\text{N:} \qquad c = 0.84d + gY \qquad (5)$$

$$\text{energy:} \qquad aHC_S + cHC_A = dHC_C + \text{heat} + gHC_P \qquad (6)$$

where HC_S is the heat of combustion of methanol (kcal/mol), HC_A is the heat of combustion of ammonia (kcal/mol), HC_C is the heat of combustion of cells (kcal/mol), and HC_P is the heat of combustion of products (kcal/mol). When the composition of the extracellular product(s) is known, there are eight unknowns and five equations. The measurement of oxygen uptake rate (OUR), carbon dioxide evolution rate (CER), and heat evolution provides a soluble system of equations. Since the heat measurement is the least accurate of the three [8], it is estimated by assuming a constant heat-to-oxygen ratio, which allows it to be calculated from OUR measurements.

A sensitivity analysis [9] was then used to analyze the expected accuracy of the indirect measurement of cell mass concentration and the rates of methanol accumulation and product formation. The accuracies of the primary measurements (OUR and CER) were first evaluated experimentally and the standard deviations of the measurements were then multiplied by the sensitivity factors determined for each indirect measurement. The analysis indicated that errors in the primary measurements could cause unacceptable errors in the indirect assessments.

As a second alternative, we assumed $G = 0$; effective process control can avoid methanol accumulation, which prevents formaldehyde and formic acid excretion. With the additional assumption, the system of equations is overspecified, and this allows the use of only two measurements to solve the balance equations. A sensitivity analysis was again used to suggest that the best on-line estimates of culture parameters could be obtained using the OUR and CER measurements with the elemental balances to calculate the methanol utilization and cell formation rates. The sensitivity analysis also showed that the accuracies of the final indirect assessments of methanol accumulation and cell concentration were sufficient as a basis for computer control. Therefore, the computer was programmed to use 15-min averages of the primary measurements to calculate rates of methanol utilization and cell formation. Material balances, also conducted at 15-min intervals, were then used to estimate methanol accumulation and cell concentration. Finally, the specific growth rate was calculated from the rate of cell formation and the cell concentration.

The performance of the indirect monitoring system is shown in Figure 3. For a culture originally operating at steady state ($D = 0.10$ hr^{-1}), the dilution rate was stepped up to 0.12 hr^{-1} at point A. When washout began, the dilution rate was returned to 0.10 hr^{-1}. The process was repeated at points C and D. Although the washout was unexpected, according to the sensitivity analysis, the experiment

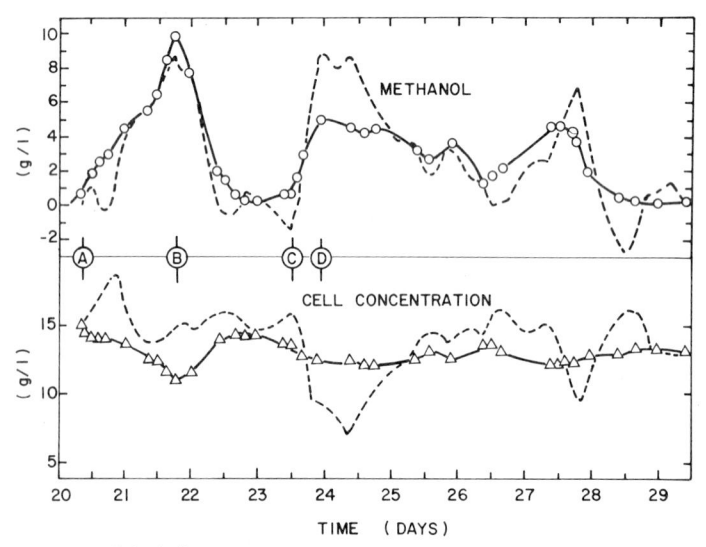

Fig. 3. Accuracy of the indirect on-line monitoring of residual methanol and cell concentrations. (—) Actual concentrations as determined by gas chromatography for methanol and by dry cell weight measurements. At point A the culture was stepped up from $D = 0.10$ hr^{-1} to 0.12 hr^{-1}; at point B, the dilution rate was returned to 0.10 hr^{-1}. Same procedure was followed at points C and D.

provides a good test for the indirect monitoring. Its accuracy is summarized in Table I and compared to the expected accuracy as suggested by the sensitivity analysis. Actual performance is better than predicted, possibly because errors of resolution in load cell measurements of methanol feed rate are reduced by integration over time. The observed error in cell yield may partially reflect actual changes in yield during the periods of culture washout and recovery since a constant yield was assumed in the error evaluation.

The most important number from Table I is the standard deviation in estimating the rate of methanol accumulation, 0.32 g/liter hr. Since the gas analyzers are recalibrated automatically every 12 hr, for the period between calibrations, measurement error could only cause a false indication of 3.8 g/liter methanol accumulation or would only mask an actual accumulation of that magnitude. Since methanol does not begin to slow down the growth rate until a concentration of 6.5 g/liter is reached, the indirect monitoring is sufficiently accurate for effective computer control.

Example of Computer Control of Continuous Culture

A control scheme was then developed to avoid methanol accumulation. If the residual methanol concentration (S) exceeds 8 g/liter, substrate addition is stopped. However, if the indicated methanol concentration is between 2 and 8 g/liter, the feed rate is only slowed down. At the end of each monitoring interval with $2 < S < 8$, the feed rate is reduced by an amount proportional to $S - 2$ and to the current feed rate.

TABLE I
Accuracy of Using Material Balancing for On-Line Monitoring

	Expected Std. Dev. By Sensitivity Analysis	NINE-DAY PERIOD (Fig. 3)		
		Ave. Error	Max. Error	Std. Dev.
Cell Mass Estimation (% Error)	30	14.5	42	17.1
Estimation of Rate of Methanol Accumulation (gr/l-hr)	0.45	0.20	0.70	0.32
Estimation of Cell Yield (% Error)	20	23	81	30

If methanol has not accumulated, the system first checks to see if the dilution rate (D) is greater than 80% of the dilution rate set point (D_{sp}). If so, the feed rate is adjusted to control at D_{sp}. If not, the feed rate is instead adjusted to control a constant specific growth rate (as calculated using elemental balances, OUR, and CER). In the last case, the feed rate is changed by an amount proportional both to the difference between the measured specific growth rate and the growth rate set point ($\mu_{sp} - \mu$) and to the measured OUR. Attenuating the control with OUR allows for a slower control action at low cell densities and a faster action when the culture can respond more quickly.

To demonstrate the control action, an experiment was performed to simulate the effect of mechanical failure or poisoning of respiration. A steady-state culture ($D = 0.075$ hr^{-1}, $X = 14$ g/liter) was subjected to a 3-hr period of reduced agitation (from 1100 to 100 rpm). Figure 4 shows the culture response without control. The period of oxygen starvation causes a 40% reduction in cell concentration, an increase in the residual methanol concentration to 16 g/liter, and, over the 16-hr period, a cell yield reduction of 20% due to methanol loss in exit streams. Also, formaldehyde and formic acid were excreted and then reutilized during the 4-hr period following the resumption of agitation.

In Figure 5, we show a case with computer control. The computer quickly stopped feed as S rose above 2 g/liter. The 3-hr period of oxygen starvation only caused a 10% reduction in concentration while methanol accumulation was limited to 5 g/liter. This reduced the methanol loss to only 3% of the loss in the uncontrolled experiment. No detectable formic acid or formaldehyde was excreted.

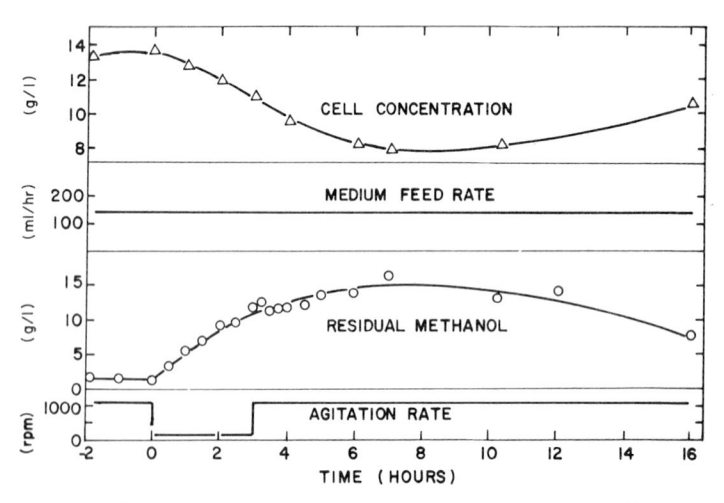

Fig. 4. Response of a continuous culture to a 3-hr period of reduced agitation (from 1100 to 100 rpm). Data points and lines represent off-line measurements. Dilution rate was constant.

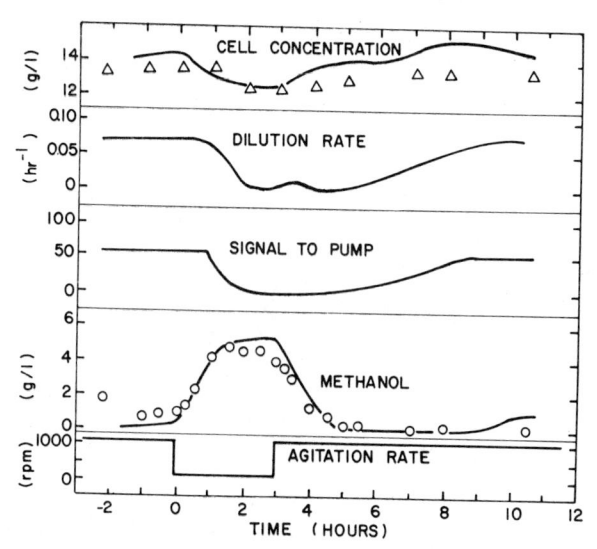

Fig. 5. Response of a continuous culture under computer control to a 3-hr period of reduced agitation (from 1100 to 100 rpm). (——) On-line computer measurements and the data points, off-line measurements. Dilution rate was controlled by the computer as shown.

CONCLUSIONS

Sensitivity analyses are valuable in helping to select an indirect monitoring technique and to determine *a priori* if that technique would be sufficient for control.

In this case, computer control based on indirect monitoring was shown to be effective in reducing the deleterious effects of process upsets. The techniques can be used in the laboratory or in large-scale process control. In addition to improving overall productivities and yields, computer control is valuable in preventing accumulation of products such as formate or formaldehyde, which could lower the quality of the SCP produced.

References

[1] L. B. Evans, *Science, 195,* 1146 (1977).
[2] D. D. Dobry and J. L. Jost, in *Annual Reports on Fermentation Processes,* D. Perlman and G. T. Tsao, Eds. (Academic, New York, 1977), pp. 95–114.
[3] J. F. Andrews, *Biotechnol. Bioeng., 10,* 707 (1968).
[4] V. H. Edwards, *Biotechnol. Bioeng., 12,* 679 (1970).
[5] P. Pilat and A. Prokop, *J. Appl. Chem. Biotechnol., 26,* 445 (1976).
[6] J. R. Swartz, "Computer monitoring and control of continuous culture: Yeast from methanol," D.Sc. thesis, M.I.T., Cambridge, Massachusetts, 1978.
[7] D. W. Levine and C. L. Cooney, *Appl. Microbiol., 26,* 982 (1973).
[8] J. R. Swartz and C. L. Cooney, *Proc. Biochem., 13*(2), 3 (1978).
[9] M. Kishimoto, T. Yamane, and F. Yoshida, *J. Ferment. Technol. 54,* 891 (1976).

Sensors and Instrumentation: Steam-Sterilizable Dissolved Oxygen Sensor and Cell Mass Sensor for On-Line Fermentation System Control

M. OHASHI, T. WATABE, T. ISHIKAWA, and Y. WATANABE

Oriental Yeast Co., Ltd., Tokyo, Japan

K. MIWA

Oriental Electric Co., Ltd., Niiza-City, Saitama, Japan

M. SHODA

Nagoya University, Nagoya, Japan

Y. ISHIKAWA

Ishikawa Works Co., Ltd., Tokyo, Japan

T. ANDO, T. SHIBATA, T. KITSUNAI, N. KAMIYAMA, and Y. OIKAWA

The Intitute of Physical and Chemical Research (RIKEN), Wako-City, Saitama, Japan

INTRODUCTION

Computer control has been an attractive topic to the fermentation industry. However, one of the most difficult problems which is impeding the implementation of computer control is a deficiency in reliable environmental sensors that can withstand repeated steam sterilizations. Therefore we have improved the dissolved oxygen (DO) sensor, and developed a new cell mass sensor that can be used *in situ* in the fermentor. Since the oxygen and cell mass concentrations in the broth are closely related to microbial growth, both sensors will provide useful information in monitoring the fermentation, and make control of the fermentor more accurate and practical.

OXYGEN ELECTRODE

It is still difficult to obtain a reliable oxygen electrode for a microbial fermentation systems since it must fulfill several demanding requirements. The electrode must: 1) be stable for a long period of operation; 2) have a simple construction; 3) have a short response time; 4) be steam sterilizable; and 5) constructed of materials that are not toxic to the microorganisms. The requirement of resistance to steam sterilization is especially important. Needless to say, several electrodes have been made available on the market. Clark [1] first devised a membrane polarographic oxygen sensor. Later, several types of electrodes based on the Galvanic cell method were developed [2-6]. The electrodes described by Johnson et al. [3], Brookman [5], and Bühler and Ingold [6] are claimed to be steam sterilizable.

Generally, conventional oxygen probes are complicated assemblies. An insulation of the anode from the cathode has been made by using adhesive. Therefore, their durability is not sufficient for long operations. This paper describes a simple membrane electrode that fulfills the requirements mentioned above. In addition, this electrode has several structural advantages.

Construction

First, the developmental process of the steam-sterilizable electrode is mentioned briefly (see Fig. 1). The mechanical and material improvements that were developed are discussed in various steps.

First Step

PTFE (polytetrafluoroethylene) was used as the initial material of the electrode cell body. The platinum (cathode)-lead (anode) cell was chosen because the residual current at a zero oxygen concentration is small. The anode was fixed to a glass tube and dipped into the electrolyte solution, whereas the platinum

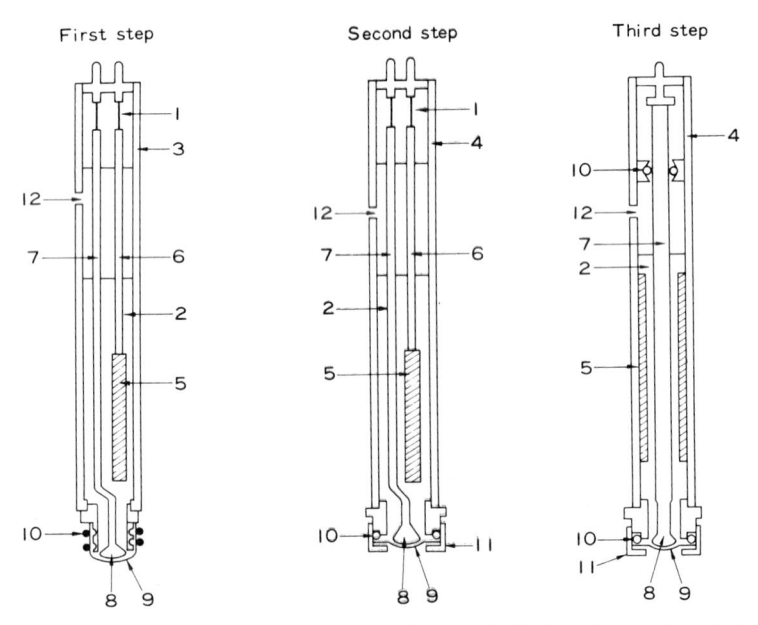

Fig. 1. Developmental process of the electrode. 1, Epoxy resin; 2, electrolyte; 3, electrolytic cell (Teflon); 4, electrolytic cell (stainless steel); 5, anode metal (Pb); 6, glass tube for anode; 7, glass tube for cathode; 8, cathode metal (Pt); 9, membrane; 10, O-ring; 11, membrane holding cap; 12, vent hole.

cathode was connected to a cathode terminal by a lead wire through a glass tube.

A fluorinated ethylene propylene copolymer (FEP) membrane was used due to its favorable heat resistance and moderate oxygen permeability.

In this step, several troubles were found. PTFE lost its plasticity when heated during steam sterilization, and the fixing of the membrane with an O-ring was not satisfactory for repeated sterilizations or long operations.

Second Step

The body of the electrode was changed to a metal tube of stainless steel. The membrane, which was glued to a washer beforehand, was fitted over the cathode surface with a stainless-steel membrane holding cap. This electrode design was fairly stable after five sterilization cycles in an autoclave at 120°C, if treated carefully.

However, when sterilized at 130°C, the insulation between the anode and the cathode failed. As shown in Figure 1, the glass tube of the cathode was partly bent, and the anode fixed on the anode glass tube was rather bulky. Therefore, the glass tubes were sensitive to outside mechanical shocks. The glass tube was especially susceptible to breakage during the process of fixing a membrane to the probe, mainly because the upper part of the electrode was fixed with epoxy resin to support the glass tubes.

Third Step

Taking the previous results into account, the lead anode was deposited as a lining on the internal wall of the electrolyte cell from the bottom to a suitable height. This allowed the cathode glass tube to be straight and the internal cell space to be rather large. The problem of cathode fragility to external shocks during membrane exchange could be eliminated by supporting the glass tube with a stainless-steel body–O-ring assembly instead of the epoxy resin.

Fourth Step

Additional improvements were attempted to enhance the resistance to steam sterilization. Figure 2(a) is a side view and a cross-sectional side view of the electrode (14 o.d. or 16 mm, 145 mm length), and Figure 2(b) is a photograph of the sensor. The anode is extended, and the cathode surface area remains unchanged, compared with the former versions. Air bubbles or silicone antifoam agents are not likely to stick to the membrane because of the grooves on the membrane-holding cap, 14 in Figure 2(a). The vent hole adjusts the inner and outer pressures of the electrode to avoid membrane rupture, especially during steam sterilization. The hole is also used for the addition of electrolyte to the probe. The protection cover (7) screwed to the electrode keeps the medium from flowing in through the vent hole when the probe is submerged into a broth. Since the electrolyte is not sealed in this assembly as it is in some other electrode de-

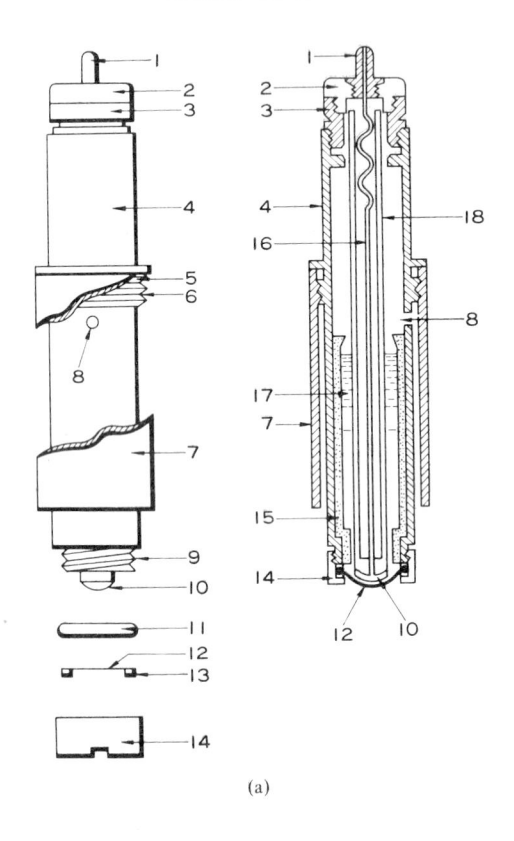

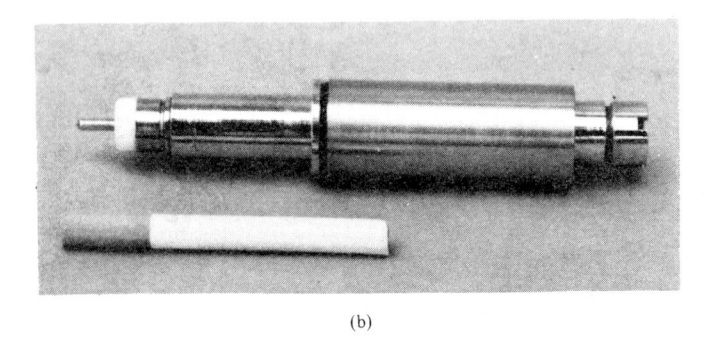

Fig. 2. (a) Construction of the DO sensor. 1, Cathode terminal; 2, insulation; 3, anode terminal; 4, electrolytic cell; 5, O-ring; 6, external screw thread for protection cover; 7, protection cover; 8, vent hole; 9, external screw thread for membrane holding cap; 10, cathode metal; 11, O-ring; 12, oxygen permeable membrane; 13, washer; 14, membrane holding cap; 15, anode metal; 16, lead wire; 17, electrolyte; 18, glass tube. (b) Photograph of the DO sensor.

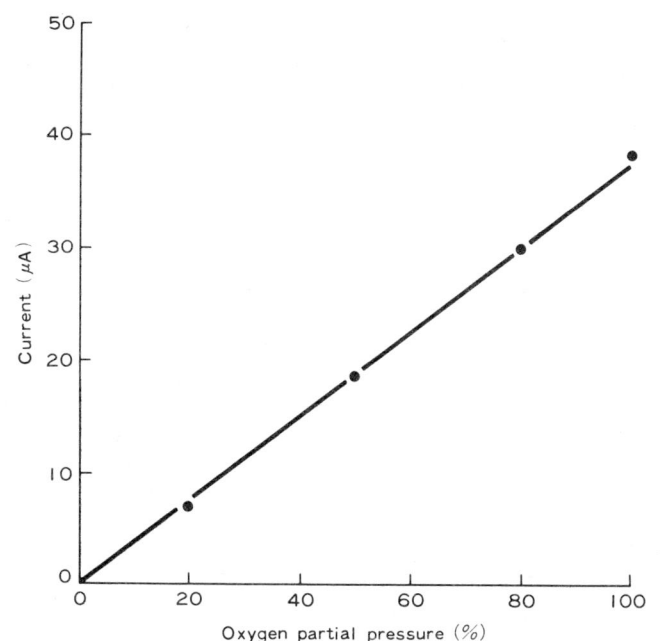

Fig. 3. Calibration curve of the DO sensor at different oxygen partial pressures. This calibration curve was obtained by mixing air, nitrogen, and oxygen in various proportions and passing the resultant mixture through a fermentor. Temperature = 30°C, total pressure = 760 mm Hg, load resistance = 2000 Ω.

signs, oxygen diffuses into the cell through the vent hole and dissolves in the electrolyte. However, this does not affect the measurement since the probe assembly prevents the oxygen that is dissolved in the electrolyte from diffusing onto the cathode surface. Only the dissolved oxygen in the medium which is outside of the membrane is allowed to reach the anode surface. Most of the electrode is constructed of stainless steel and glass, although PTFE was used for insulation (part 2) because of its resistance to heat and its chemical inertness. Furthermore, the electrode avoids using adhesives by employing threaded connections.

These improvements simplified the construction of the electrode and enhanced its resistance to chemical and thermal damage. Problems such as a loosening of integrated parts or insulation failure were not experienced throughout repeated heating and cooling.

The reaction of the electrolytic cell is as follows:

Cathode reaction:
$$O_2 + 2H_2O + 4e^- \rightarrow 4OH^-$$

Anode reaction:
$$2Pb + 4OH^- \rightarrow 2Pb(OH)_2 + 4e^-$$

Overall reaction:

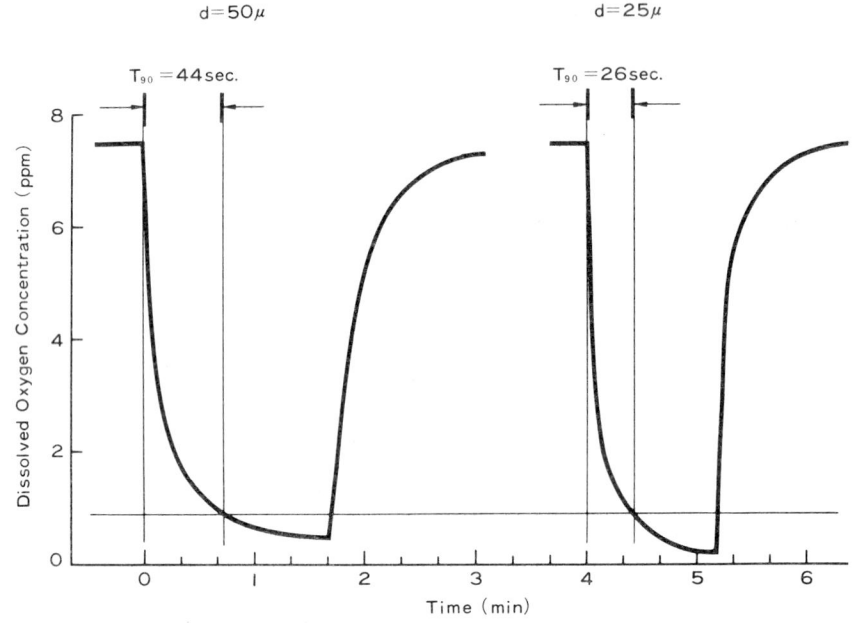

Fig. 4. Response curve of the DO sensor. T_{90} (90% response time, sec), d (membrane thickness, μm), temperature = 30°C.

$$2Pb + O_2 + 2H_2O \rightarrow 2Pb(OH)_2$$

Lead is oxidized to $Pb(OH)_2$ so that the current produced is proportional to the oxygen tension. A solution of 2% sodium hydroxide was chosen as the electrolyte so that the product, $Pb(OH)_2$ deposited on the surface of the anode is solubilized in the solution, and new anode surface are always exposed to the electrolyte.

Electrode Performance

The relationship between the partial pressure of oxygen and the diffusion current of the electrode at 30°C is shown in Figure 3.

Air, nitrogen, and oxygen were mixed in various proportions, keeping the total pressure at 760 mm Hg. Gas prepared in this manner was passed through the fermentor. The current of the electrode was found to be proportional to the oxygen tension.

The response time of the electrode is an important characteristic, especially when the dissolved oxygen concentration has to be controlled. Figure 4 shows the response curve of the electrode when the electrode was placed alternately in air-saturated water at 30°C and 5% Na_2SO_3 solution at 30°C. The T_{90} (90% response time) was 44 sec, when a membrane thickness of 50 μm was used, while a 25 μm thick membrane electrode exhibited a faster response time (T_{90} = 26 sec).

TABLE I
Sterilizing Test in the Autoclave[a]

Times of sterilization	Diffusion current (μA)	90% Response time (sec)
1	8.0	30
2	8.0	31
3	7.5	28
4	8.5	29
5	8.0	30
6	8.5	29
7	8.0	34
8	8.5	29
9	8.5	29
10	8.5	28
11	8.5	30
12	8.5	29
13	8.0	32
14	10.0	31
15	9.5	32
16	10.0	30
17 [a]	9.5	29
18	9.0	30
19	9.0	27
20	9.0	29
21	8.5	31
22	9.0	30
23	8.0	28
24	8.5	32
25	8.5	30
26	8.5	29
27	8.5	28
28	8.5	26
29	8.0	30
30 [b]	8.0	32
31	10.0	33
32	10.0	36
33	10.0	39

[a] Autoclaving conditions: autoclave was HA-30 (Hirayama Seisakusho, Japan); gauge pressure = 1.2 kg/cm^2; temperature = 120°C; sterilization = 30 min; cool-down time = 30 min. Measurements: Diffusion current was measured in the air-saturated water (p_{O_2} = 0.21 atm) at 30°C; the response time was transferred from air-saturated water (p_{O_2} = 0.21 atm) to 5% Na$_2$SO$_3$ solution at 30°C.

[b] Membrane and electrolyte were changed.

[c] Membrane and electrolyte were changed and the cathode was repolished.

Steam Sterilization Test

The 50 μm thick membrane electrode was steam sterilized in an autoclave at a pressure of 1.2 kg/cm^2 gauge, and a temperature of 120°C for 30 min. Table I gives the diffusion current and the response time measured in air-saturated water at 30°C after each sterilization.

Clearly, the electrode was stable for both diffusion current and response time,

TABLE II
Sterilizing Test in the Autoclave[a]

Times of sterilization	Diffusion current (μA)	90% Response time (sec)
1	14	33
2	10	54
3	13	40
4	12	29
5	12	33
6	14	27

[a] Autoclaving conditions: autoclave is the ZM-CU (Hirayama Seisakusho, Japan); gauge pressure = 1.7 kg/cm², temperature = 130°C; sterilization time = 60 min; cool-down time = 30 min. Measurements: diffusion current was measured in air-saturated water (p_{O_2} = 0.21 atm) at 30°C; the response time was transferred from air-saturated water (p_{O_2} = 0.21 atm) to 5% Na_2SO_3 solution at 30°C.

although the membrane and electrolyte were exchanged, respectively, for safety after the 15th sterilization. No significant changes in diffusion current or 90% response time were noted after repeated sterilization tests when the electrode was installed in the side wall of a 30 liter jar fermentor (either empty or charged with medium).

It is also confirmed that the electrode worked satisfactorily even after more severe sterilization conditions, i.e., at 1.7 kg/cm² gauge, a temperature of 130°C, for 60 min as shown in Table II.

Estimation of Culture Variables Using the Oxygen Electrode

As an example of the application of the oxygen electrode to the fermentation system, the specific respiration rate (Q_{O_2}) and volumetric oxygen transfer coefficient ($K_L a$) were estimated with the dynamic method during the exponential growth of batch culture of the yeast *Saccharomyces cerevisiae* at 30°C. The apparatus used for the yeast cultivation was a minijar fermentor (nominal volume of 2 liter, working volume of 1.3 liter). The aeration rate and the agitation speed were fixed at 330 ml/hr and 500 rpm, respectively. When the air supply was turned off, the dissolved oxygen concentration in the medium decreased as shown in Figure 5. From the slope of this figure, the specific respiration rate (Q_{O_2}) was calculated as 1.73 mmol O_2/hr mg dry cell, whereas the value of the cell concentration (x) was measured to be 0.64 g dry weight/liter. When the supply of oxygen was restarted, the DO concentration increased rapidly. From

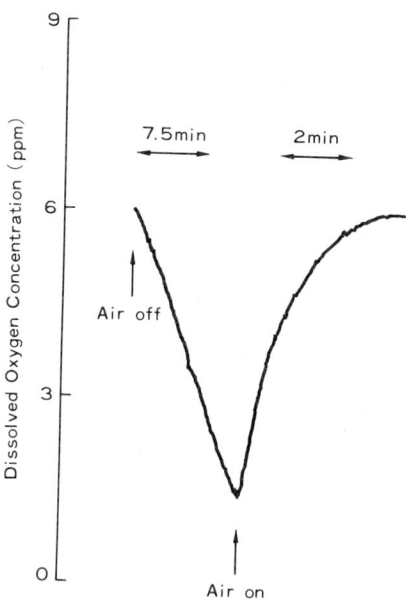

Fig. 5. DO change in the "dynamic method." Jar fermentor from Iwashiya K. Sawada Co., Ltd., Japan, type MB (nominal volume is 2 liter). Operation conditions: aeration rate = 330 ml/min; agitation speed = 500 rpm; operation volume = 1.3 liter; temperature = 30°C; the strain used was *S. cerevisiae;* the cell concentration = 0.64 g/liter.

the data on oxygen dissolution and the Q_{O_2} value calculated above, the volumetric oxygen transfer coefficient (K_La) could be estimated (see Fig. 6). Using a value of 7.5 ppm for the saturated concentration of oxygen in the culture medium, a K_La value of 0.91 min^{-1} was computed from these dynamic data. The sulfite oxidation method was also applied to this system under the same operation conditions, although the medium was replaced by distilled water. The K_La value determined by this procedure was 0.99 min^{-1}.

The traditional procedures to estimate the Q_{O_2} value, e.g., by the Warburg manometer, or K_La values by the sulfite oxidation method, are rather tedious, and the measuring conditions are often different from those occurring in the actual fermentation. It is well known that the medium properties change during the progress of a fermentation, so the K_La or Q_{O_2} values also vary according to changes in the culture conditions.

Consequently, it is of particular significance from the viewpoint of process control that the Q_{O_2} or K_La values be estimated *in situ* and continuously in the fermentor using an oxygen electrode. This oxygen probe has been successfully used in a continuous cultivation of a bakers' yeast for 500 hr. During this experiment in a minijar fermentor (working volume of 0.8 liter) at a dilution rate of 0.1 hr^{-1} for a glucose-limited chemostat, replacements of the membrane or electrolyte were not needed. This probe was also found to respond properly in a continuous fermentation with a phenol-utilizing yeast for five days when the phenol solution (250 ppm) was fed at a rate of 250 ml/hr.

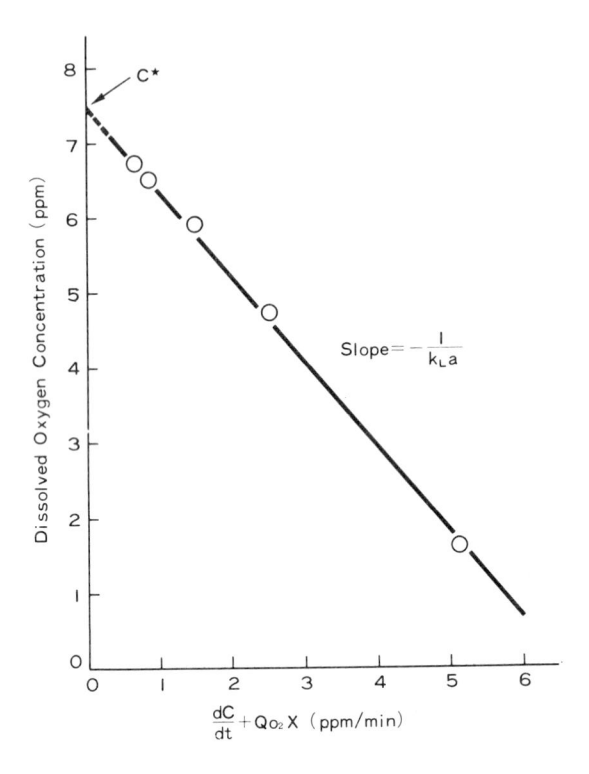

Fig. 6. Estimation of values (C^* and K_La) by the "dynamic method." Jar fermentor: Iwashiya K. Sawada Co., Ltd., Japan, type MB (nominal volume 2 liter). Operation conditions: aeration rate = 330 ml/min; agitation speed = 500 rpm; operation volume = 1.3 liter, temperature = 30°C. Strain used was *S. cerevisiae;* the cell concentration = 0.64 g/liter. K_La = volumetric oxygen transfer coefficient; C^* = saturation concentration of oxygen in the medium.

Since the maximum current caused by the oxidation–reduction reaction shown in Table I was 10 µA, we could easily measure the oxygen concentration as a voltage drop in a recorder when a proper shunt resistance (e.g., 2 kΩ) was inserted into the anode–cathode circuit.

CELL MASS SENSOR [7]

Air bubbles produced in great quantities are unavoidable in the aerobic culture vessel. Accordingly, the optical density (OD) measurement *in situ* is generally complicated.

This paper proposes a submersible colorimeter probe that permits accurate, easy, and continuous measurement of OD of the solution by removing air bubbles from the solution.

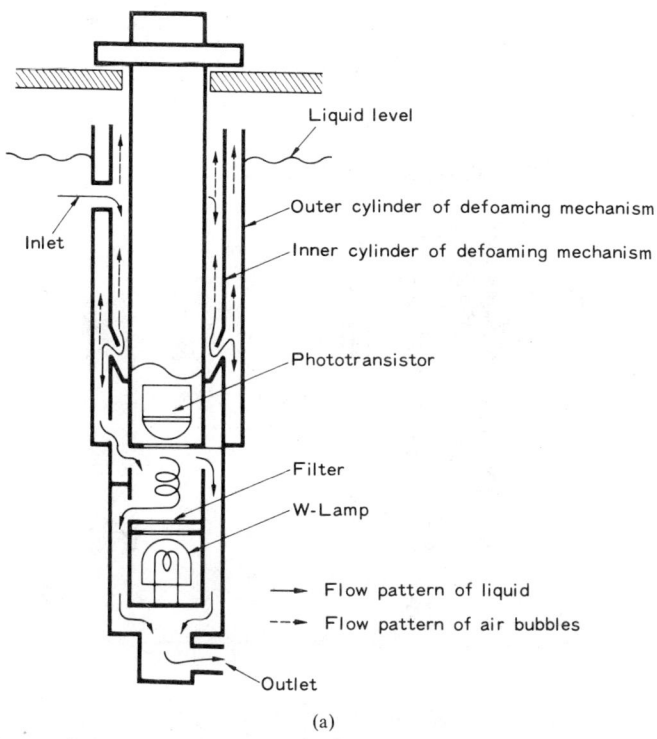

Fig. 7. (a) Construction of cell growth meter (throw-in type colorimeter with defoaming mechanism). (b) Photograph of cell growth meter. A: setting view of the head plate to the fermentor; B: colorimeter rod; C: inner cylinder of defoaming mechanism; D: outer cylinder of defoaming mechanism; E: cap.

Construction

Figure 7(a) is a schematic sectional view and Figure 7(b) is a photograph of a submersible colorimeter probe, the "cell mass sensor," attached to a fermentor. This sensor (62 mm in diam and 490 mm in length) consists of a double cylinder for defoaming, a measuring chamber with a light source, a photocell or phototransistor, and a discharge port. When the fermentation broth is agitated, the whirlpool motion caused by the impeller forces the broth into the opening at the upper part of the sensor, and the broth flows into the inner cylinder. The downward flow in the inner cylinder is reversed in its direction upon reaching the lower end of inner cylinder. During the flow entrained air bubbles in the broth are completely purged and discharged. The bubble-free culture medium flows into the measuring chamber provided at the lower part to determine the optical density. The broth is finally discharged into the fermentor from a discharge port. A residence time of the broth in the colorimeter is approximately 1 to 3 min.

In the measuring chamber, the light emitted from the light source (e.g., W-lamp) passes through the broth and is projected on a photocell or a photo-

(b)

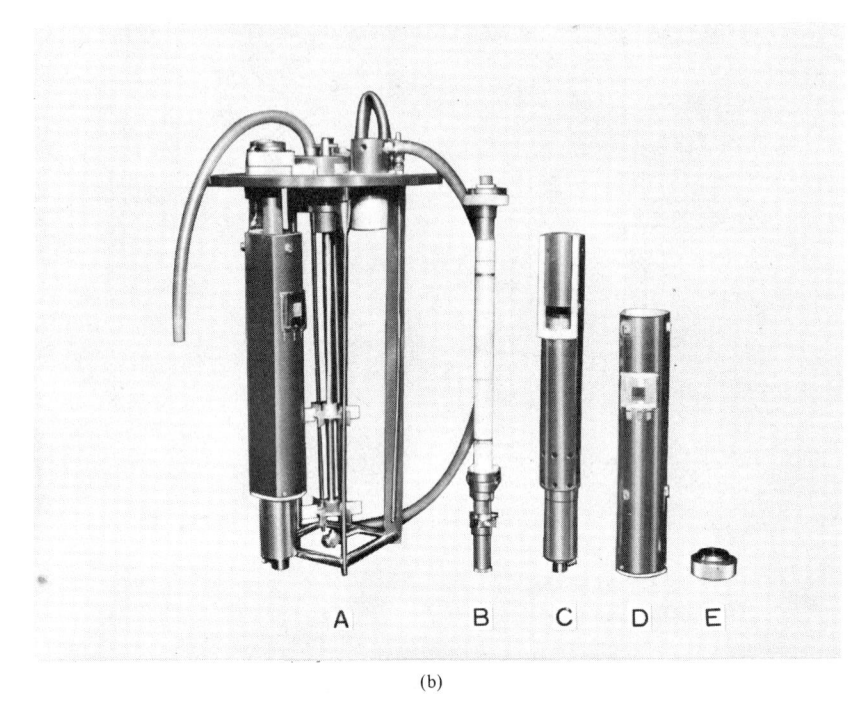

Fig. 7. (*Continued from previous page*).

transistor. The optical density of the broth can be determined from the photoelectromotive force given by a dc voltmeter connected to the photocell or phototransistor. The light source may be of any type that provides light of a desired wavelength. Either a photocell or a phototransistor may be used for the measurement of transmitted light.

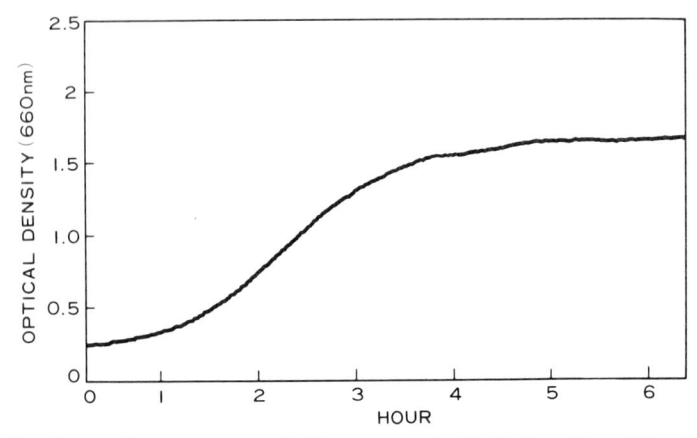

Fig. 8. Automatic measurement of cell mass concentration in the culture of *B. subtilis*.

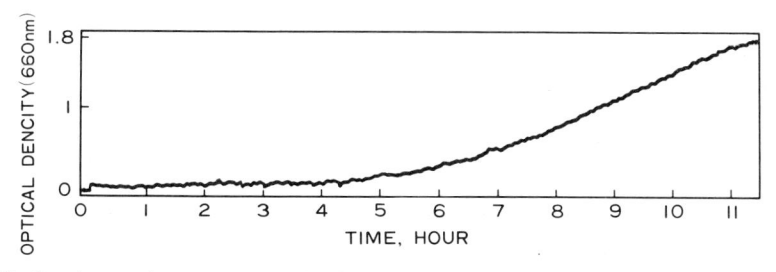

Fig. 9. Automatic measurement of cell mass concentration in the culture of *S. cerevisiae*.

Example 1

First *Bacillus subtilis* was cultured in a jar fermentor (14 liter in capacity, 22 cm i.d., 45 cm in height: Microferm Fermentor MF-114 manufactured by New Brunswick Scientific Co., Inc.). The sensor was inserted into the fermentor and 10 liter culture medium (containing 100 g polypeptone, 25 g yeast extract, and 25 g sodium chloride and adjusted to pH 7.0) was charged into the fermentor. Then the setup was subjected to a high-pressure steam sterilization under 1 kg/cm^2 for 30 min and, after cooling, it was maintained at 37°C. The fermentor was inoculated with 400 ml of a culture of *B. subtilis 20* Marburg GSY 1026 grown for 20 hr in a shake flask at 37°C. The aeration rate was 6 liter/min, agitation was 500 rpm, and a temperature of 37°C. The liquid level dropped by approximately 10 mm during the fermentation period. The optical density was measured at 660 nm wavelength. The results are shown in Figure 8.

Example 2

Saccharomyces cerevisiae was also cultivated by using a culture medium containing 3% malt extract, 0.5% yeast extract, and 0.5% glucose at 30°C, using

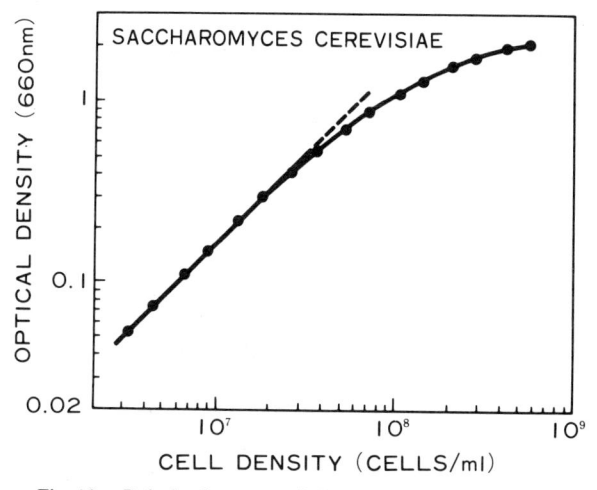

Fig. 10. Relation between cell density and optical density.

an agitation rate of 800 rpm and an aeration rate of 14.0 liter/min, as shown in Figure 9. These results indicate that the measurement was accomplished without any interference by the air bubbles in the fermentor. This permits an accurate, easy, and continuous measurement of the optical density of the broth. The relationship between the measured optical density and the cell concentration (cells/ml) is presented in Figure 10. The linearity of the sensor with respect to cell concentration was observed up to 1.0 in optical density at 660 nm.

The authors wish to thank Professor S. Aiba of Osaka University for his helpful suggestions in the course of this work.

References

[1] L. C. Clark, Jr., *Trans. Am. Soc. Artificial Internal Organs, 2,* 41 (1956).
[2] F. J. H. Mackereth, *J. Sci. Instrum., 41,* 38 (1964).
[3] M. J. Johnson, J. Borkowski, and C. Engblom, *Biotechnol. Bioeng., 6,* 457 (1964).
[4] D. S. Flynn, D. G. Kilburn, M. D. Lilly, and F. C. Webb, *Biotechnol. Bioeng., 9,* 623 (1967).
[5] J. S. G. Brookman, *Biotechnol. Bioeng., 11,* 323 (1969).
[6] H. Bühler and W. Ingold, *Process. Biochem., 11*(3), 19 (1976).
[7] T. Ando, T. Shibata, T. Kitsunai, N. Kamiyama, and Y. Oikawa, U.S. patent 4,075,062 (1978).

Use of Culture Fluorescence for Monitoring of Fermentation Systems

DANE W. ZABRISKIE
BioChem Technology, Inc., 66 Great Valley Parkway, Malvern, Pennsylvania 19355

INTRODUCTION

Living cells produce compounds that fluoresce under appropriate experimental conditions. One of these substances is the reduced form of nicotinamide adenine dinucleotide (NADH), which fluoresces at 460 nm when irradiated with 366-nm wavelength light. This phenomenon suggests that the continuous monitoring of the fluorescence of a culture using the conditions that cause NADH to fluoresce could provide important information on the status of a fermentation and aid in the control of the process.

EXPERIMENTAL METHODS

The fluorometer used in these experiments was originally described by Harrison and Chance [1] and is shown in Figure 1. The instrument was mounted on an observation port of the fermentor located beneath the surface of the culture. The culture was irradiated with light near 366 nm using an ultraviolet lamp and a narrow bandpass optical filter. This light excited the culture in the vicinity of the observation port, causing it to fluoresce near 460 nm. The fluorescent light intensity was measured using a photomultiplier, which was filtered to screen out the exciting light and other fluorescence wavelengths.

The fermentation experiments to be described produced bakers' yeast in a complex medium composed of 2% corn steep liquor, 8.0 g $(NH_4)_2HPO_4$/liter, and glucose. A glucose syrup was added on demand to the fed-batch aseptic fermentor to prevent glucose repression effects and excessive conversion to ethanol. Instrumentation was available for measuring pH, temperature, dissolved oxygen concentration, gas flow rates, and CO_2 and O_2 concentrations in the exhaust gas. Biomass concentration was determined by a dry weight analysis. The experimental details have been discussed elsewhere [2,3].

RESULTS

The fluorometer response to variations in fluorophore concentrations was tested using solutions of quinine or NADH, and fermentation broth [2,3]. In all cases, a nonlinear relationship was obtained as illustrated in Figure 2 for an

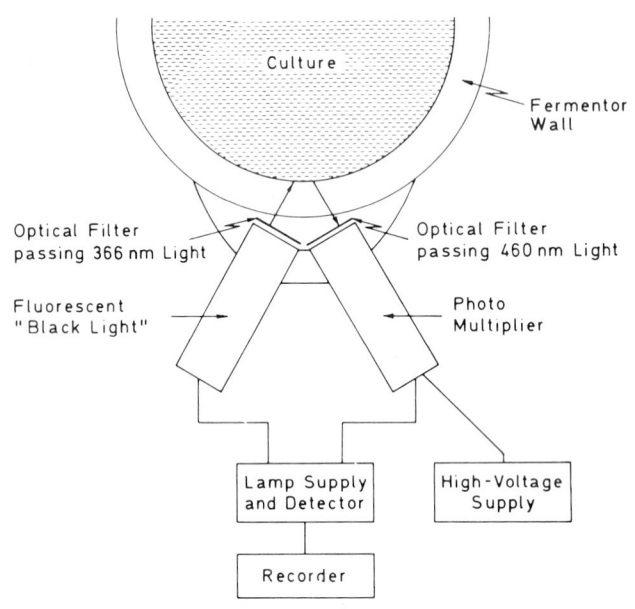

Fig. 1. Fermentor fluorometer assembly.

experiment involving the broth from a bakers' yeast fermentation. Since log–log coordinates have been known to linearize fluorescence measurements from other instruments with surface detector designs [4], these data were replotted as shown

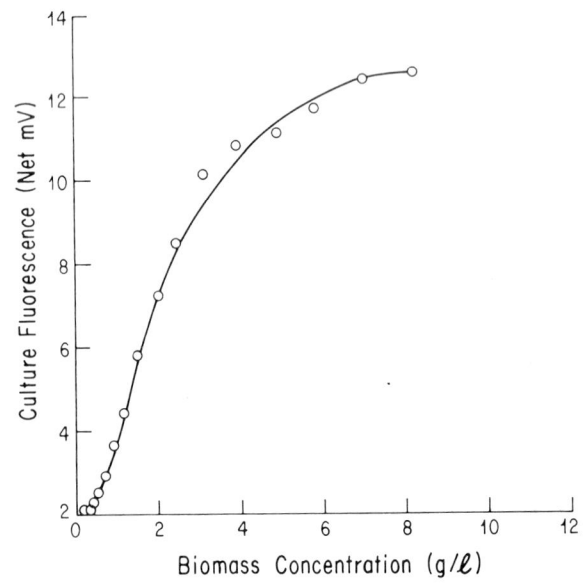

Fig. 2. Fermentor fluorometer response to the concentration of bakers' yeast biomass.

in Figure 3. This transformation was effective in linearizing the data and suggested that the concentration of biomass could be estimated directly using culture fluorescence measurements and an elementary correlation equation of the form

$$X = \exp(-m/b) \, F^{1/m}$$

Parameters m and b are the slope and intercept of the line drawn on Figure 3, respectively. Results of this estimation technique for a fed-batch bakers' yeast fermentation are shown in Figure 4.

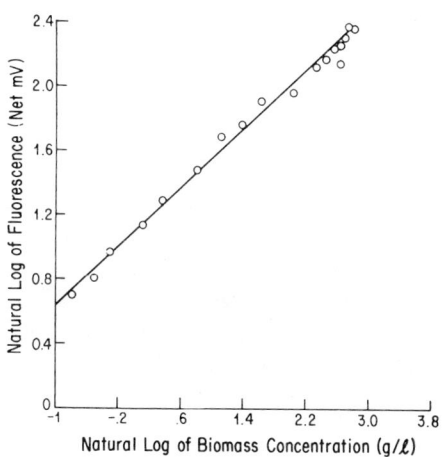

Fig. 3. Fermentor fluorometer response to the concentration of bakers' yeast biomass (log–log coordinates).

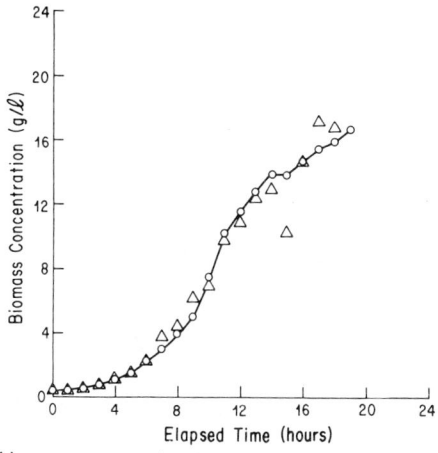

Fig. 4. Bakers' yeast biomass concentration data and estimates derived from culture fluorescence measurements. (O) Biomass concentration data (dry weight), (△) biomass concentration estimates (fluorescence).

Accurate estimates of biomass concentration were only obtained when the culture conditions were carefully controlled so that pH and temperature were constant and all nutrients were available in excess. This observation can be explained by recognizing that the culture fluorescence measurement is sensitive to the number of cells containing NADH, and the level of NADH content within the cells. Since NADH is an important high-energy intermediate, factors that influence cellular metabolism and energy production can influence the fluorescence measurements without any change in cell number. The chemical and physical environment of the fluorescent compound also affect the fluorescence efficiency of the molecule. Variations in culture pH and temperature, for example, affect culture fluorescence by both mechanisms, causing changes in cell metabolism and molecular fluorescence efficiency.

Several experiments were performed to expose some of the culture conditions that cause culture fluorescence changes by altering the internal NADH content of the microorganisms [2,3]. A withholding of oxygen to aerobically growing bakers' yeast caused culture fluorescence to increase as shown in Figure 5. An oxygen deficiency blocks the oxidation of NADH to nonfluorescent NAD through the oxidative phosphorylation process which causes the NADH fluorophore to accumulate. Culture fluorescence decreased when the carbon source (glucose) was depleted as shown in Figure 6. This nutrient limitation prevented the cells from replenishing their internal energy supply, which caused a depletion of the high-energy NADH reserves during the starvation period. A similar effect was obtained when iodoacetate, which blocks the Embden–Meyerhof–Parnas (EMP) metabolic pathway, was added to the culture medium as shown in Figure 7. Based on this experiment, it was possible to estimate that approximately 50% of the culture fluorophore content in bakers' yeast is intracellular NADH [3].

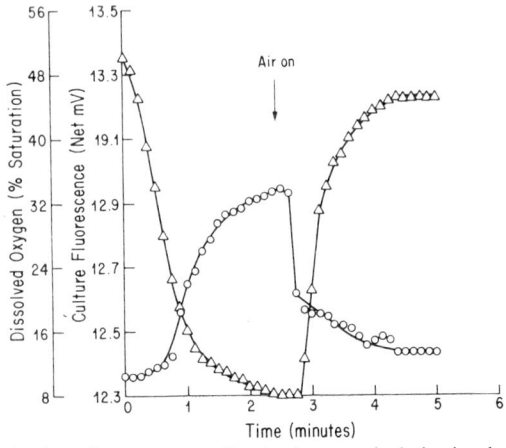

Fig. 5. Response of culture fluorescence to dissolved oxygen depletion in a bakers' yeast fermentation. (O) Culture fluorescence; (△) dissolved oxygen concentration.

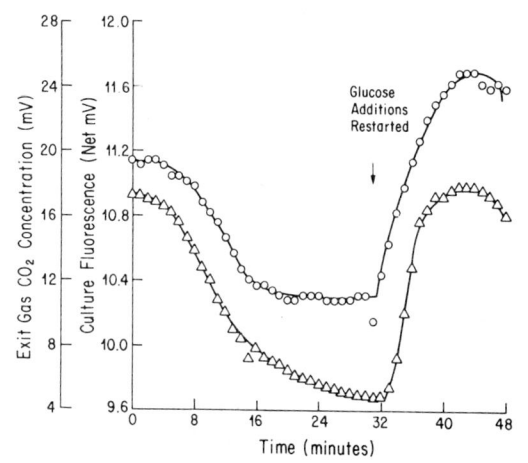

Fig. 6. Response of culture fluorescence to depletion of the carbon source (glucose) in a bakers' yeast fermentation. (O) Culture fluorescence; (△) CO_2 concentration.

DISCUSSION

It can be concluded that culture fluorescence is a complex function of biomass concentration, cellular metabolic activity, and a variety of environmental factors. Although this complexity makes a detailed understanding of the behavior of these data impossible at this time, culture fluorescence appears to provide a cumulative index of culture activity and may therefore have importance in the control of a variety of fermentation processes.

One study [5] that adopted this interpretation used culture fluorescence measurements to control the addition of ethanol to a fed-batch single-cell protein

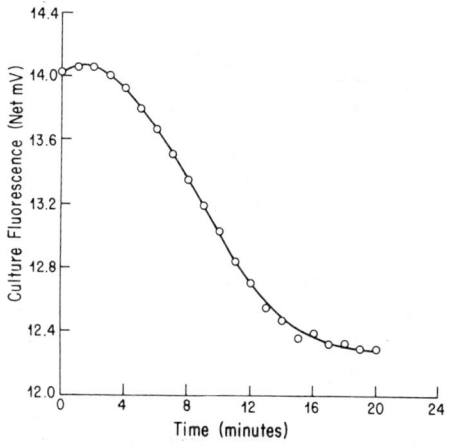

Fig. 7. Response of culture fluorescence to the inhibition of the Embden-Meyerhof-Parnas metabolic pathway by iodoacetate in a bakers' yeast fermentation.

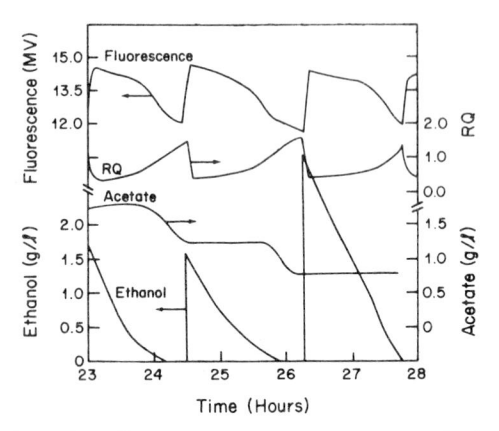

Fig. 8. Basis for using culture fluorescence measurements to control substrate additions to a fed-batch fermentation for SCP production [5].

(SCP) fermentation (*Candida utilis*). In this process, a underfeeding of ethanol leads to low productivity while an overfeeding of ethanol leads to the production of acetate as a by-product and a corresponding reduction in SCP yield. The experimental basis for the control policy is shown in Figure 8. Initially, the concentration of ethanol and the fluorometer signal are high and the respiratory quotient is low near its theoretical value of 0.29 for growth on ethanol [5]. As the ethanol becomes limiting, the signal drops sharply. Soon some of the acetate is assimilated by the yeast using a diauxic growth mechanism, and the respiratory quotient increases, reflecting this transition. The cycle repeats with each addition of ethanol. The fluorometer can be used to indicate the proper time for ethanol addition in this manner. Experiments that employed this procedure were effective in maintaining a high SCP productivity and yield from ethanol [5].

CONCLUSIONS

This evidence indicates that culture fluorescence measurements may provide essential information that may be useful in the control of fermentation processes. The measurement has a very rapid response, is continuous, and does not require the generation of discrete samples. It has been used in a variety of fermentations including those involving cellulose suspensions, complex medium, and mycelial morphology without any measurement problems [2,3]. Further refinements in the stability of the fluorometer should enable a more complete understanding of the significance of this promising on-line measurement.

References

[1] D. E. F. Harrison and B. Chance, *Appl. Microbiol., 19,* 446 (1970).

[2] D. W. Zabriskie, "Real-time estimation of aerobic batch fermentation biomass concentration by component balancing and culture fluorescence," Ph.D. dissertation, University of Pennsylvania, Department of Chemical and Biochemical Engineering, Philadelphia, PA, 1976.
[3] D. W. Zabriskie and A. E. Humphrey, *Appl. Environ. Microbiol.*, 35 (2), 337 (1978).
[4] S. Udenfriend, *Fluorescence Assay in Biology and Medicine* (Academic, New York, 1969).
[5] D. L. Ristroph, C. M. Watteeuw, W. B. Armiger, and A. E. Humphrey, *J. Ferment. Technol.*, 55 (6), 599 (1977).

Use of an Immobilized Penicillinase Electrode in the Monitoring of the Penicillin Fermentation

J. W. HEWETSON, T. H. JONG, and P. P. GRAY

School of Biological Technology, University of New South Wales, Kensington, N.S.W. 2033 Australia

INTRODUCTION

The use of computer-linked fermentation systems has expanded rapidly in recent years. However, the lack of adequate sensors for key metabolic parameters has limited considerably the use of computers for closed-loop control. In the development of a fermentation for any microbial product the most important parameters are the concentration and rate of accumulation of the product. These parameters will be the most critical in any optimization strategy and so a continuous readout of their values would be of considerable help. The development of enzyme electrodes [1] offers the potential for the continuous readout of product concentration, which would allow closed-loop control of the fermentation based on this parameter. Initial reports described the construction of penicillin-sensing electrodes [2,3]. These electrodes trapped the enzyme penicillinase in a layer of polyacrylamide gel next to the surface of a glass pH electrode. When the electrode was placed in a solution of penicillin, there was a change in the pH reading of the electrode proportional to the concentration of penicillin. Papariello et al. [2] observed that monovalent cations also affected the response of the electrode, and subsequent papers [4,5] overcame this effect by adsorbing or covalently linking the enzyme to various glass supports. A more recent report [6] describes an electrode where a solution of the enzyme is retained around a pH electrode by dialysis membrane and is not sensitive to monovalent cations in buffered solutions with relatively high ionic strength. When these electrodes have been used to assay for penicillin present in fermentation broths, some form of conditioning, such as filtration, has been carried out first. To date there have been no reports on the use of these electrodes in a fermentor for the on-line analysis of penicillin during a fermentation. This paper describes the development of a penicillin-sensing electrode suitable for use in a fermentation environment and initial results from its use to monitor penicillin accretion during the fermentation.

EXPERIMENTAL

Preparation of Immobilized Penicillinase and Construction of Electrode

Penicillinase was covalently attached to the support [controlled pore glass (CPG) (ZIR-CLAD™/CPG-550, Corning)] by the method described by Messing [7], which involves silanization of the CPG, followed by glutaraldehyde coupling of the enzyme. One gram of CPG was silanized by refluxing it in a solution of 10% (by weight) 3-aminopropyltriethoxysilane (Aldrich) in distilled water. The pH of the solution was adjusted to 2.5 with glacial acetic acid, and the refluxing was carried out for 6 hr. The CPG was then rinsed with distilled water and dried at 105°C overnight.

A similar method to that reported by Weetall [8] for the linking of glucoamylase to alkylamine glass by glutaraldehyde, was used to prepare penicillinase derivatives. To 1 g silanized CPG were added 10 ml 2.5% aqueous glutaraldehyde solution. The reaction mixture was placed in a vacuum dessicator, which was evacuated with an aspirator to remove air and gas bubbles from the pores. The reaction was allowed to proceed for approximately 30 min at reduced pressure and room temperature, and was then continued at atmospheric pressure for 60 min. Finally, the glutaraldehyde solution was decanted, and the CPG washed at least three times with distilled water. Protein coupling was carried out at room temperature. The penicillinase solution consisted of 800,000 Riker units in 3 ml $0.05M$ phosphate buffer (pH 7.0). This penicillinase solution was added to the CPG-glutaraldehyde so that the liquid just covered the glass, and the mixture was slurried and evacuated for approximately 30 min, before being left at atmospheric pressure for a further 2 hr. The resulting product, penicillinase immobilized onto CPG, was then washed several times with distilled water, followed by phosphate buffer. This final product was stored in phosphate buffer at 5°C.

The activity of the preparations was determined by pumping a solution of penicillin V in phosphate buffer, pH 7.0, through a column of the CPG-penicillinase (approximately 0.4 g) and assaying for penicillin before and after hydrolysis. To determine the pH-activity profile, the same procedure was followed using phosphate buffers of differing pH values to dissolve the penicillin V.

A Riker unit of penicillinase activity is defined as that amount of penicillinase that can inactivate 1 unit of penicillin/min, at pH 7.0 and temperature 25°C. In this study international units are used and 1 unit of penicillinase activity is defined as that amount of penicillinase that can inactivate 1 μmol penicillin/min, at pH 7.0 and temperature 30°C.

The penicillin electrode was constructed by placing approx. 0.5 g wet weight of the CPG-penicillinase preparation in a thin layer in front of a flat surface pH electrode (Markson Science, Inc., model No. 1205). A thin threaded Perspex sleeve was attached to the electrode with Silastic (Dow Corning). The CPG-penicillinase was held against the electrode by a membrane stretched over a threaded Perspex sleeve which was screwed tightly onto the matching thread attached to the electrode.

Sterilization of the Electrode

Two approaches were developed to ensure the electrode was sterile before insertion into the fermentor. First, the electrode could be assembled using sterile components. The CPG was autoclaved after silanization and the glutaraldehyde and the penicillinase solutions filter sterilized before immobilization, allowing the production of a sterile CPG-penicillinase preparation. The electrode and ancillary fittings were sterilized with formaldehyde and the presterilized components were assembled in a laminar flow hood. Alternatively, the penicillin electrode could also be sterilized after assembly by immersing in chloroform-saturated water at 37°C for 3 hr.

Calibration of the Electrode

The response of the penicillin electrode (ΔmV) depends on the buffering capacity (β = mequiv H^+/liter/pH-unit) of the solution. The penicillin electrode was used to construct calibration curves (ΔmV vs. mg/ml penicillin V) at several buffering capacities. The slopes of these lines [(ΔmV/(mg/ml) penicillin V] represent the sensitivity (k) of the electrode at the different buffering capacities, and were plotted against β, which was calculated by titrating each buffer over the range pH 7-6. When determining calibration curves, penicillin solutions were maintained at pH 7.0 using a proportional pH controller (Radiometer Titrator TTTlc).

Penicillin Assay

Samples were analyzed for penicillin content according to the hydroxylamine method [9]. Broth samples were centrifuged at 1500g for 10 min to remove the mycelia and suspended solids. Assays were carried out on the same day as sampling. Media blanks were created by treating duplicate broth samples with Labpenase (Commonwealth Serum Laboratories, Australia).

Organism and Medium

A strain of *Penicillium chrysogenum* (ATCC 26818) was used and was stored in freeze-dried ampuls. The sporulation and germination medium were as described by McCann and Calam [10]. The production medium contained: 40 g/liter corn steep liquor; 40 g/liter lactose; 6 g/liter $CaCO_3$; 8 g/liter soybean oil; 1 g/liter Na_2SO_4; 0.5 g/liter $(NH_4)_2SO_4$; 1.5 g/liter phenoxyacetic acid; 2.04 g/liter KH_2PO_4.

Fermentor Operation

A freeze-dried ampul was used to inoculate sporulation slants which were incubated at 25°C for 9-12 days. Spores from these slants were used to inoculate 50 ml germination medium in a 250-ml flask shaker at 250 rpm at 25°C. After 30 hr growth, 200 ml of the germination medium were used to inoculate the fermentor, containing 3.5 liter production medium. The temperature during the fermentation was controlled at 25°C and the dissolved oxygen level was maintained above 25% by varying the stirrer speed. Foaming was automatically controlled by the addition of Antifoam A (Dow Corning). The pH was allowed to drift for approximately 20 hr, usually in the range 5.2-6.5 before being brought gradually to pH 7.0 with 5M NaOH.

Initially the pH in the fermentor was controlled with a two-way ON-OFF controller. However, this control was not sufficiently accurate and a system employing a constant bleed of acid and proportional control on alkali addition was developed.

Data Logging

The signals from the pH and penicillin-sensing electrodes were amplified and fed into the computer (EAI Pacer 500). The signals were averaged over a 10-min period and then the mean of each signal was printed out.

RESULTS AND DISCUSSION

Properties of CPG-Penicillinase Preparations

The average apparent activity of the CPG-penicillinase preparations was found to be 320 IU/g CPG. This represented a tenfold increase in specific activity over that obtained when the enzyme was immobilized onto a sintered glass disk [11]. The pH activity profile of the enzyme was changed upon immobilization and is shown in Figure 1. The free enzyme displayed a rather broad peak of maximum activity between pH 6.6 and 7.1, while the CPG-penicillinase preparation displayed a much lower percentage activity, compared with the free enzyme, in the pH range 5.0-6.8. Furthermore, the immobilized enzyme displayed a narrow peak of maximum activity at pH 7.2, which represented a slight shift in the maximum activity toward alkaline conditions. Imsande et al. [12] have shown that the penicillinase used in this study consists of three distinct species, each with a slightly different pH optimum, and it is possible that the ratio of the species after immobilization was different from that present in the free enzyme. In order for the electrode to function satisfactorily, there must always be an excess of enzyme activity when compared with the amount of penicillin diffusing into the electrode. The preparation used in this study was found to be satisfactory, and the problem of having insufficient enzyme loading in the electrode was not experienced; however, the possibility exists of selecting a support that would cause the pH optimum to be shifted to more acidic conditions, allowing more concentrated penicillin solutions to be assayed and a larger mV reading to be obtained from the electrode.

To test the stability of the CPG-penicillinase, a solution of penicillin V was pumped continuously through a column of the preparation and the percentage hydrolysis of the penicillin V measured daily. The results, shown in Figure 2, indicate that the CPG-penicillinase was very stable. The percentage of hydrolysis only decreased by 1.8% over a 10-day period.

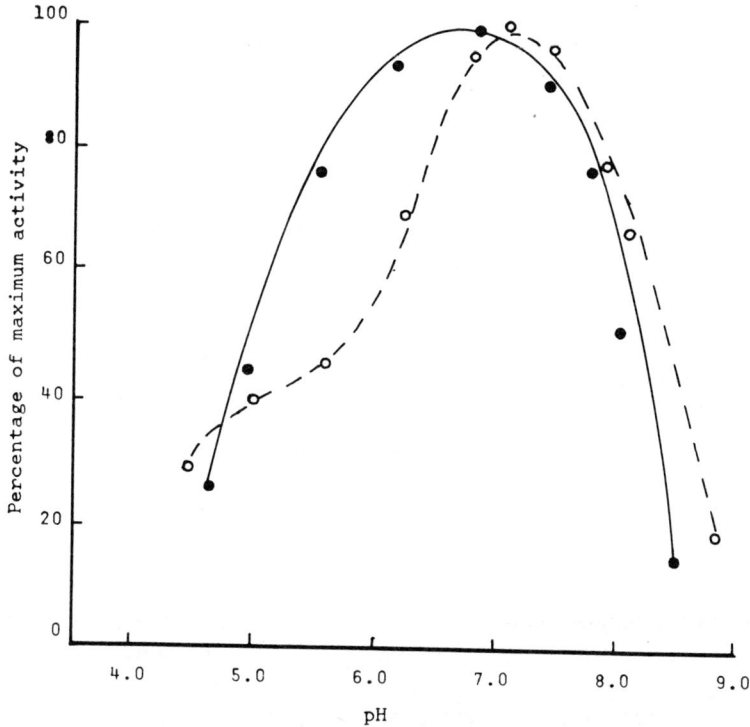

Fig. 1. pH-activity profiles for (●) soluble and (○) immobilized penicillinase.

Calibration of Electrode

When the electrode was used to measure penicillin V concentrations in buffered solutions it was found that, a series of linear calibration curves was obtained, as has been reported by other workers [5,6] (Fig. 3). As the buffering capacity of the solution is increased, the slope of the calibration curve (ΔmV/mg/ml penicillin V) decreases, owing to the fact that the hydrolysis of a given concentration of penicillin V is producing less of a pH change within the electrode. This offers the possibility of extending the upper limit of concentrations that the probe can assay by increasing the buffering capacity, although a limit will be reached when the sensitivity will be too low to allow accurate readings from the electrode.

In Figure 4 the slopes of calibration curves shown in Figure 3 are plotted against buffering capacity. The results obtained when the calibration is carried out in filtered fermentation broth supplemented with different amounts of phosphate are also plotted in Figure 4. It can be seen that there is good agreement between calibrations carried out in broth and in phosphate buffer. Nilsson et al. [6] obtained different β. vs. k curves for broth and phosphate buffer. They attributed this behavior to different buffering capacities inside the electrode

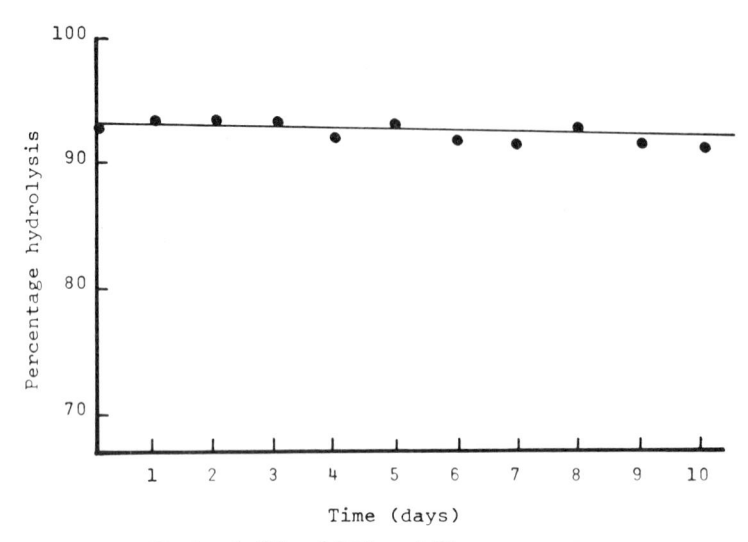

Fig. 2. Stability of CPG–penicillinase preparation.

caused by the dialysis membrane they used to retain the enzyme while excluding macromolecular proteolytes. This was not observed in our study, probably because the 2-μm filtration membrane used to retain the CPG would allow the diffusion of high-molecular-weight molecules. It can be seen that at low values of β, the response changes markedly for only small changes of β, whereas at high buffering capacities, the inverse is true. In order to operate in a region where the electrode's response is not too sensitive to changes in β, the buffering capacity during the fermentation was increased by adding $0.015 M$ phosphate to the medium to give a value of β of approximately 14.

The penicillin electrode described by Cullen et al. [4] had the enzyme covalently linked to a fritted glass disk. This would not be particularly suitable in a fermentation environment in that the mycelia could penetrate and clog the disk. In order to retain the CPG–penicillinase as a layer in front of the pH electrode, a membrane was needed that had small enough pores to allow rapid diffusion of penicillin into the CPG preparation. A range of membranes was tested and the response times that were obtained following step changes in penicillin V concentrations and pH were determined. In Table I are shown the results obtained for a step change of 4 mg/ml penicillin V; similar curves were obtained following a step change in pH. The general trend observed was that the response time decreased with increasing pore size and water flow rates. For each membrane, the final mV reading of the electrode was the same. The 2-μm poly(vinyl chloride) (PVC) membrane was chosen as it represented a compromise between response time and pore size. By altering the configuration of the electrode by stretching the membrane over a curved stainless-steel mesh, it has now been possible to obtain response times of 4–5 min with the PVC membrane.

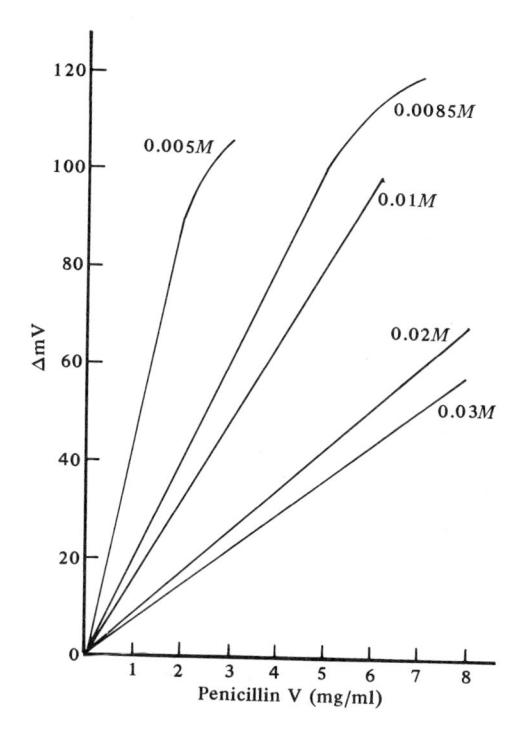

Fig. 3. Response of the penicillin electrode (ΔmV) in phosphate buffers.

The results for chloroform sterilization of a contaminated CPG preparation are shown in Table II. The viable count dropped rapidly and was zero after 1 hr. The effect of chloroform on the activity of the penicillinase was determined by pumping penicillin V solutions, saturated with chloroform, through a column of CPG–penicillinase and determining the percentage hydrolysis as a function of time. From Table II it can be seen that after 3 hr the activity had dropped by 11%. This drop in activity could be tolerated as the price necessary to achieve a sterile preparation.

The results from a fermentation where the electrode was used to measure penicillin accretion are plotted in Figure 5. Penicillin synthesis started at 24 hr and proceeded in a roughly linear fashion up to 96 hr. The electrode was inserted into the fermentor at 30 hr and the signal increased to a final reading of 24 mV after initial fluctuations caused by problems of pH control and electrical interference. The buffering capacity of the broth increased slightly from 12.2 to 15.5 during the period the electrode was in the fermentor. From Figure 4 it can be seen that this would only represent a decrease in k from 8.5 to 7.5 mV/mg/ml. An average value of $k = 8$ mV/mg/ml penicillin V was used to convert the mV reading to penicillin concentration. As can be seen from Figure 5, there is good agreement between the concentration calculated from the electrode and that obtained by the hydroxylamine assay. Thus, one of the main aims of the study,

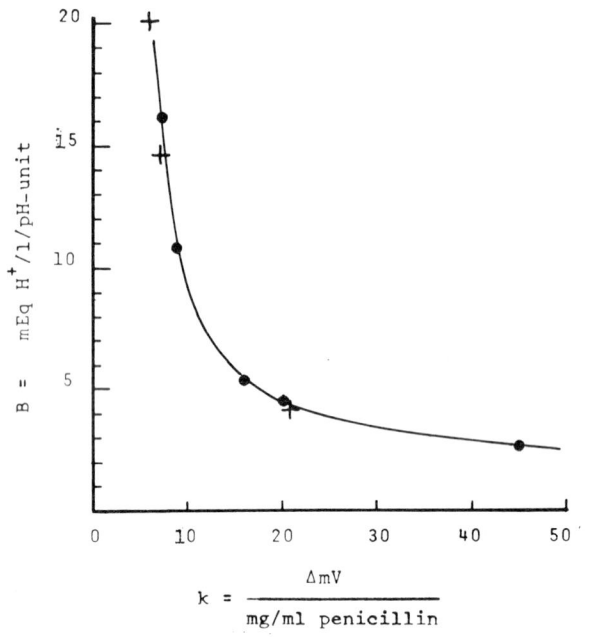

Fig. 4. Calibration curve of electrode sensitivity as a function of buffering capacity for (●) phosphate buffers and (+) fermentation broth.

$$k = \frac{\Delta mV}{mg/ml\ penicillin}$$

$B = mEq\ H^+/1/pH\text{-unit}$

to see whether it was possible to get a continuous readout from the electrode over a period of days, has been achieved. Runs where the electrode was still monitoring penicillin accretion after four days in the fermentor have been achieved. The problems of electrical interference and pH variations causing inaccuracies and noise in the electrode's output are being minimized in the following ways. At the buffering capacity existing in the fermentation, the value of k is approximately 8 mV/mg/ml penicillin V. This means that the output of the probe is very sensitive to changes in pH. As well as trying to get accurate control of pH in the vessel, we are also developing a technique to compensate for the vessel pH. It is possible to construct a second electrode identical to the penicillin electrode except that CPG without penicillinase attached is used in its construction. This electrode has the same dynamic response as the penicillin electrode and is used to monitor vessel pH. The two electrodes are inserted in the fermentor and the output of each logged. The difference between the electrodes' readings is also calculated and logged, and is used to calculate penicillin concentration. In dealing with small mV signals for the various electrodes, problems are encountered with respect to electrical interference. All of the electrode signals are amplified at the fermentor before being relayed to the computer while a single reference electrode is used for pH and penicillin electrodes. As far as possible, floating or differential input instrumentation has been employed.

After three to four days in the fermentor, the electrode's response time usually increased from the 5–10 min observed initially with the PVC membrane to up

TABLE I

Effect of the Membrane Retaining CPG–Penicillinase on the Response Time of the Electrode Following a Step Input of 4 mg/ml Penicillin V in 0.03M Phosphate Buffer

Membrane		Water flow rate (ml/min.cm^2; 25°C,) (Δp = 52 cm Hg)	Response time (mins) (63% of final value)
Polyamide (Sartorius)	8μm	1100	2.5
	3μm	420	12.0
	1.2μm	300	18.0
	0.8μm	225	36.0
PVC (Sartorius)	8μm	1100	2.0
	0.8μm	225	20.0
PVC (Millipore)	2μm	230	9.5
Nylon (Millipore)	14μm	750	6.0
	7μm	450	19.0
	1μm	130	24.0

to 40 min. It has not yet been determined whether this increase is gradual throughout the fermentation or is only occurring at the end of the fermentation when some cell lysis occurs due to lactose exhaustion.

As the penicillin electrode gives a continuous readout, its potential value for control purposes is greatly increased by linking its signal to a computer. At the moment we are only logging the electrode's signal, but the next stage would be to use the signal for control purposes, e.g., for the control of carbohydrate feed rates. The computer is currently being used to control pH, allowing the possibility of using it to carry out titration over small pH increments, e.g., 0.1–0.2 pH units

TABLE II

Chloroform Sterilization of CPG–Penicillinase Preparation

Exposure time (hr)	Viable Count No/ml	% loss in activity
0.0	5 x 10^5	0
0.5	300	4
1.0	0	6
2.0	0	9
3.0	0	11

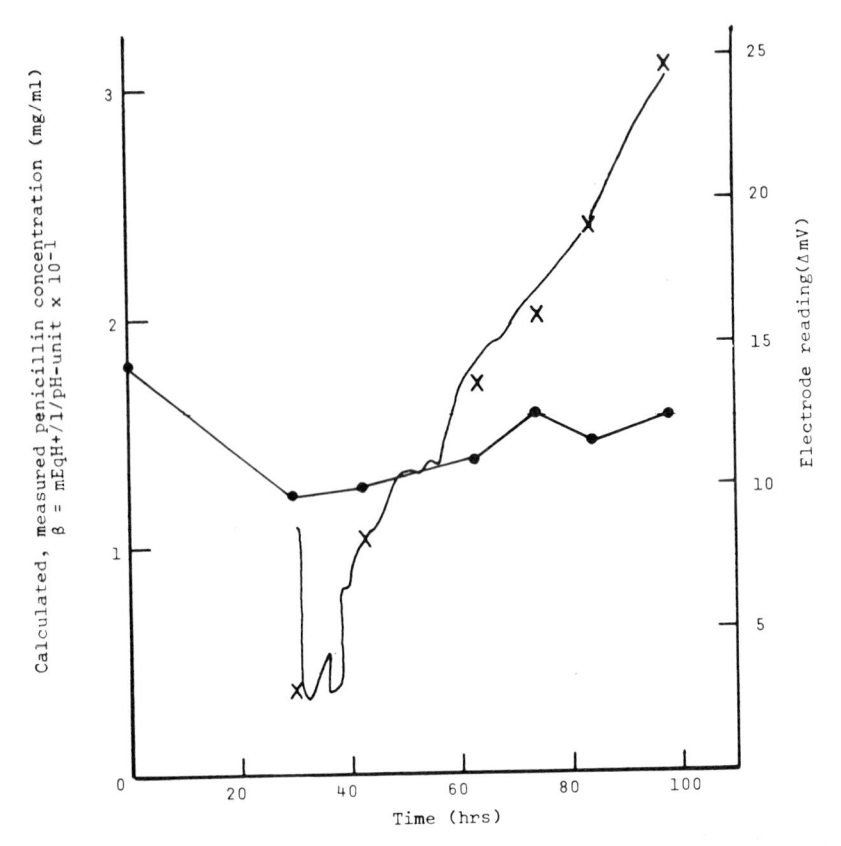

Fig. 5. Penicillin concentration of batch culture (—) as calculated from ΔmV output of penicillin electrode; (×) as assayed by the hydroxylamine method; (●) buffering capacity.

in order to calculate β and hence provide a corrected value of k throughout the run.

CONCLUSION

It was shown that it was possible to measure penicillin accretion in a fermentation lasting four days using a penicillin sensing electrode constructed by retaining a layer of penicillinase covalently linked to CPG in front of a flat-surface pH electrode. Using a previously determined relationship between the buffering capacity of the medium and the sensitivity of the electrode, it was possible to calculate from the ΔmV readings of the electrode penicillin concentrations throughout the fermentation which agreed well with those determined externally by the hydroxylamine assay. Even in a fermentation environment, the immobilized enzyme displayed good stability and the loss of enzyme activity was not a problem. The main problems experienced were pH variations and electrical interference upsetting the readings from the electrode, and an increase in response time throughout the fermentation.

The signal from the electrode would seem to have potential for future closed-loop control.

References

[1] G. G. Guilbault, in *Methods in Enzymology,* K. Mosbach, Ed. (Academic, New York, 1976), Vol. 44.
[2] G. J. Papariello, A. K. Mukherji, and C. M. Shearer, *Anal. Chem., 45,* 790 (1973).
[3] H. Nilsson, A.-D. Åkerlund, and K. Mosbach, *Biochim. Biophys. Acta, 320,* 529 (1973).
[4] L. F. Cullen, J. F. Rusling, A. Schleifer, and G. J. Papariello, *Anal. Chem., 46,* 1955 (1974).
[5] J. F. Rusling, G. H. Luttrel, L. F. Cullen, and G. J. Papariello, *Anal. Chem., 48,* 1211 (1976).
[6] H. Nilsson, K. Mosbach, S.-O. Enfors, and N. Molin, *Biotechnol. Bioeng. 20,* 527 (1978).
[7] R. A. Messing, *J. Am. Chem. Soc., 91,* 2370 (1969).
[8] H. H. Weetall and N. B. Havewala, *Biotechnol. Bioeng. Symp., 3,* 241 (1972).
[9] G. E. Boxer and P. M. Everett, *Anal. Chem.,* 670 (1949).
[10] E. P. McCann and C. T. Calam, *J. Appl. Chem. Biotechnol., 22,* 1201 (1972).
[11] T. H. Jong, M.Sc. thesis, University of New South Wales, Sydney, Australia, 1977.
[12] J. Imsande, F. D. Gillin, R. J. Tanis, and A. G. Atherly, *J. Biol. Chem., 245,* 2205 (1970).

Mathematical Models and Computer Simulations of Dialysis and Nondialysis Continuous Processes for Ammonium-Lactate Fermentation

R. W. STIEBER and PHILIPP GERHARDT

Department of Microbiology and Public Health, Michigan State University, East Lansing, Michigan 48824

INTRODUCTION

Of the various separation processes that involve membranes (dialysis, microfiltration, ultrafiltration, and reverse osmosis), dialysis is the most easily applied to the production of microbes and their metabolites. Microbial cultivation often can be enhanced by introducing a dialysis membrane between the culture chamber and a medium-containing reservoir. Not only can nutrients thereby diffuse from the reservoir to the culture, but metabolic products can also escape from the culture to the reservoir so that the effects of inhibitory metabolic products are minimized. Such a system of dialysis culture can be operated batchwise, continuously, or with a combination of these modes [1].

We have undertaken a fundamental study of dialysis continuous culture using the ammonium-lactate fermentation of whole or deproteinized whey as a model process [2,3]. The fermentation can be managed without sterilization or asepsis as a result of the restrictive conditions of low pH, high concentration of undissociated acid, high temperature, and anaerobiosis. The substrate whey is a residue of cheese manufacture and has become a pollution problem as well as an economic and nutrient loss, so that useful disposal of whey is important. The rationale of ammoniated organic acid fermentation is applicable also to other carbohydrate-rich agricultural residues. The ammonium lactate produced can be used effectively as a feed supplement for ruminant animals, or can be converted into lactic acid for use as an industrial chemical and ammonium sulfate for use as a fertilizer. The lactobacillus cells can provide high-quality single-cell protein for animal or human consumption. Thus the fermentation is useful for its commercial potential as well as for fundamental study of dialysis continuous-culture processes.

In the dialysis continuous process for ammonium-lactate fermentation, the substrate is fed into a fermentor and the fermentor contents are dialyzed through a membrane against water. Thereby the small molecular products are removed from the immediate environment of the bacterial cell, thus relieving the inhibition by a product that normally regulates its production. As more product is withdrawn by dialysis, more substrate is consumed and more product is made, i.e., the fermentation becomes more efficient. The dialysis separation of much of the small molecular products additionally represents a purification step toward alternative uses.

In a review of the principles and applications of dialysis culture, Schultz and Gerhardt [1] showed the predictive value of mathematical modeling and computer simulation, especially for a completely continuous dialysis culture system. Afterwards, Coulman et al. [4] developed a generalized mathematical model of a dialysis continuous process for application to the steady-state ammonium-lactate fermentation of whey and then [5] confirmed the model predictions with experimental tests using whole whey as the substrate. Stieber and Gerhardt [6] improved the model by incorporating separate terms for substrate limitation and product inhibition into the equation describing the rate of cell growth, further validated the model, and extended the applicability by using deproteinized whey as the substrate. Stieber and Gerhardt [7] also showed the usefulness of the model for nondialysis single- and two-stage continuous processes for the fermentation.

In this paper, we summarize the improved mathematical model and computer simulations for the ammonium-lactate fermentation applied to nondialysis as well as dialysis continuous processes. We also report on the validity, use, and potential of the model.

MODEL OF AMMONIUM-LACTATE FERMENTATION

A set of rate relationship equations was selected for modeling the ammonium-lactate fermentation. The equation for the rate of substrate utilization includes terms for cell growth and cell maintenance [8]:

$$-r_s = \alpha r_g + \beta X_f \tag{1}$$

The symbols in this and subsequent equations are listed in Table I.

The rate of product formation is proportional to that of substrate utilization [9]:

$$r_p = -\gamma r_s \tag{2}$$

The equation for the rate of cell growth is based on competitive inhibition kinetics and includes terms for both substrate limitation and product inhibition [6]:

$$r_g = \mu_m (S_f X_f)/(K_s + S_f + K_p P_f) \tag{3}$$

TABLE I
Glossary of Mathematical Symbols

Symbol	Description	Units
A_m	Area of membrane available for dialysis	cm^2
F_d	Flow rate into and out of dialysate circuit	ml/h
F_f	Flow rate into and out of fermentor circuit	ml/h
K_p	Product inhibition constant	mg/mg
K_s	Substrate limitation constant	mg/ml
P_d	Product concentration in dialysate circuit	mg/ml
P_f^o	Product concentration in fermentor feed	mg/ml
P_{mp}	Permeability of membrane to product	mg/cm^2-h
P_{ms}	Permeability of membrane to substrate	mg/cm^2-h
r_g	Rate of cell growth	mg/ml-h
r_p	Rate of product formation	mg/ml-h
$-r_s$	Rate of substrate utilization	mg/ml-h
S_d	Substrate concentration in dialysate circuit	mg/ml
S_f^o	Substrate concentration in fermentor feed	mg/ml
S_f	Substrate concentration in fermentor circuit	mg/ml
t	Time	h
V_d	Volume of liquid in dialysate circuit	ml
V_f	Volume of liquid in fermentor circuit	ml
X_f	Cell-mass concentration in fermentor circuit	mg/ml
α	Substrate/cell ratio	mg/mg
β	Specific maintenance rate	h^{-1}
γ	Product/substrate ratio	mg/mg
μ_m	Maximum specific growth rate of cells	h^{-1}
T_f	Cell-retention time in fermentor circuit	h

MODEL AND SIMULATIONS OF DIALYSIS CONTINUOUS PROCESS

Design of Fermentation System

Figure 1 shows a schematic of the dialysis continuous fermentation system. For purposes of mathematical modeling, the following assumptions are made: high rates of mixing and circulation ensure homogeneity throughout the system; liquid turbulence and excess membrane surface ensure insignificant fouling of

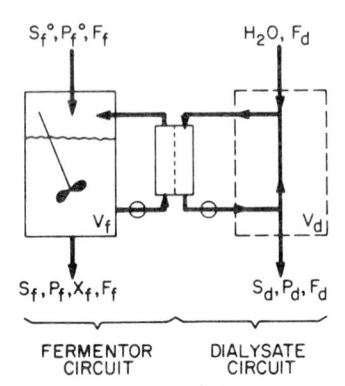

Fig. 1. Schematic of dialysis continuous fermentation system [4]. Symbols are described in Table I.

the dialyzer membranes for a useful period of time; bacterial metabolic values (μ_m, K_s, K_p, α, β) remain unchanged; the volume of the dialysate circuit is negligible relative to that of the fermentor; the rate of ammonia solution addition is negligible relative to that of substrate feed; and hydraulic pressures are equalized between both circuits.

Material-Balance Equations

A set of material-balance equations was formulated for substrate, product, and cell mass in the fermentor circuit and for substrate and product in the dialysate circuit [4].

Generalized Model

The rate equations were combined with the material-balance equations and the variables were defined in dimensionless parameters (Table II) to obtain a generalized model.

The resulting equations for the fermentor circuit are as follows:

$$\frac{d\overline{S}_f}{dt} = \left[-(1+\Pi)\overline{S}_f - \left(\theta \frac{\overline{S}_f}{\overline{K}_s + \overline{S}_f + \overline{K}_p \overline{P}_f} + \frac{\beta T_f}{\alpha} \right) \overline{X}_f + \Pi \overline{S}_d + 1 \right] / T_f \quad (4)$$

$$\frac{d\overline{P}_f}{dt} = \left[-(1+R\Pi)\overline{P}_f + \left(\theta \frac{\overline{S}_f}{\overline{K}_s + \overline{S}_f + \overline{K}_p \overline{P}_f} + \frac{\beta T_f}{\alpha} \right) \overline{X}_f + R\Pi \overline{P}_d + \overline{P}_f^0 \right] / T_f \quad (5)$$

$$\frac{d\overline{X}_f}{dt} = \left(\theta \frac{\overline{S}_f}{\overline{K}_s + \overline{S}_f + \overline{K}_p \overline{P}_f} - 1 \right) \overline{X}_f / T_f \quad (6)$$

where $T_f = V_f/F_f$ and is an operational parameter.

For the dialysate circuit, the corresponding equations are

$$\frac{d\overline{S}_d}{dt} = \frac{[-(\Pi + \phi)\overline{S}_d + \Pi \overline{S}_f]F_f}{V_d} \quad (7)$$

TABLE II
Glossary of Dimensionless Parameters

Type	Symbol and definition	Description
Material parameters	$\overline{P}_d = P_d/(\gamma S_f^o)$	Product factor in dialysate circuit
	$\overline{P}_f = P_f/(\gamma S_f^o)$	Product factor in fermentor circuit
	$\overline{P}_f^o = P_f^o/(\gamma S_f^o)$	Product factor in fermentor feed
	$\overline{S}_d = S_d/S_f^o$	Substrate factor in dialysate circuit
	$\overline{S}_f = S_f/S_f^o$	Substrate factor in fermentor circuit
	$\overline{X}_f = \alpha X_f/S_f^o$	Cell factor in fermentor circuit
Operational parameters	$R = P_{mp}/P_{ms}$	Ratio of product/substrate membrane permeabilities
	$\Pi = P_{ms} A_m/F_f$	Membrane permeability factor
	$\phi = F_d/F_f$	Flow-rate ratio
Kinetic parameters	$\overline{K}_s = K_s/S_f^o$	Substrate-limitation factor
	$\overline{K}_p = \gamma K_p$	Product-inhibition factor
	$\theta = \mu_m T_f$	Time factor

$$\frac{d\overline{P}_d}{dt} = \frac{[-(R\Pi + \phi)\overline{P}_d + R\Pi\overline{P}_f]F_f}{V_d} \tag{8}$$

Generalized Steady-State Solution

The equations of the generalized model were rearranged and combined and the time derivatives were set at zero to obtain a generalized solution for the steady state.

Predictions by Computer Simulation

The generalized steady-state solution was used to simulate the fermentation by programming on a digital computer. When the simulations were performed before actual experimental fermentations were conducted, approximate values were used for the various conditions [4].

Such predictive simulations are exemplified by Figure 2, which shows the simulated effects of changes in the cell-retention time (T_f) on the concentration of substrate in the fermentor circuit (S_f) and the dialysate circuit (S_d) at constant flow rates. The simulations predicted that satisfactorily low levels of residual substrate (<5 mg/ml) could be attained by use of a reasonably short retention time in the fermentor (<20 hr).

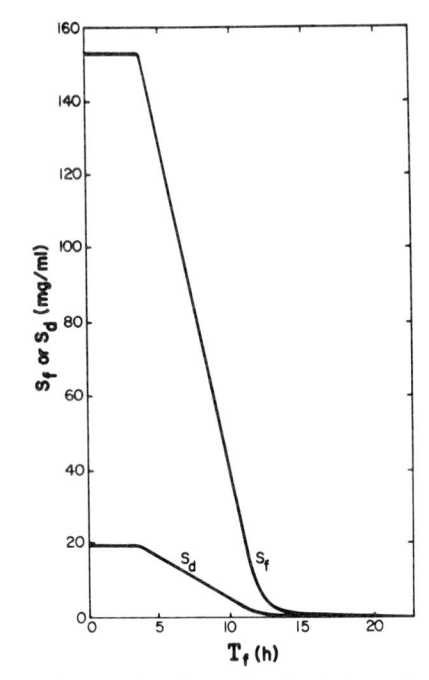

Fig. 2. Simulated effects of changes in cell retention time (T_f) on substrate concentrations in the fermentor circuit (S_f) and the dialysate circuit (S_d) at constant flow rates [4].

Simulations also were computed to predict the effects of changes in the dialysate flow rate (F_d) on the concentration of substrate in the fermentor circuit (S_f) and the dialysate circuit (S_d) at constant cell-retention time, and to illustrate the interrelationships of the various dimensionless operating parameters [4].

Validation of the Model

Computer simulations were performed after experimental fermentations were conducted, using actual results, to validate the model and to evaluate the experimental results.

The values in Table III were used in these computer simulations, which agreed with the experimental tests using whole whey as the substrate [5]. The values

TABLE III
Values Used for Computer Simulation of Experimental Dialysis Fermentations with Whole Whey [5]

Figure (Days)	μ_m (h^{-1})	$\bar{K}_p$	$\bar{K}_s$	ϕ	Π	R
3 (19-24)	.145	.6	.0004	3.7	.66	3.0
3 (50-75)	.25	.6	.0004	3.4	.62	3.0

for ϕ and Π were calculated directly from the experimental data. The values for μ_m, K_s, and K_p were obtained by successive curve fitting of the simulated with the experimental results. A side effect of the dialysis process was a large osmotic influx of water from the dialysate into the fermentor, diluting its contents. The dilution was accounted for in the simulations by assuming that the diluting water entered the fermentor with the feed stream rather than from the dialysate and by correcting S_f^0 accordingly.

Figure 3 shows the computer-simulated curves obtained by using the mathematical model with the values in Table III to describe the correlation between one important operating parameter and the conversion efficiency, e.g., T_f vs. S_f and S_d. A similar comparison between the simulated and the experimental results was also made for a second operating parameter, e.g., F_d vs. S_f and S_d (not shown). The curves all fit closely with the superimposed points of experimental results. Thus the results demonstrated the validity of the mathematical model.

The values in Table IV were used in further validating simulations, which agreed with experimental tests of the dialysis process using deproteinized whey as the substrate [6]. The values for ϕ and γ were calculated directly from the experimental data. The values for μ_m, K_s, and K_p were obtained from the results

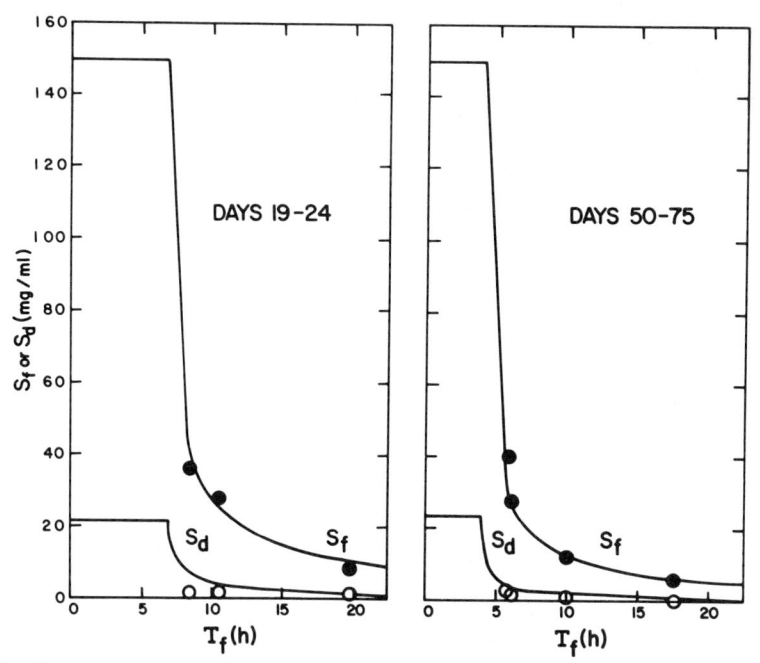

Fig. 3. Computer-simulated effects of cell retention time (T_f) on residual lactose in the fermentor circuit (S_f) and the dialysate circuit (S_d) during two periods of dialysis continuous fermentation. Curves were plotted by use of the mathematical model with the values in Table III. Points were replotted from Figure 2 in Stieber et al. [5] to demonstrate the close fit between experimental results and the computer simulations [6].

TABLE IV
Values Used for Computer Simulation of Experimental Dialysis Fermentations with Deproteinized Whey [6]

Figure (Days)	μ_m (h^{-1})	$\bar{K}_p$	$\bar{K}_s$	ϕ	Π	R	γ
4 (26–36)	.35	2.2	.0004	1.0	.50	3.0	.96
4 (40–49)	.35	2.2	.0004	2.5	.40	3.0	.96

of a nondialysis continuous process for the fermentation [7]. S_f^0 was corrected (160.3 mg/ml) as above to account for osmotic dilution of the fermentor contents.

Figure 4 shows the computer-simulated curves obtained by using the mathematical model with the values in Table IV to describe the relation between the two principal operating parameters (T_f and ϕ) and the accumulation of product (P_f and P_d). Similar comparisons also were made between the same operating parameters and the conversion efficiency (S_f and S_d) and the cell mass accumulation (X_f) (not shown). All of the simulation curves fitted well with the superimposed points of experimental data. Furthermore, the kinetic constants (μ_m, K_s, K_p), used to correlate the simulated results with the experimental data in the dialysis system [6], were evaluated in a nondialysis system [7]. Altogether, these results additionally demonstrated the validity of the mathematical model.

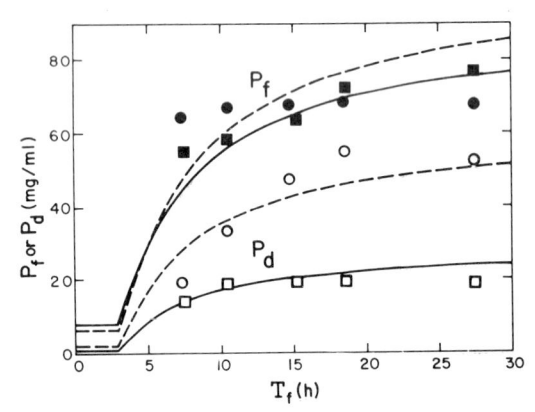

Fig. 4. Computer-simulated effects of cell retention time (T_f) on accumulated lactate in the fermentor circuit (P_f) and the dialysate circuit (P_d) at two flow rate ratios (ϕ) and during two time periods of dialysis continuous fermentation [6]. Curves were plotted by use of the mathematical model and the values in Table IV. Points are experimental data and demonstrate the fit between the experimental results and the computer simulations. Dashed curves and circle points were obtained at $\phi = 1.0$ and during days 26–36. Smooth curves and square points were obtained at $\phi = 2.5$ and during days 40–49.

Process Monitoring

On-line sensors are becoming increasingly important in the operation of fermentations. In the present process, the dialysate effluent is a relatively pure solution of ammonium lactate and so was measurable by its electrical conductivity. Measurements showed a positive correlation between the concentration of ammonium lactate and the conductance in the dialysate [6]. If coupled with the model, on-line monitoring of product concentration could indicate other parameters of the fermentation, e.g., cell concentration in the fermentor and membrane permeability in the dialyzer.

MODEL AND SIMULATIONS OF NONDIALYSIS CONTINUOUS PROCESSES

Design of Fermentation System

The nondialysis single-stage continuous fermentation system consisted of an ordinary continuous fermentor for which the flow rate of the effluent equalled that of the feed [7]. For purposes of mathematical modeling, the appropriate assumptions are the same as those made for modeling the dialysis continuous process.

Generalized Model

A set of material-balance equations was formulated for substrate, product, and cell mass in the fermentor. The rate equations for the ammonium-lactate fermentation were combined with the material-balance equations and the variables were defined in dimensionless parameters (Table II) to obtain a generalized model for nondialysis continuous fermentation.

The resulting equations are as follows:

$$\frac{d\overline{S}_f}{dt} = \left[-\left(\theta \frac{\overline{S}_f}{\overline{K}_s + \overline{S}_f + \overline{K}_p \overline{P}_f} + \frac{\beta T_f}{\alpha}\right)\overline{X}_f - \overline{S}_f + 1\right] / T_f \qquad (9)$$

$$\frac{d\overline{P}_f}{dt} = \left[\left(\theta \frac{\overline{S}_f}{\overline{K}_s + \overline{S}_f + \overline{K}_p \overline{P}_f} + \frac{\beta T_f}{\alpha}\right)\overline{X}_f - \overline{P}_f + \overline{P}_f^0\right] / T_f \qquad (10)$$

$$\frac{d\overline{X}_f}{dt} = \left(\theta \frac{\overline{S}_f}{\overline{K}_s + \overline{S}_f + \overline{K}_p \overline{P}_f} - 1\right)\overline{X}_f / T_f \qquad (11)$$

Generalized Steady-State Solution

The equations of the generalized model were rearranged and combined and the time derivatives were set at zero to obtain a generalized solution for the steady state.

Validation of the Model

The generalized steady-state solution was used to simulate the fermentation by programming on a digital computer. Values for μ_m, K_s, K_p, α, and β, which were obtained by successive curve fitting of the simulated with the experimental results (Table V), proved best for correlating the computer simulations with the fermentation results using deproteinized whey as the substrate. Values for S_f^0 (72.8 mg/ml), P_f^0 (2.8 mg/ml), and γ (0.96) were obtained from the experimental results.

Figure 5 shows the computer-simulated curves obtained by using the mathematical model with the values in Table V to describe the relation between the main operating parameter (T_f) and the cell mass accumulation (X_f). Similar comparisons also were made between T_f and the conversion efficiency (S_f) and the accumulation of product (P_f) (not shown). All of the simulation curves agreed with the superimposed points of experimental results and thereby demonstrated the validity of the mathematical model.

A two-stage operation for the nondialysis continuous fermentation also was computer-simulated at various retention-time ratios [7]. In this system the effluent from the first fermentor was used as the feed for the second fermentor, with the assumption that there was no lag in growth of the cells entering the second fermentor. The results predicted that the combined retention time could be reduced to 17 hr, with conversion efficiency as good as that with the single-stage operation.

TABLE V
Values Used for Computer Simulation of Experimental Nondialysis Continuous Fermentations with Deproteinized Whey [7]

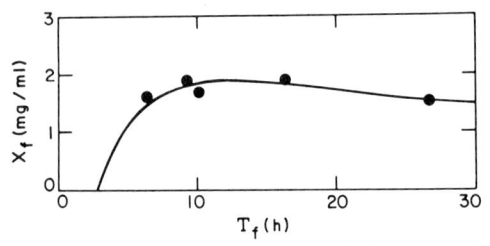

Figure	μ_m (h^{-1})	$\overline{K}_p$	$\overline{K}_s$	R	γ (mg/mg)	α (mg/mg)	β (h^{-1})
5	.35	2.2	.001	3.0	.96	11.0	1.0

Fig. 5. Computer-simulated effects of cell retention time (T_f) on accumulated cell mass in the fermentor (X_f) during nondialysis single-stage continuous fermentation [7]. Curve was plotted by use of the mathematical model and the values in Table V. Points are experimental data and demonstrate the fit between the experimental results and the computer simulations.

DISCUSSION

Fermentation Analysis by Use of the Model

The experimental and simulated results also were correlated to evaluate the fermentation processes. Figure 3 shows that the conversion efficiency increased substantially as time progressed during the fermentation of whole whey [6]. Values for the bacterial metabolic variables (Table III), obtained by use of the model, showed that the adaptation after day 50 resulted in μ_m increasing from 0.145–0.25 hr^{-1}, while K_s and K_p remained unchanged. Thus, although the maximum concentration of the product tolerated by the culture did not change, the culture was able to grow and metabolize at a faster rate after the adaptation than before.

The values for Π (Table IV: 0.5 at days 26–36 and 0.4 at days 40–49) showed that the permeability of the membrane decreased as the fermentation of deproteinized whey progressed. Calculations of Π using a P_{ms} of 0.06 mg/cm^2-hr showed that Π should have been about 0.9. Thus, by four weeks of continuous fermentation, the permeability factor decreased by 50%. Coulman et al. [4] showed that a Π of 2.0 should be used for the dialysis process to obtain suitable relief from product inhibition.

A comparison of the values for μ_m, K_s, and K_p (Tables IV and V) showed that the metabolic variables remained unchanged, indicating no change in the bacterial culture in these fermentations, which were run sequentially and with deproteinized whey. The values for K_s and K_p also showed that the fermentations were not affected by substrate limitation but were greatly limited by increasing concentration of product.

The relative importance of growth and maintenance metabolism in the fermentation can be calculated from knowledge of the values for α and β [9]. Calculations showed that, at conditions which allowed maximal substrate utilization, maintenance metabolism accounted for more than 70% of the total energy of catabolism.

Potential of the Model

The model for the ammonium-lactate fermentation is simple and agreed with experimental tests of both dialysis and nondialysis processes for the fermentation over a wide range of operating parameters, with high substrate and product concentrations, and with both whole and deproteinized whey as substrate. Specifically, the model agreed with experimental results of substrate utilization, product formation, and cell-mass accumulation. Moreover, values for the bacterial metabolic variables (μ_m, K_s, and K_p) remained unchanged over the whole range of conditions with deproteinized whey as substrate. The model also contains substrate-limitation and product-inhibition terms that have a biological basis [7], and the values used for these terms are realistic.

A shortcoming of the model is that the product-inhibition effect is not strong enough at lactate concentrations greater than 70 mg/ml. Furthermore, eq. (1)

implies that substrate utilization can continue after all of the substrate has been utilized. A solution to both problems may lie with the incorporation of substrate-limitation and product-inhibition terms into the maintenance term of eq. (1), as well as in eq. (3). Since a dialysis culture is able to tolerate greater lactate concentrations than is a nondialysis culture [5], there may be an unknown dialyzable factor that also has an inhibition effect.

Altogether, the mathematical model gives a good description of the ammonium-lactate fermentation and should be useful for on-line monitoring whether managed as a dialysis or nondialysis continuous process.

This work was supported by grants No. ENG 76-17260 and DAR 79-10236 from the National Science Foundation. Illustrations from original articles are reprinted with permission of the publisher and authors.

References

[1] J. S. Schultz and P. Gerhardt, *Bacteriol. Rev., 33*, 1 (1969).
[2] P. Gerhardt and C. A. Reddy, *Dev. Ind. Microbiol., 19*, 71 (1978).
[3] F. W. Juengst, Jr., *J. Dairy Sci., 62*, 106 (1979).
[4] G. A. Coulman, R. W. Stieber, and P. Gerhardt, *Appl. Environ. Microbiol., 34*, 725 (1977).
[5] R. W. Stieber, G. A. Coulman, and P. Gerhardt, *Appl. Environ. Microbiol., 34*, 733 (1977).
[6] R. W. Stieber and P. Gerhardt, *Appl. Environ. Microbiol., 37*, 487 (1979).
[7] R. W. Stieber and P. Gerhardt, *J. Dairy Sci.,* (1979) in press.
[8] A. G. Marr, E. H. Nilson, and D. J. Clark, *N.Y. Acad. Sci., 102*, 536 (1963).
[9] A. K. Keller and P. Gerhardt, *Biotechnol. Bioeng., 17*, 997 (1975).

Modeling and Optimal Control of an SCP Fermentation Process

J. ALVAREZ and J. RICAÑO

Electrical Engineering Department, Centro de Investigaciones del I. P. N., Mexico City, Mexico

INTRODUCTION

There are several works dealing with the modeling, identification, and control aspects of biomass fermentation processes [1-4]. One of the main problems of those studies is the lack of statistical analysis on the estimated parameters. In addition, it is very difficult to apply these models for the optimal control of the process.

This research deals with a batch culture fermentation process for producing single cell protein (SCP) from yeast grown on methanol. Nonlinear programming methods are employed to estimate the growth kinetic parameters. These methods obtain the parameter values of a nonlinear equation by an iterative procedure that minimizes the sum of the least squares errors. The influence of the control variables on the growth kinetic parameters can be represented by a polynomial expression. A maximum likelihood method was used for the identification of the parameters in this expression.

In order to analyze the success of the estimation method, it is necessary to make several statistical tests. A Z test is made on the residuals to prove their randomness, and a Student-Fischer test is used to detect the meaningless parameters of the polynomial regression.

It is well known that closed-loop control improves the performance of a process. However, in order to design an optimal control strategy that could be applied in a closed-loop system, a dynamic programming method is used for maximizing a productivity cost function.

MATHEMATICAL MODEL AND PARAMETER ESTIMATION

A Monod [5] type model was used for a set of experiments. These experiments covered the range of conditions where the best operating points were believed to be.

The Monod equation can be written

$$\frac{dx}{dt} = \mu_m \frac{s}{s + K_s} x \qquad (1)$$

The substrate equation is expressed

$$\frac{ds}{dt} = -\frac{1}{Y}\frac{dx}{dt} \qquad (2)$$

where x is the biomass concentration, s the substrate concentration, μ_m is the maximum specific growth rate, Y is the overall yield factor; K_s is the Michaelis–Menten constant, and t is time.

By considering the yield factor as a constant, eq. (2) can be integrated and combined with eq. (1):

$$\frac{dx}{dt} = \mu_m \frac{x_0 + Ys_0 - x}{x_0 + Ys_0 + YK_s - x} x \qquad (3)$$

with x_0 and s_0 being the initial concentrations of biomass and substrate, respectively.

It was found in the literature [2,6] and by experimentation that the temperature (T) and pH influence the maximum specific growth rate and the overall yield. The nature of this influence suggested polynomial relationships as follows:

$$\mu_m = a_0 + a_1 T + a_2 \text{pH} + a_3 T^2 + a_4 \text{pH}^2 + a_5 T\text{pH} \qquad (4)$$

$$Y = b_0 + b_1 T + b_2 \text{pH} + b_3 T^2 + b_4 \text{pH}^2 + b_5 T\text{pH} \qquad (5)$$

Examining eqs. (3)–(5), it was observed that it is first necessary to make estimates of μ_m, Y, and K_s before one can identify a_i and b_i. Equation (3) is nonlinear; hence, a nonlinear programming iterative procedure was utilized. This method combined the high-speed convergence property of the gradient method and the high-accuracy property near the optimum of the Gauss–Newton [7]. Both methods use an iterative algorithm to obtain the minimum. This can be expressed as

$$\theta_{k+1} = \theta_k + \Delta\theta_k \qquad (6)$$

where θ_{k+1} is the parameter vector at iteration $k + 1$ and $\Delta\theta_k$ is a correction factor that is different in each method.

For the gradient method

$$\Delta\theta_k = -A \left.\frac{\partial J}{\partial \theta}\right|_k \qquad (7)$$

and for the Gauss–Newton method

$$\Delta\theta_k = -G_k^{-1} \left.\frac{\partial J}{\partial \theta}\right|_k \qquad (8)$$

where J is the sum of square errors.

G is defined as follows:

$$G_k = \begin{bmatrix} g_{11}(k) & g_{12}(k) & \cdots & g_{1n}(k) \\ g_{21}(k) & g_{22}(k) & \cdots & g_{2n}(k) \\ g_{n1}(k) & g_{n2}(k) & \cdots & g_{nn}(k) \end{bmatrix} \qquad (9)$$

with

$$g_{ij}(k) = \sum_{\lambda=1}^{N} \left[\frac{\partial e(\lambda)}{\partial \theta_i} \frac{\partial e(\lambda)}{\partial \theta_j} \bigg|_k \right] \quad (10)$$

where e is an error between an experimental point and the one estimated by the iterative procedure.

Table I shows some results obtained from the set of experiments along with standard deviations for each estimated parameter. It is observed that the standard deviations of μ_m and Y are small, which indicates a good estimation. However, the standard deviations for K_s indicate that the estimated value is not very accurate. This inaccuracy is due to the inadequate number of experimental points around the inflection zone, where the numerical value of the substrate concentration is very small. Nevertheless, Figure 1 shows a reasonable fit of the model curve to the experimental points.

TABLE I
Some Results Obtained for the Estimation of μ_m, Y, and K_s with their Respective Standard Deviations

Temp. (°C)	pH	μ_m (hr^{-1})	Y (g/g)	K_s (g/liter)
37.5	4.5	0.2102 ± 0.0056	0.3309 ± 0.0029	0.055 ± 0.095
37.5	4.0	0.1715 ± 0.0009	0.3485 ± 0.0036	0.0123 ± 0.0014
33.0	5.0	0.1322 ± 0.0013	0.3339 ± 0.0049	0.0055 ± 0.0004
45.0	5.0	0.0863 ± 0.0060	0.2734 ± 0.0062	0.0068 ± 0.0005
42.5	3.5	0.1786 ± 0.0032	0.2943 ± 0.0055	0.0067 ± 0.0036

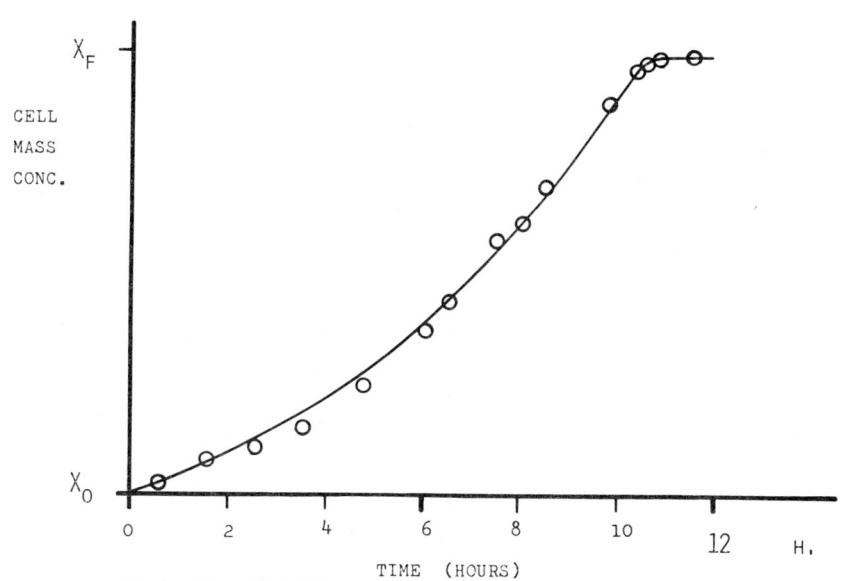

Fig. 1. Curve fitting between the experimental points and the model.

In order to avoid a biased estimation of a_i and b_i the maximum likelihood method was utilized [8]. On a previous identification, it was found that some parameters had very high standard deviation. It was also suspected that some of the parameters had no physical meaning. A Student–Fisher [9] test was used to find such parameters, and after a new estimation of T and pH, the final results were obtained (see Table II).

STATISTICAL ANALYSIS

To test the accuracy of the parameter estimation, it is necessary to examine the residuals. The Z test was applied to examine the randomness of the residuals. This test basically consists of counting the number of runs, where a run is defined as a sequence of residuals with the same sign. If the number of runs is out of the range determined by the test, a lack of randomness is suspected. For a set of 17 experiments, 16 of them passed this test, so the residuals could be explained as a measurement error.

The mean and standard deviation values with their respective confidence intervals were calculated. Since the absolute numerical values (Table III) were very small compared to the measured values, it appears to be a nonbiased estimate.

TABLE II
Results Obtained in the Estimation of the Parameters of the Regression for μ_m and Y

Parameter	Value
a_0	-0.2658 ± 0.0774
a_1	0.6440 ± 0.0739
a_2	0.9551 ± 0.2363
a_3	-0.4129 ± 0.0411
a_4	-0.6744 ± 0.1890
a_5	-0.2989 ± 0.0820
b_2	1.1826 ± 0.0395
b_4	-1.0690 ± 0.0622

TABLE III
Values Obtained of the Mean, Variance, and their Respective Confidence Intervals for Some of the Experiments

Mean value	Confidence interval of the mean value	Variance	Confidence interval of the variance
-0.0088	$-0.0290 < \mu < 0.0114$	0.0011	$0.0007 < \sigma^2 < 0.0038$
-0.0137	$-0.0397 < \mu < 0.0123$	0.0020	$0.0013 < \sigma^2 < 0.0065$
-0.0087	$-0.0257 < \mu < 0.0084$	0.0010	$0.0006 < \sigma^2 < 0.0029$
-0.0074	$-0.0185 < \mu < 0.0036$	0.0004	$0.0003 < \sigma^2 < 0.0012$
0.0018	$-0.0191 < \mu < 0.0227$	0.0008	$0.0005 < \sigma^2 < 0.0036$

OPTIMAL CONTROL STRATEGIES

Pontriagyn's maximum principle has been used several times to construct an optimal control strategy [2,10]. This method gives an open-loop control law $u^0(t)$, because the process is nonlinear as expressed in eq. (3) [11]. To obtain a closed-loop control law $u^0(x;t)$, the dynamic programming method was employed.

This method is based on the Bellman optimality principle, which states that: "An optimal policy has the property that whatever the initial state and initial decision are, the remaining decisions must constitute an optimal policy with regard to the state resulting from the first decision [12]."

To apply this method, a discretization of the working field is necessary, so time, biomass, temperature, and pH ranges were divided into small intervals. The method proceeds with an opposite sense of time, calculating for each state point (biomass-time) the best control point (temperature-pH) for maximizing the productivity cost function. At the end of the calculations, when time equals zero, a table of optimal control values that are functions of the process state is obtained.

Since eq. (3) does not have an analytical solution, an approximate discrete model was utilized. For obtaining this model, a first-order approximation of a derivative is applied, which gives the following:

$$x_{k+1} = \left(1 + \Delta t \, \mu_m \frac{x_0 + Ys_0 - x_k}{x_0 + Ys_0 + YK_s - x_k}\right) x_k \quad (11)$$

with Δt as the time interval.

If productivity is defined as the maximum biomass obtained in minimum time, the cost function is as follows:

$$J = (x_f - x_0)/t_f \quad (12)$$

where x_f is the final biomass concentration, x_0 is the inoculum concentration, and t_f is the fermentation time. It t_f is constant, maximization of Jt_f gives the same result as maximizing J. The optimal control law is

$$J_k^0 = \max(x_k - x_{k-1} + J_{k-1}^0) \quad (13)$$

When the method is applied, a set of control values that are functions of the process state is obtained (Table IV).

Incorporating this control strategy into the process resulted in an improvement of about 1%. This small quantity can be explained by examining the control values where the maximum specific growth rate and the maximum yield were obtained:

$$\mu_m^0 = 0.215; \quad T_{\mu m}^0 = 38.7°C; \quad pH_{\mu m}^0 = 4.2$$
$$Y^0 = 0.349; \quad T_Y^0 = 38.8°C; \quad pH_Y^0 = 4.1$$

TABLE IV
Relationship between the Process State Variables and the Control Variables

Biomass (g/liter)	Temp. (°C)	pH
0.5	38.5	4.25
1.0	38.5	4.25
1.5	38.5	4.25
2.0	39.0	4.00
2.5	39.0	4.00

This effect of optimization is reduced because of the small trajectory necessary to move the control variables.

CONCLUSION

For all experiments, the model structure with three parameters gives the best description of the system. The aid of nonlinear programming for estimating parameters led to successful results, as it is observed in the statistical analysis.

The optimal control scheme was not so favorable because the operation conditions of this yeast were almost unimprovable; however, on a simulated process the results shown were much better.

The most important remark about this work is that a general method of identification and optimal control for biomass batch fermentation processes was implemented.

This research was sponsored by the Organization of American States.

References

[1] A. Lukasik, "Identification de procédés de fermentation discontinus," Institut National Polytechnique de Grenoble, D. 3éme. Cycle (1974).
[2] D. Bourdaud and C. Foulard, "Identification and optimization of batch culture fermentation processes," on 1st. European Conference on Computer Process Control in Fermentation (1973).
[3] H. Y. Wang, "Computer control of yeast fermentation," Massachusetts Institute of Technology, Ph.D. thesis, 1977.
[4] M. Nihtilä and J. Virkkunen, *Biotechnol. Bioeng.*, 19, 1831 (1977).
[5] J. Monod, *Recherche sur la Croissance des Cultures Bactériennes* (Hermann et Cie, Paris, 1941).
[6] V. K. Eroshin, I. S. Utkin, S. V. Ladynichev, V. V. Samoylov, V. D. Kuyshinnkov, and G. K. Skryabin, *Biotechnol. Bioeng.*, 18, 289 (1976).
[7] R. Boudarel, J. Delmas, and P. Guichet, *Commande Optimal des Processus* (Dunod, Paris, 1968) Tome 2.
[8] P. Eykhoff, *System Identification* (Wiley, New York, 1974).
[9] P. Hoel, *Introduction to Mathematical Statistics* (Wiley, New York, 1971).
[10] A. Constantinides, J. L. Spencer, and E. L. Gaden, *Biotechnol. Bioeng.*, 12, 1081 (1970).
[11] M. Athans and P. L. Falb, *Optimal Control* (McGraw-Hill, New York, 1966).
[12] R. Bellman and S. Dreyfus, *Applied Dynamic Programming* (Princeton U. P., Princeton, 1962).

Growth Kinetics and Antibiotic Synthesis during the Repeated Fed-Batch Culture of *Streptomycetes*

M. BOŠNJAK, V. TOPOLOVEC, and VERA JOHANIDES

PLIVA–Pharmaceutical and Chemical Works and University of Zagreb, Zagreb, Yugoslavia

INTRODUCTION

Upon the introduction of the term "fed-batch culture" [1] (FBC) a series of papers concerning the study of FBC appeared [2–11]. However, only some of these papers [2,3,8] include discussions concerning the theory and application of a repeated fed-batch culture (RFBC). RFBC seems to be a very attractive continuous-cultivation method. The reasons for suggesting the study of RFBC include the ease of applying this technique of continuous cultivation in industrial production of some products (e.g., those termed as "secondary metabolites") and the convenience of the method for the study of kinetics of microbial processes. RFBC can be applied as one stage in multistage systems. Two-stage semicontinuous cultivation (TSSC) of *Streptomyces erythreus* [12] is an example. The recent report concerning RFBC as the first stage of TSSC [13] can be considered as an introduction to the theory of RFBC of *Streptomycetes*. The subject of this article is the further development of the theory discussing its applicability to experimental data. The model describing growth kinetics during the batch process [14–16] has to be used as a basis in developing the model for RFBC.

THEORY

Growth Kinetics

In the batch process the growth kinetics of *Streptomycetes* can be expressed by the equation [14]

$$\frac{dx}{dt} = k_1 x^{2/3} - k_2 x \tag{1}$$

or more generally, including the decline phase, by the equation [15,16]

$$\frac{dx}{dt} = k_1 x^{2/3} - k_2 x - k_3 xt \tag{2}$$

Since FBC is defined as a batch culture provided with a feed of complete medium, the rate of change of biomass concentration is a consequence of growth rate and dilution rate, and it can be expressed as follows:

$$\frac{dx}{dt} = k_1 x^{2/3} - k_2 x - D_{FBC} x \tag{3}$$

The dilution rate varies during FBC, and in the case of a constant flow rate it can be defined as

$$D_{FBC} = v/V = v/(V_0 + vt) \tag{4}$$

It follows that

$$\frac{dx}{dt} = k_1 x^{2/3} - k_2 x - \frac{v}{V_0 + vt} x \tag{5}$$

If a portion of the culture is withdrawn from time to time, then FBC becomes RFBC and the equation

$$\frac{dx}{dt} = k_1 x^{2/3} - k_2 x - \frac{v_i x}{V_{0i} + v_i(t - t'_{i-1})} \tag{6}$$

appears to be an adequate expression for growth kinetics during RFBC.

Adhesion of microorganisms on the inner surfaces of the fermentor is a phenomenon that must be taken into account when growth kinetics is considered. Microbial films are a feature of every fermentor [17] but the thickness of films and their properties depend upon cultivation conditions and upon character of the microorganisms. Usually mycelial microorganisms, molds, and *Streptomycetes* form films that interfere with the process kinetics. The thickness of the microbial layer within a given fermentor configuration can only be determined by experimentation [17] and the study of the influence of this layer on process kinetics is a complex problem. However, in the limited range of RFBC one can presume that this influence on process kinetics, in bulk culture, increases with time. Suppose that the consequent decrease of biomass concentration can be represented by the equation

$$\left(\frac{dx}{dt}\right)_{ML} = -k'_3 x t \tag{7}$$

it follows for RFBC that

$$\frac{dx}{dt} = k_1 x^{2/3} - k_2 x - k'_3 x t - \frac{v_i x}{V_{0i} + v_i(t - t'_{i-1})} \tag{8}$$

Kinetics of Antibiotic Biosynthesis

The biosynthesis of secondary metabolites is a complex function of many factors, and it is difficult to express exactly the connection between growth rate and biosynthesis rate. In secondary metabolism, the physiological state of the culture plays a very important role. In addition, special quality biomass must be produced in the most appropriate way in order to make the production of secondary metabolites as high as possible. Usually, the specific rate of product formation is not constant, and often the relationship between specific growth rate and specific rate of product formation is unknown [8]. In recent report [13] biosynthesis of oxytetracycline and erythromycin were discussed, and a relationship expressing the connection between growth and biosynthesis was applied. Similarly, the simple empirical equation

$$p/x = kt + b \qquad (9)$$

can be used in order to express the biosynthesis of oxytetracycline [14] and erythromycin [18] as a function of biomass concentration during the batch process. This equation is also compatible with the structural model that was developed for qualitative explanation of the batch process of oxytetracycline biosynthesis [19]. Taking into account that the specific rate of antibiotic synthesis is a function of many factors, the rate of antibiotic synthesis in the batch process can be defined by the equation

$$\frac{dp}{dt} = k_p x \qquad (10)$$

According to this equation, the rate of change of antibiotic concentration during RFBC can be expressed as

$$\frac{dp}{dt} = k_p x - \frac{v_i p}{V_{0i} + v_i(t - t'_{i-1})} \qquad (11)$$

Since there is almost always a range of cultivation conditions for which k_p will be constant, the rate of antibiotic synthesis [eq. (9)] can be expressed as

$$\frac{dp}{dt} = kx + \frac{p}{x}\frac{dx}{dt} \qquad (12)$$

Upon substituting

$$\frac{dp/dt}{x} = k_p \qquad (13)$$

it follows that

$$k_p = k + \frac{p}{x^2}\frac{dx}{dt} \qquad (14)$$

If the second term of eq. (14) includes the influence of the physiological state on the value of the specific rate of antibiotic synthesis, then the derivative dx/dt in eq. (14) can be considered as a rate of change of biomass concentration rather than as a growth rate. According to this assumption the specific rate of antibiotic synthesis is a constant when there is no change of biomass concentration. This can be expected in the stationary phase of the batch process, in the steady state of the continuous process, and in the "quasi-steady-state" of FBC and RFBC.

COMPUTER SIMULATION

The best way to solve eqs. (6), (8), and (12) is by computer simulation. In previous works [15,16] digital-analog simulation has been successfully applied to experimental data for the growth kinetics of batch cultures. A similar method can be utilized for solving eqs. (6), (8), and (12) by extending the computer program. The simplification of eqs. (6), (8), and (12) would be useful. For

$$V_{0i}/v_i = K_{vi} \tag{15}$$

eqs. (6), (8), and (12) become

$$\frac{dx}{dt} = k_1 x^{2/3} - k_2 x - \frac{x}{K_{vi} + (t - t'_{i-1})} \tag{16}$$

$$\frac{dx}{dt} = k_1 x^{2/3} - k_2 x - k'_3 xt - \frac{x}{K_{vi} + (t - t'_{i-1})} \tag{17}$$

and

$$\frac{dp}{dt} = k_p x - \frac{p}{K_{vi} + (t - t'_{i-1})} \tag{18}$$

respectively.

When FBC and RFBC are considered, it is important to know the relationship between the specific growth rate and the dilution rate. Since for a batch process the specific growth rate

$$\mu = \frac{1}{x}\frac{dx}{dt} \tag{19}$$

eqs. (1) and (2) becomes

$$\mu = k_1/x^{1/3} - k_2 \tag{20}$$

or

$$\mu = k_1/x^{1/3} - k_2 - k_3 t \tag{21}$$

By defining the specific growth rate in such a way, the change of biomass concentration during RFBC can be expressed as

$$\frac{dx}{dt} = \mu x - D_{RFBC} x \tag{22}$$

The value

$$\mu/D = \mu x/D_{RFBC} x$$

can be calculated by the computer so that the changes during RFBC can be analyzed. Figure 1 presents the block diagram for simulation of growth kinetics during RFBC according to eqs. (16) and (17).

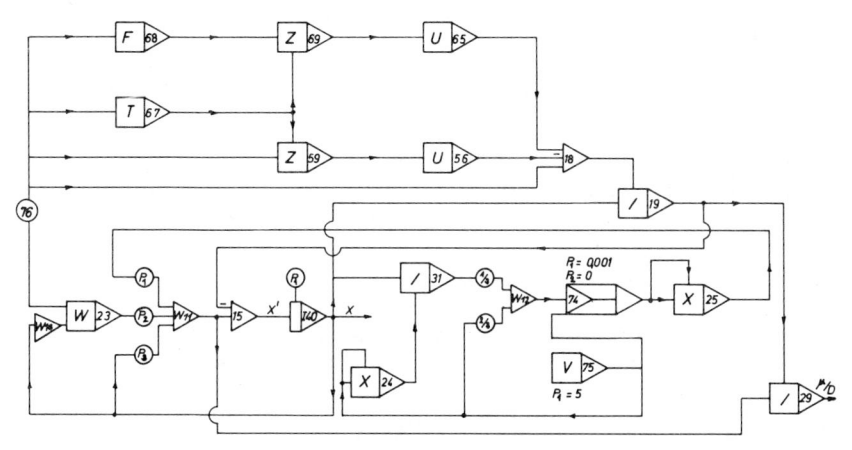

Fig. 1. Repeated fed-batch culture of *Streptomyces* sp.: digital–analog simulator–block diagram. F, function generator; Z, zero-order hold; T, time pulse generator; U, unit delay; W, weighted summer; X, multiplier; /, divider; I, integrator; Y, wye; V, vacuous.

COMPARISON OF SIMULATED AND EXPERIMENTAL DATA

In a recent report [13] a comparison of simulated results to experimental data for the RFBC of *S. rimosus* was investigated. The simulated curves of biomass concentration agreed well with experimental data. However, since antibiotic synthesis was not simulated, the objectives of this work are the further analysis of this simulation and the simulation of antibiotic synthesis during RFBC according to the system of eqs. (16) [or (17)] and (18). Table I illustrates some results of statistical analysis. Since the value of the specific rate of antibiotic synthesis is influenced by the physiological state of the culture as well as by the biomass layer adhering to the inner wall of the fermentation vessel [13], it is reasonable (for the time being) to consider only two cycles of RFBC. It is assumed that an analysis of two cycles will provide useful information for further

TABLE I
Growth Kinetics of *S. rimosus* T_6 during RFBC; Fitting Simulated to Experimental Data

Experiment No.	k_1	k_2	k_3'	a	b_x	Correlation	Biomass (mg/ml) Expressed as:
1	0.315	0.0775	0.00013	1.111	−1.672	0.99203	dry weight
	0.114	0.0775	0.00010	1.078	−0.067	0.96736	nitrogen in dry weight
2	0.315	0.0775	0.00025	1.143	−2.541	0.87622	dry weight
	0.114	0.0775	0.00010	1.125	−0.166	0.97083	nitrogen in dry weight
Both				1.122	−1.981	0.95592	dry weight
	0.114	0.0775	0.00010	1.091	−0.100	0.95865	nitrogen in dry weight

[a] Simulation according to eq. (17) and statistical analysis of $x_E = ax_S + b_x$; x_E = experimental and x_S = simulated values of biomass concentration.

study of RFBC. The values of constants (k_1, k_2, k, k_p) along with the biomass and antibiotic concentrations are given in Table II.

Based on the model, the response of the process to different conditions can be simulated so that the influence of medium flow rates upon process kinetics can be predicted. Some examples of the simulation utilizing constants calculated from experimental data are illustrated by Figures 2 and 3. Data presented in Table III illustrate the influence of flow rates on productivity of RFBC.

DISCUSSION

Data from Table I show that the model is acceptable, especially for the first experiment. It is evident from the data that the same values of constants k_1, k_2, and k'_3 can be applied to the simulation of growth kinetics during both experiments of RFBC if biomass concentration is expressed as dry weight nitrogen. Regardless of the possibility to choose more precise values of constants in order to improve the fit to experimental data, it can be considered that the fit is quite acceptable. It seems that the lower value of the correlation coefficient in the case of second experiment (biomass expressed as dry weight) is the consequence of some experimental error. In addition, it is not so reasonable to expect a perfect fit since ideal cultivation conditions during RFBC cannot be obtained. The conditions concerning mass transfer change during each cycle of RFBC even in the case of perfect control of temperature, impeller velocity, and air flow rate. Experiments were performed varying K_v values in individual cycles [13] and so discrepancies can be expected.

In this light, the results presented in Table II can be discussed. Moreover, the presence of the microbial layer adhering to the vessel wall (its quantity was not measured) is a fact that must not be neglected. The influence of this layer on antibiotic synthesis is not expressed in the model and an explanation of the difference between experimental value and simulated value in the second cycle cannot (for the time being) be complete. Nevertheless, it seems that the model can be accepted if one takes into account some variability of k_p values during

TABLE II
Comparison of Experimental and Simulated Data for Two Cycles of RFBC

| \multicolumn{4}{c|}{Value of Constants} | Cycle No. (simulated) | Simulated Time hr | \multicolumn{4}{c}{Concentration (mg/ml)} |
k_1	k_2	k	k_p			Biomass Sim.	Expt.	Antibiotic Sim.	Expt.
			k_{p1}	1	0	25.0	25.2	1.96	1.96
				1	24	27.1	28.4	3.38	3.42
0.315	0.0775	0.005	k_{p2}	1	24	27.1	28.4	3.61	3.42
			k_{p1}	2	48	25.5	24.5	3.71	4.48
			k_{p2}	2	48	25.5	24.5	3.78	4.48

[a] Simulation according to the system of eqs. (16) and (18); $k_{p1} = k$; $k_{p2} = k + (p/x^2)(dx/dt)$. Duration of cycle = 24 hr.

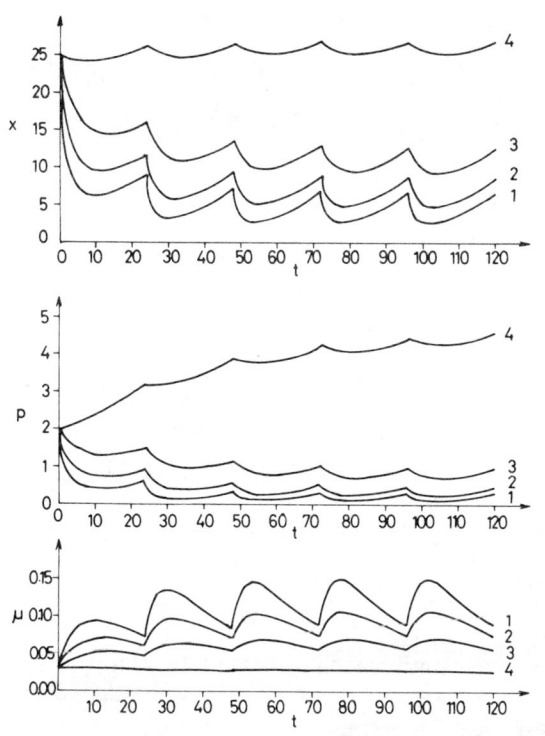

Fig. 2. Simulation of the influence of medium flow rates on process kinetics of RFBC. Medium flow rate $v = (V_1 - V_0)/t_{cycle}$; $t_{cycle} = 24$ hr; $x_0 = 25$ (g/liter); $p_0 = 1.96$ (g/liter); $k_1 = 0.315$; $k_2 = 0.0775$; $k_p = k_{p_1} = k = 0.005$; dilution rate ranges: (1) $V_1/V_0 = 20$; (2) $V_1/V_0 = 10$; (3) $V_1/V_0 = 5$; (4) $V_1/V_0 = 2$. Abscissa: cultivation time (hr); ordinates: x = biomass concentration; p = antibiotic concentration; μ = specific growth rate (hr^{-1}).

RFBC. Based on the presented data, one cannot recommend what to choose for the values of k_p-k_{p_1} or k_{p_2}. For a chosen value of k, both k_{p_1} and k_{p_2} can explain the antibiotic concentration at the end of first simulated cycle, but this number gives somewhat low values at the end of the second cycle in comparison with the experimental value. To some extent, an increase in the k value would increase the disagreement with the experimental results for the first cycle, but it would predict a better agreement between the simulated and experimental values of the second cycle.

If we accept the model, we can apply it for different purposes. As can be observed from Figure 2, medium flow rate has a very significant influence on the course of the process. For constants chosen on the basis of their fit to experimental data (see Table II), the simulated data demonstrate clearly differences in the course of the fermentation when different ranges of dilution rates (V_1/V_0 ratios) are applied. Comparing the presented examples, the following can be seen.

1) At a high medium flow rate (case 1), biomass and antibiotic concentrations

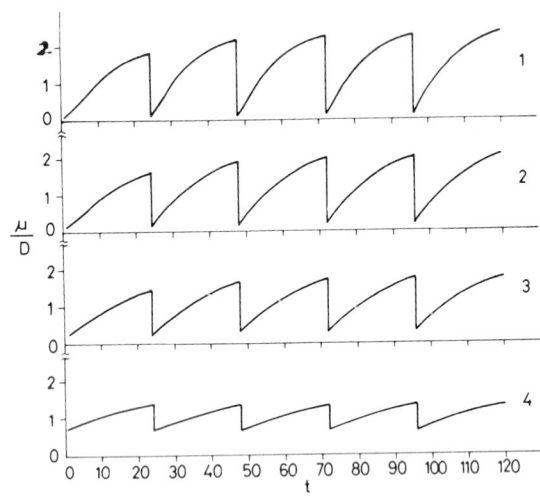

Fig. 3. Changes in the values of ratio μ/D during RFBC and shift up of dilution rates. Abscissa: cultivation time (hr); ordinate: μ/D (dimensionless). Curves 1–4 correspond to flow rates 1–4 as indicated in Figure 2.

decrease successively until reaching an oscillatory level of very low concentrations. Relatively very high and significantly varying values of the specific growth rate can be observed. The amplitudes of μ/D values (Figure 3) are very high.

2) The decrease of medium flow rate (cases 2–4) leads to higher levels of biomass and antibiotic concentrations, lower specific growth rates, and lower amplitudes for oscillating μ/D values.

3) At the lowest flow rate (case 4), the levels of biomass and antibiotic concentrations are the highest. The amplitudes of oscillations are very small and the specific growth rate is practically constant.

The further analysis of simulated data led to the results in Table III. Evidently, better productivities can be obtained by applying lower medium flow rates. However, it must be emphasized that antibiotic productivity increases faster (relatively) than biomass productivity when flow rate is decreasing. Such a phenomenon was also observed in experiments [13]. Comparing the consequences of different expressions for the specific rate of antibiotic synthesis, it can be seen that the influence of these differences is more evident at higher rather than at lower medium flow rates. This is as expected. The data suggest detailed experimental investigation of RFBC in the range of high medium flow rates. Furthermore, the investigation of the applicability of other expressions for the specific rate of antibiotic synthesis which take into account the influence of such factors (e.g., phosphorus compounds [20]) on the biosynthesis and properties of microorganisms and on the lower values of the specific rate of antibiotic synthesis can be tested for the range of higher dilution rates. On the other hand, at cultivation conditions similar to those presented by case 4, the constant value of the specific rate of antibiotic synthesis can be applied. In order to appreciate

TABLE III
Biomass and Antibiotic Productivities in Individual Cycles as a Function of Medium Flow Rates[a]

Example No.	Cycle No.	Biomass Produced in Indicated Cycle (g)	Value of k_p	Antibiotic Produced in Indicated Cycle (g)
1	1	78.70	k_{p1}	4.76
			k_{p2}	5.13
	5	63.22	k_{p1}	2.83
			k_{p2}	4.19
2	1	92.00	k_{p1}	7.03
			k_{p2}	6.69
	5	77.79	k_{p1}	4.29
			k_{p2}	5.70
3	1	108.70	k_{p1}	11.09
			k_{p2}	10.00
	5	100.47	k_{p1}	7.66
			k_{p2}	9.05
4	1	136.29	k_{p1}	22.29
			k_{p2}	23.96
	5	134.98	k_{p1}	23.38
			k_{p2}	24.29

[a] Function of the ratio V_1/V_0; conditions as in Figure 2; calculation based on $V_1 = 10$ liter.

the perspective of the application of the model, it is useful to mention that eqs. (1) and (2) satisfactorily describe growth kinetics of different *Streptomycetes* cultivated in different media [14–16,21]. In addition, eqs. (1), (16), and (17) can be applied to modeling the growth kinetics of *Aspergillus niger* [22]. Therefore its applicability can be extended to other microbial processes.

Nomenclature

a	straight-line direction coefficient(dimensionless)
b	constant(dimensionless)
b_x	constant(ML^{-3})
dp/dt	derivative of p with respect to t; antibiotic synthesis rate or rate of change of antibiotic concentration ($ML^{-3}T^{-1}$)
dx/dt	derivative of x with respect to t; growth rate or change of biomass concentration ($ML^{-3}T^{-1}$)
$(dx/dt)_{ML}$	rate of decrease of biomass concentration due to the presence of microbial layer($ML^{-3}T^{-1}$)
D	dilution rate(T^{-1})
D_{FBC}	dilution rate in FBC(T^{-1})
D_{RFBC}	dilution rate in RFBC(T^{-1})
i	ordinal number of the cycle(dimensionless)
k_1	cubic growth rate constant($M^{1/3}L^{-1}T^{-1}$)

k_2	constant of growth rate suppression(T^{-1})
k_3	constant of growth rate retardation(T^{-2})
k'_3	constant of growth rate retardation in RFBC(T^{-2})
k, k_p, k_{p_1}, k_{p_2}	specific rates of antiobiotic synthesis as defined in the text (T^{-1})
K_{v^i}	constant; reciprocal dilution rate at the beginning of the cycle i(T)
L	length
M	mass
p	product(antibiotic) concentration(ML^{-3})
t	cultivation time(T)
t'_i	cultivation time at the end of cycle i, in RFBC(T)
T	time
v, v_i	volumetric medium flow rates (L^3T^{-1})
V	fermentation broth volume(L^3)
V_0, V_{0i}	fermentation broth volumes at the beginning of the cycle(L^3)
V_1	fermentation broth volume at the end of the cycle(L^3)
x	biomass concentration (ML^{-3})
x_E	experimental value of the biomass concentration (ML^{-3})
x_S	simulated value of the biomass concentration (ML^{-3})
μ	specific growth rate (T^{-1})

References

[1] F. Yoshida, T. Yamané, and K. I. Nakamoto, *Biotechnol. Bioeng., 15,* 257 (1973).
[2] S. J. Pirt, *J. Appl. Chem. Biotechnol., 24,* 415 (1974).
[3] S. J. Pirt, *Principles of Microbe and Cell Cultivation* (Blackwell, Oxford, 1975), pp. 211–218.
[4] I. J. Dunn, and J. R. Mor, *Biotechnol. Bioeng. 17,* 1805 (1975).
[5] S. Nagai, Y. Nishizawa, and T. Yamagata, in *Abstracts of Fifth International Fermentation Symposium,* H. Dellweg, Ed. (Verlag, Berlin, 1976), p. 30.
[6] R. C. Jones and R. M. Anthony, *Eur. J. Appl. Microbiol., 4,* 87 (1977).
[7] H. C. Lim, B. J. Chen, and C. C. Creagan, *Biotechnol. Bioeng., 19,* 425 (1977).
[8] A. Trilli, V. Michelini, V. Mantovani, and S. J. Pirt, *J. Appl. Chem. Biotechnol., 27,* 219 (1977).
[9] T. Yamané and S. Hirano, *J. Ferment. Technol., 55,* 156 (1977).
[10] T. Yamané and S. Hirano, *J. Ferment. Technol., 55,* 380 (1977).
[11] T. Yamané, T. Kume, E. Sada, and T. Takamatsu, *J. Ferment. Technol., 55,* 587 (1977).
[12] M. Bošnjak, M. Holjevac, and V. Johanides, *J. Appl. Chem. Biotechnol., 26,* 333 (1976).
[13] M. Bošnjak, V. Topolovec, and V. Johanides, "The repeated fed-batch culture as the first stage of the two-stage semicontinuous process: growth kinetics," reported at the *7th International Symposium on Continuous Culture of Microorganisms, Prague 10–14 July 1978* (Czechoslovak Academy of Science, Prague, 1978).
[14] M. Bošnjak and V. Johanides, *Mikrobiologija, 10,* 179 (1973).
[15] V. Topolovec and M. Bošnjak, Proceedings of the 11th Yugoslav International Symposium on Information Processing—Informatica 76,5 109, Bled, Yugoslavia (October 1976).
[16] M. Bošnjak, V. Topolovec, and M. Holjevac, Proceedings of the 12th Yugoslav International Symposium on Information Processing—Informatica 77, 6 219, Bled, Yugoslavia (October 1977).
[17] B. Atkinson, *Biochemical Reactors* (Pion, London, 1974), pp. 153–175.
[18] M. Bošnjak, M. Vrana, and V. Topolovec, *Abstracts of Fifth International Fermentation Symposium,* H. Dellweg, Ed. (Berlin, 1976), pp. 219.
[19] M. Bošnjak, N. Šerman, and V. Johanides, Proceedings of the 10th Yugoslav International

Symposium on Information Processing—Informatica 75, 5-18, Bled, Yugoslavia (October 1975).
[20] J. F. Martin, in *Advances in Biochemical Engineering,* K. Ghose, A. Fiechter, and N. Blakebrough, Eds. (Springer-Verlag, Berlin, 1977), Vol. 6, pp. 105-127.
[21] R. Nožinić, G. Marinković, M. Dražić, and M. Bošnjak, "Growth of *S. bambergiensis* and glucose isomerase biosynthesis," report in *1st European Congress on Biotechnology* (Interlaken, Switzerland, Sept. 25-29, 1978) (Dechema, Frankfurt/Maine, 1978).
[22] J. Beljak, M. Bošnjak, V. Topolovec, and R. Valinger, "Growth kinetics of *Aspergillus niger* and glucoamylase biosynthesis in RFBC," report in *12th International Congress of Microbiology* (Munich, Sept. 3-8, 1978) (IAMS, Munich, 1978).

Kinetic Model for the Control of Wine Fermentations

ROGER BOULTON

University of California, Davis, California 95616

INTRODUCTION

A wine fermentation can be classified as the batch fermentation of a mixture of D-glucose and D-fructose under anaerobic conditions by a pure culture of a strain of *Saccharomyces cerevisiae*. The fermentation medium is the juice of mature grapes of the *Vitis* species, in general *Vitis vinifera*. The fermentation is conducted under batch conditions because the complex mixture of natural components and fermentation by-products developed by this method has not been duplicated under continuous fermentation conditions.

Wine fermentations are characterized by a combination of conditions rarely found in fermentation systems, even those producing other alcoholic beverages. These conditions are as follows:

a) An unusually low pH, typically between 3.0 and 3.6, caused mainly by the levels of tartaric and malic acids in the juice.

b) Levels of bisulfite ions, in the range 50 to 150 ppm, which have been added to inhibit wild yeasts and oxidative enzymes.

c) Concentrations of D-glucose and D-fructose, typically between 90 and 130 g/liter, which cause substrate inhibition [1,2].

d) A mixture of hexose isomers which causes competitive inhibition [1,2].

e) Ethanol concentrations, typically between 90 and 130 g/liter, which cause product inhibition [1,2] and significant reductions in the viable fraction of the yeast population.

Evidence of substrate inhibition of strains of *S. cerevisiae* can be found in the studies by Hopkins and Roberts [3]. Similar results have also been reported with the strains used in wine making [4–6]. As early as 1922, the competitive inhibition between D-glucose and D-fructose had been observed [7] and is also found in studies with wine yeasts [8]. Investigations of ethanol inhibition are reported with general strains [9,10] and with wine strains [11,12]. Finally, the effect of ethanol on the viability of the yeast population has long been recognized in wine fermentations [11,13].

THE MODEL

The proposed model has been developed on the assumption that the rate of fermentation is controlled by the rate at which substrates are consumed by the growth and maintenance activities of the yeast. The way in which maintenance activities are expressed follows the proposal by Pirt [2] rather than Herbert [14].

The expression for growth rate is of the Monod-type with a term included for the product inhibition by ethanol. Additional relationships are included for yeast viability, medium temperature, and ethanol formation.

Yeast Growth and Substrate Utilization

Metabolic activity is considered to occur only in the viable yeast cells. The rate of increase in biomass (dX/dt) can then be written

$$\frac{dX}{dt} = \mu X_v \tag{1}$$

where μ is the specific growth rate of the yeast and X_v is the viable biomass concentration. Since dual substrates are available and their relative uptake is a function of the yeast strain, two expressions for utilization are written:

$$\frac{dG}{dt} = -\left(\frac{\mu_G X_v}{Y_M} + \frac{mX_v G}{F+G}\right) \tag{2}$$

and

$$\frac{dF}{dt} = -\left(\frac{\mu_F X_v}{Y_M} + \frac{mX_v F}{F+G}\right) \tag{3}$$

where G and F are the concentrations of D-glucose and D-fructose and μ_G and μ_F are the growth rates based on glucose and fructose, respectively. Y_M is the maximum growth yield [2] and m is the specific maintenance rate. It has been assumed that both of these properties are independent of the substrate. The maintenance activity is considered to consume the substrates in the same ratio that they exist in the medium.

The total substrate concentration (S) and the total specific growth rate (μ) can be written

$$S = G + F \tag{4}$$

and

$$\mu = \mu_G + \mu_F \tag{5}$$

Inhibition Effects

In the presence of high substrate concentrations, the specific growth rate can be expressed [2]

$$\mu_G = \frac{\mu_M G}{K_G + G + G^2/K_{GI}}$$

where μ_M is the maximum specific growth rate, K_G is the saturation constant for glucose, and K_{GI} is the substrate inhibition constant for glucose. A similar expression can be written for fructose.

In the presence of an inhibiting competitor, the specific growth rate can be expressed [2]

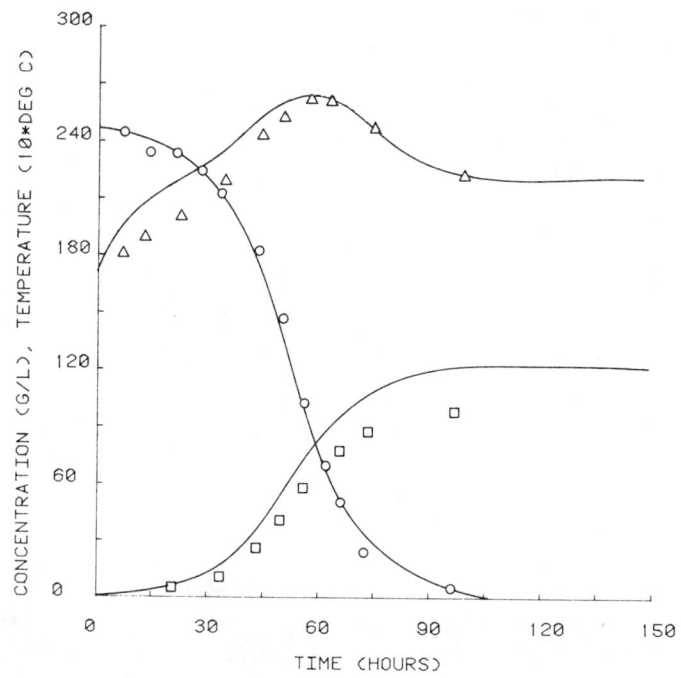

Fig. 1. Comparison of model predictions with the experimental data of Castor and Archer [24]. (△) Temperature; (○) total sugars; (□) ethanol.

$$\mu_G = \frac{\mu_M G}{K_G(1 + F/K_{CI}) + G}$$

where F is the inhibitor (fructose) concentration and K_{CI} is the competitive inhibition constant. A similar expression can be written for fructose.

Although a model incorporating these expressions has been fitted to wine fermentation data [15], the inhibition constants are not reliable because they have been deduced from results with general strains of *S. cerevisiae* in other media. Fortunately, simpler expressions of the type

$$\mu_G = \mu_M G/(K_G + G)$$

where K_G is a lumped approximation of the substrate inhibition contributions, have been successfully used to simulate wine fermentations and are employed in the proposed model.

Ethanol inhibition is incorporated into the specific growth relationship as [2]

$$\mu_G = \frac{\mu_M G}{(K_G + G)(1 + E/K_{EI})} \qquad (6)$$

where E is the concentration of ethanol and K_{EI} is the ethanol inhibition constant.

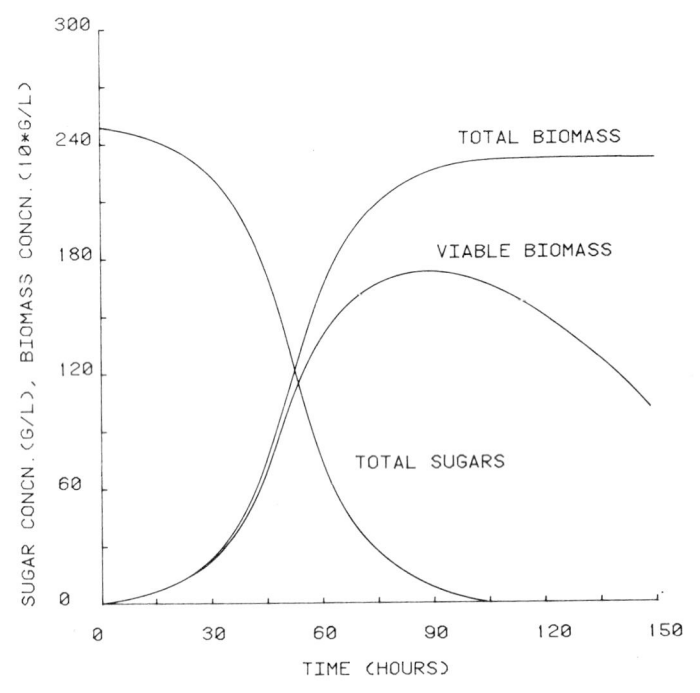

Fig. 2. Predicted total and viable biomass for the data of Castor and Archer [24].

Product Formation

The rate of ethanol formation (dE/dt) can be related stoichiometrically to the rate of total sugar consumption (dS/dt). In the ideal conversion, 180 mass units of sugar will produce 92 mass units of ethanol. Then

$$\frac{dE}{dt} = -\frac{92}{180} \cdot \frac{dS}{dt} \tag{7}$$

Heat Transfer Effects

The rate of change of medium temperature (dT/dt) can be related to the rate of heat generation by the fermentation and the rate of heat removal by the cooling system in an energy balance:

$$\frac{dT}{dt} = \frac{\Delta H}{\rho C_p} \frac{dS}{dt} - \frac{UA}{\rho C_p V}(T - T_c) \tag{8}$$

ΔH is the heat liberated by the fermentation, and C_p are the medium density and heat capacity, and T_c is the coolant temperature. U is the overall heat transfer coefficient, A is the heat transfer area, and V is the volume of juice. Estimates of ΔH [16–19] vary between 22.5 [19] and 28 [17] kcal/mol glucose. A value of 22.5 has been chosen in this study.

TABLE I
Values of Model Parameters at 20°C for the Fermentation of Castor and Archer [24]

Parameter	Value
Maximum specific growth rate, μ_M	0.11 hr^{-1}
Specific maintenance rate, m	1.0×10^{-2} hr^{-1}
Lumped substrate constant, for Glucose, K_G	112 g. liter^{-1}
for Fructose, K_F	112 g. liter^{-1}
Ethanol inhibition constant, K_{EI}	40 g. liter^{-1}
Maximum yield coefficient, Y_M	0.1
Overall heat transfer coefficient, U	0.27 Kcal cm^{-2} sec^{-1} °k^{-1}

Temperature Effects

The growth and maintenance terms must be expressed as functions of temperature if the model is to be useful in temperature control applications. The maximum specific growth rate can be expressed as a combination of two terms [2,20]. The first, an exponential increase in activity with temperature and the second an exponential increase in denaturation or death with temperature. Studies with the Montrachet strain [21,22] indicate an activation energy of 14.2 kcal/mol for activity and a value of 121 kcal/mol for denaturation:

$$\mu_M = 0.18 \exp[14200(T - 300)/300 \, RT] \\ -0.0054 \exp[121000(T - 300)/300 \, RT] \quad (9)$$

R is the universal gas constant, and T is the absolute temperature.

Activation energies have not yet been determined for the maintenance coefficient m, the saturation constants K_G, K_F, and K_{EI}, the inhibition constant for ethanol, with strains of *S. cerevisiae*. As a first approximation the values reported for *Aerobacter aerogenes* [20] have been used. The resulting expressions are

$$m = m_0 \exp[-9000(T - 293.3)/293.3 \, RT] \quad (10)$$

and

$$K = K_0/\exp[-11\,000(T - 293.3)/293.3 \, RT] \quad (11)$$

K_0 and m_0 are reference values of the constants K_G, K_F, K_{EI}, and m.

Yeast Viability

Since the growth and maintenance rates depend on the viable biomass, it is important to attempt to account for the changes in viability as the fermentation proceeds. Schanderl [13] has shown the reduction in viability of wine yeasts at various ethanol concentrations. Portno [23] reported that yeast viability decreased almost linearly with time in the continuous culture of a brewing yeast. In this study, the following empirical expression has been used:

$$X_v = X(1 - Et/42\,000) \tag{12}$$

MODEL VERIFICATION

The set of eqs. (1)–(12) together with additional forms of eqs. (6), (10), and (11) constitute the proposed model. Their solution was obtained using a Euler constant step-size integration scheme with time steps of 1 hr. The integration was carried out on a Tektronix 4051 computer.

The model was tested using the data of Castor and Archer [24]. This is one of the few reports in which sugar and ethanol concentration and temperature are recorded in the absence of temperature control. They used the Montrachet strain with a French Colombard juice in open ceramic vessels in an ambient temperature of 22°C.

The predictions of the model are compared with their results (Fig. 1) for total sugar and ethanol concentrations and juice temperature. The predicted rise in biomass and viable biomass for this case are also presented (Fig. 2). The constants used for this simulation are given in Table I.

DISCUSSION

The data of Castor and Archer [24] have been chosen because they permit the model to be tested under nonisothermal conditions. The predicted sugar consumption is in good agreement with the measurements. The poor agreement between predicted and measured ethanol concentrations is primarily due to the open fermentor and relatively high fermentation temperature causing significant ethanol losses by evaporation. The early temperatures are not well described because no allowance has been made for the heat capacity of the ceramic fermentor. The maximum temperature and later cooling are well described by the model. Values of 112 g/liter were deduced for the lumped substrate constants (K_G and K_F) with a specific maintenance coefficient of 1×10^{-2}/hr. An evaluation of the extent to which the model fits the data was not conducted because the ratio of glucose to fructose was not reported, but assumed to be 1:1. The predicted biomass and viable biomass (Fig. 2) indicate that the final cell concentration is approximately 23 g/liter and that between 65 and 75% of this is viable.

The proposed model will need to be used in an adaptive control scheme whereby the initial estimates of the lumped substrate constants can be modified as the fermentation proceeds. This will be necessary because various species

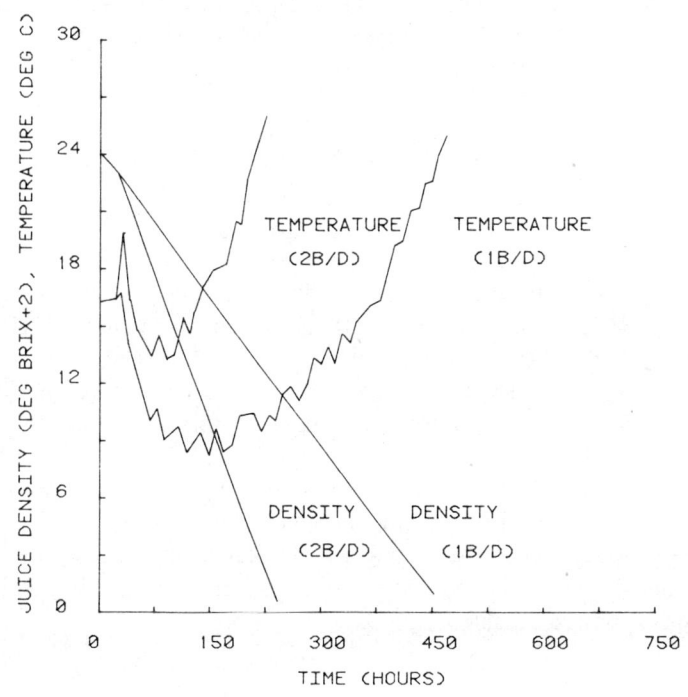

Fig. 3. Predicted temperature patterns for uniform fermentation rates of a Chenin blanc juice.

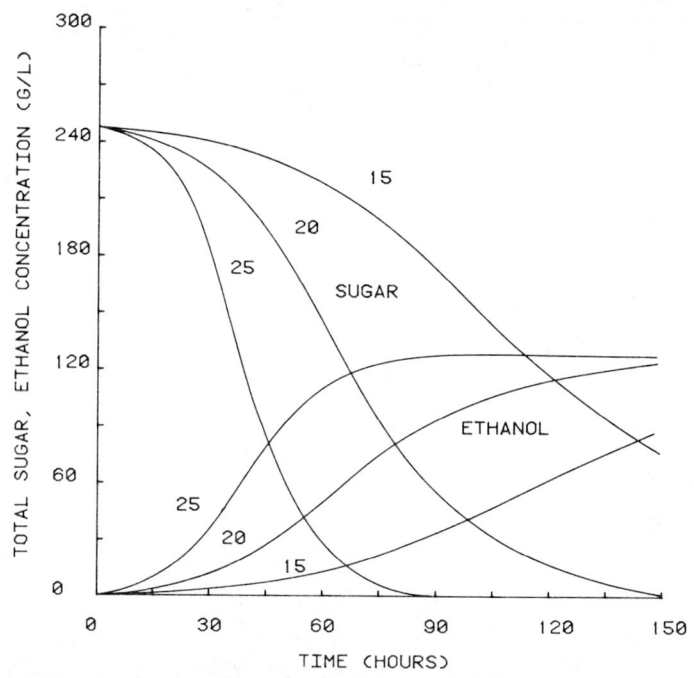

Fig. 4. Predicted fermentation patterns of a French Colombard juice at 15, 20, and 25°C.

found in grape juices but not included in the model are known to stimulate the fermentation rate. In particular, nitrogen and phosphate levels will vary between juices as will the pH, glucose to fructose ratio, and ionic strength.

APPLICATIONS TO FERMENTATION CONTROL

Uniform Fermentation Rate

There is considerable interest in controlling white wine fermentations so that the rate of reduction in sugars is essentially constant. The interest stems from the work of Saller [25], who reported that higher quality wines were produced under these conditions. The subject has not been pursued because of the difficulty in achieving the required control either in laboratory studies or commercial operations. Using the proposed model, in conjunction with an iteration routine, predictions of the required temperature patterns were made for two uniform fermentations (Fig. 3). The iteration was conducted with steps of 1°C, and the adjustments were chosen to be 10 hr apart. No allowance has been made in these predictions for lags in cooling and cell response, but they could be added into an adaptive control scheme.

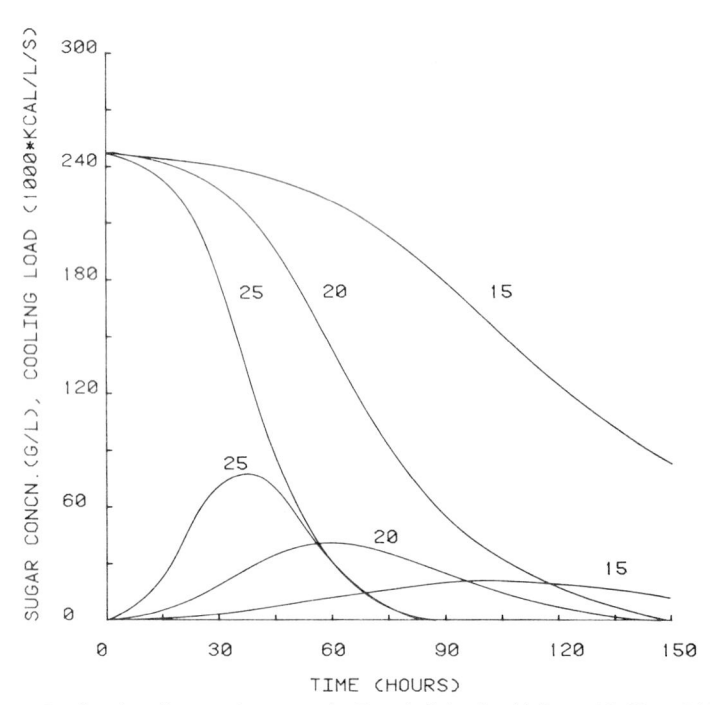

Fig. 5. Predicted cooling requirements of a French Colombard juice at 15, 20, and 25°C.

Optimum Cooling Demand

Commercial wine fermentations are presently conducted under setpoint temperature control. At the setpoint temperature, the batch fermentation will cause a cooling load in proportion to the rate of heat release by the fermentation. Predicted fermentation patterns at constant temperatures of 15, 20, and 25°C are shown in Figure 4. The predicted cooling loads required for setpoint temperature control are shown in Figure 5. Low-temperature situations, typical of white wine fermentations, will have their heat release spread over several days, while the warmer situations, typical of red wine fermentations, will be more rapid, requiring a larger cooling load. It can be seen that a fermentation at 25°C requires almost four times the cooling rate for temperature control than does a similar fermentation at 15°C. The application of the model to the optimum control of this cooling demand for several simultaneous fermentations lies in the faster-than-real-time simulation of the daily (or even hourly) cooling loads and the making of adjustments in order to minimize the overall cooling load and prevent overload situations.

For fermentations that would normally be inoculated at the same time, the model can be used to evaluate several alternative strategies in regard to cooling and inoculation. The purpose would be to either advance or retard the fermen-

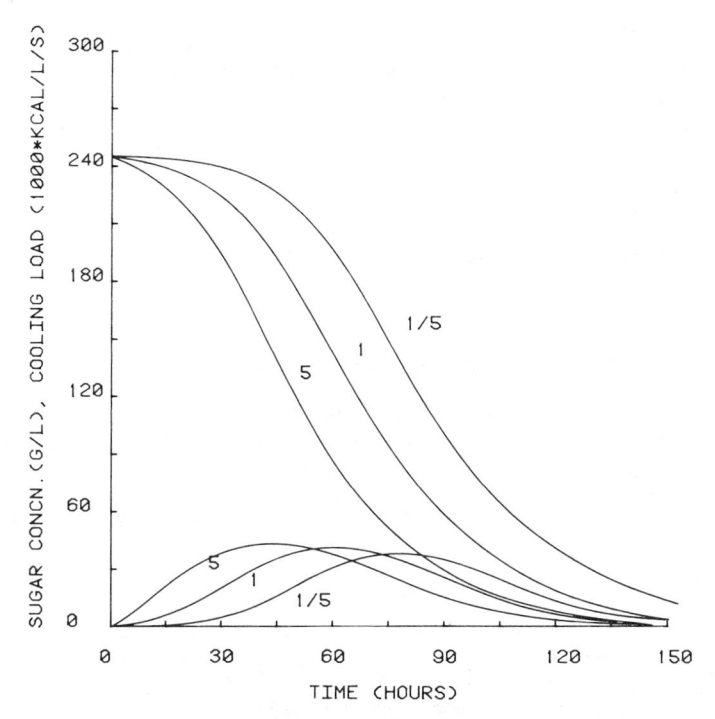

Fig. 6. Predicted cooling requirements of a French Colombard juice with $1/5$, 1, and 5% inocula at 20°C.

tation in time so that the peak cooling loads do not overlap. One approach would be by the control of the initial temperature, another would be by the control of the inoculum strength, and a third would be a combination of these two. The influence of the initial juice temperature on the early development of the fermentation can be seen in Figure 5. The influence of the inoculum strength at 20°C on the fermentation rate and refrigeration load is shown in Figure 6. Using one fifth of the usual 1% inoculum has the effect of delaying the peak fermentation rate and refrigeration load by 18 hr. Using a 5% inoculum has the effect of advancing these peaks by 18 hr. For two fermentors, this approach would separate the peak loads by up to 36 hr and reduce the combined load by approximately 20%. With more fermentors or warmer fermentations this reduction is even more pronounced. Finally, the combination of juice temperature and inoculum strength provides an even more flexible alternative for the minimization of the cooling demand.

CONCLUSION

A comprehensive model has been developed for wine fermentations. It has been verified by a comparison with a typical batch fermentation. The model has immediate application to fermentation rate control and to the optimization of refrigeration loads in commercial wineries.

References

[1] M. Dixon and E. C. Webb, *Enzymes* (Longmans, London, 1967), pp. 75-84.
[2] S. J. Pirt, *Principles of Microbe and Cell Cultivation* (Blackwell Scientific, Oxford, 1975).
[3] R. H. Hopkins and R. H. Roberts, *Biochem. J., 29*, 931 (1935).
[4] E. Vogt, in *Table Wines,* 2nd ed., M. A. Amerine and M. A. Joslyn, Eds. (University of California Press, Berkeley, 1970), p. 422.
[5] L. Benvegnin, E. Capt, and G. Pignet, *Traité de Vinification,* 2nd ed. (Librairie Payot, Lausanne, 1951), p. 584.
[6] R. E. Kunkee and M. A. Amerine, *Appl. Microbiol., 16,* 1067 (1968).
[7] R. H. Hopkins, *Biochem. J., 26,* 245 (1931).
[8] C. S. Ough and M. A. Amerine, *Am. J. Enol. Viticult., 14,* 194 (1963).
[9] S. Aiba, M. Shoda, and M. Nagatani, *Biotechnol. Bioeng., 10,* 845 (1968).
[10] C. D. Bazua and C. R. Wilke, Lawrence Berkeley Laboratory, Rept. No. 442 (1975).
[11] C. S. Ough, *Am. J. Enol. Viticult., 17,* 74 (1966).
[12] I. Holzberg, R. K. Finn, and K. H. Steinkraus: *Biotechnol. Bioeng., 9,* 413 (1967).
[13] H. Schanderl, *Die Mikrobiologie des Mostes und Weins,* 2nd ed. (Eugen Ulmer, Stuttgart, 1959), p 229.
[14] D. Herbert, in A. G. Fredrickson, R. D. Megee, III, and H. M. Tsuchiya, *Adv. Appl. Microbiol., 23,* 419 (1970).
[15] R. B. Boulton, unpublished data.
[16] A. Bouffard, *Prog. Agr. Viticol., 24,* 345 (1895).
[17] L. Genevois, *Ann. Ferment., 2,* 65 (1936).
[18] M. Rubner, in *Table Wines,* 2nd ed., M. A. Amerine and M. A. Joslyn, Eds. (University of California Press, Berkeley, 1970), p. 375.
[19] R. J. Winzler and J. P. Baumberger, *J. Cell. Comp. Physiol., 12,* 183 (1938).
[20] H. H. Topiwala and C. G. Sinclair, *Biotechnol. Bioeng., 13,* 795 (1971).
[21] J. Cahill, private communication (1976).

[22] F. C. Jacob, T. E. Archer, and J. G. B. Castor, *Am. J. Enol. Viticult.,* 15, 69 (1964).
[23] A. D. Portno, *J. Inst. Brew.,* 74, 448 (1968).
[24] J. G. B. Castor and T. E. Archer, *Am. J. Enol.,* 7, 19 (1956).
[25] W. Saller, *Am. J. Enol.,* 9, 41 (1958).

Process Modeling Based on Biochemical Mechanisms of Microbial Growth

A. R. MOREIRA,* G. VAN DEDEM,[†] and M. MOO-YOUNG[‡]

Department of Chemical Engineering, University of Waterloo, Waterloo, Ontario, Canada N2L 3G1

INTRODUCTION

There is an increasing need for the development of better control and optimization strategies for industrial processes. In fermentation technology, much research work is being directed to the development of suitable sensors to measure key process variables, quantification of specific fermentation process kinetics, interfacing of fermentation equipment to computerized systems, and implementation of sophisticated control algorithms and optimization routines.

The current availability of inexpensive minicomputers and microprocessors has boosted interest in the application of mathematical techniques to the optimization of fermentation processes. However, before this goal can be fully achieved, basic kinetic models must be formulated that are capable of quantitatively describing the actual behavior of biochemical systems during their interaction with the process microenvironment. A class of fermentations that presently challenges development of adequate process modeling is that with socalled polyauxic growth in which a microorganism, growing on mixtures of several substrates, shows preferential uptake of a given substrate. Although polyauxic growth has been reported for specific fermentation systems [1-5], it is likely to be the general pattern in most industrial fermentations due to the highly complex composition of the growth media utilized.

In this paper, the evolution of recent structured models for cell growth kinetics will be delineated. In particular, the ability of a recently proposed model [6] for diauxic growth will be assessed. Experimental results obtained with *Candida lipolytica*, growing on mixed substrates of oleic acid and glucose, and *Trichoderma viride*, growing on mixed substrates of glucose and lactose, will also be reported and compared to the model predictions.

---

* Present address: Department of Agricultural and Chemical Engineering, Colorado State University, Ft. Collins, Colorado.
[†] Present address: Biosynth BV, P.O. Box 20, Oss, Holland.
[‡] To whom correspondence should be addressed.

EVOLUTION OF STRUCTURED GROWTH MODELS

Monod's model [7] is widely used to describe cell growth limited by a single substrate. Its mathematical formulation, shown below,

$$\frac{dX}{dt} = \frac{\mu_m S}{K_s + S} X \quad (1)$$

is simple and basically reflects empirical observations. Several modifications of Monod's model have been suggested in an attempt to depict the effect of maintenance energy, inhibition, etc. For example, Yoon et al. [5] have suggested the following relationship:

$$\mu = \sum_{i=1}^{n} \left[\mu_{mi} S_i \bigg/ \left(K_i + \sum_{j=1}^{n} a_{ij} S_j \right) \right] \quad (2)$$

where each substrate exerts competitive inhibition on the utilization of the other substrates by the microorganism. Also, Tsao and Hanson [2] used the following extension to the Monod equation:

$$\frac{dX}{dt} = \left(\frac{k_{11} S_{11}}{K_{11} + S_{11}} + \cdots + \frac{k_{1i} S_{1i}}{K_{1i} + S_{1i}} \right) \left(\frac{k_{21} S_{21}}{K_{21} + S_{21}} + \cdots + \frac{k_{2j} S_{2j}}{K_{2j} + S_{2j}} \right)$$
$$\times \left(\frac{k_{m1} S_{m1}}{K_{m1} + S_{m1}} + \cdots + \frac{k_{mn} S_{mn}}{K_{mn} + S_{mn}} \right) X \quad (3)$$

suggesting the existence of several classes of substrates with growth-enhancing effects among substrates of the same class.

However, these types of unstructured models do not account for the fact that the physiological state and composition of microbial cells change in response to alterations in the environment; consequently, they have met with limited success where applied to complex situations such as polyauxic growth where metabolic regulatory mechanisms play a vital role in the cell behavior. The need for a better description of these interactions has led to the development of structured models. An important feature that has received considerable attention is related to the formulation of the rate of enzyme biosynthesis.

Jacob and Monod [8] proposed that enzyme biosynthesis is regulated by an induction–feedback repression mechanism described by the following equilibria:

$$R + nP \underset{}{\overset{k_1}{\rightleftarrows}} RP_n \quad (4)$$

$$O + R \underset{}{\overset{k_3}{\rightleftarrows}} OR \quad (5)$$

$$O + RP_n \underset{}{\overset{k_4}{\rightleftarrows}} ORP_n \quad (6)$$

where R is the cytoplasmic repressor, P is the effector, RP_n is the effector–

cytoplasmic-repressor complex, O is the operator, OR is the operator–repressor complex (induction), ORP_n is the operator–repressor complex (repression), and n is the number of binding sites for effector on R. Yagil and Yagil [9] derived a quantitative relationship between the system components as

$$\log\left(\frac{\alpha}{1-\alpha} - \alpha_b\right) = \pm n \log [P] + \log \alpha_b \mp \log k_1 \qquad (7)$$

where $\alpha = [O]/[O]_t$, $\alpha_b = k_3/[R]_t$ and the lower sign is applicable for repressible systems.

Based on the molecular equilibria previously established and on Terui's model [10] for the rate of enzyme synthesis as limited by the amount of specific messenger RNA (mRNA) available, van Dedem and Moo-Young [11] established the following equations for the (normalized) enzyme activity in batch culture: induction:

$$e = e_0 + x - x_0 - \frac{k_3'}{k_2}\left[\tan^{-1}\left(\frac{k_1'}{k_2}(x + x_0 - 1)\right) - \tan^{-1}\left(-\frac{k_1'}{k_2}\right)\right] \qquad (8)$$

repression:

$$e = e_0 + k_2'(x - x_0) - \frac{k_4}{k_1} k_2'^2 \ln \frac{k_1(x - x_0 - 1) - k_2'}{-k_1 - k_2'} \qquad (9)$$

where the symbols are explained in the Nomenclature section at the end of this paper. In a subsequent paper, van Dedem and Moo-Young [6] incorporated their initial model for enzyme induction and repression into a rather comprehensive model for diauxic growth (reviewed in the next section). Although this model was found to be compatible with a number of previously observed phenomena, the mathematical formulation for the mechanism of catabolite repression for complex situations was not fully developed.

Recently, Toda [12] offered an alternative suggestion that induction and repression are simultaneous phenomena within a "dual control" mechanism. As shown in Figure 1, Toda considered the existence of two operator genes (O_1 and O_2) next to the structural gene on the lac operon; in order to have enzyme synthesis both operator genes are freed from repressor molecules. As a consequence, the specific activity of the enzyme (Q) is given by

$$Q = Q_1 \cdot Q_2 \qquad (10)$$

where $Q_1 = [O_1]/[O_1]_t$ is the fraction of O_1 gene that is free from the action of the repressor R_1 and $Q_2 = [O_2]/[O_2]_t$ is the fraction of O_2 gene that is free from the action of the repressor $S_2^n R_2$. The suggestion of the need for the product $Q_1 \cdot Q_2$ to express enzyme biosynthesis is probably the most important concept in Toda's work.

Imanaka and Aiba in a recently proposed model for catabolite repression [13] expanded on Toda's work. Their model, summarized in Figure 2, is based on further postulations on the biochemical mechanisms of enzyme biosynthesis regulation including the possible interactions between cyclic AMP (cAMP),

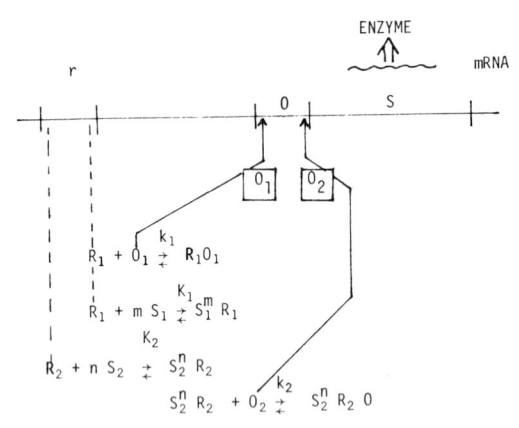

Fig. 1. Dual control model of Toda (from Ref. 12). Original nomenclature used as explained in text.

catabolite activator gene protein (CAP), and the promoter gene (P). An important difference between the Toda model and the Imanaka–Aiba model is that the latter assumes positive control for the catabolite repression mechanism [in order to have enzyme synthesis a complex $(cAMP)^n \cdot CAP \cdot P$ must be formed] while the Toda model is based on negative control (formation of $S_2^n R_2 O_2$ prevents enzyme expression). The specific activity of the enzyme is still calculated as eq. (10), but in Imanaka and Aiba's model Q_2 is the fraction of promotor gene occupied and/or impinged by the derepressing compound. In their derivation, Imanaka and Aiba obtained the same expression for Q_2, namely,

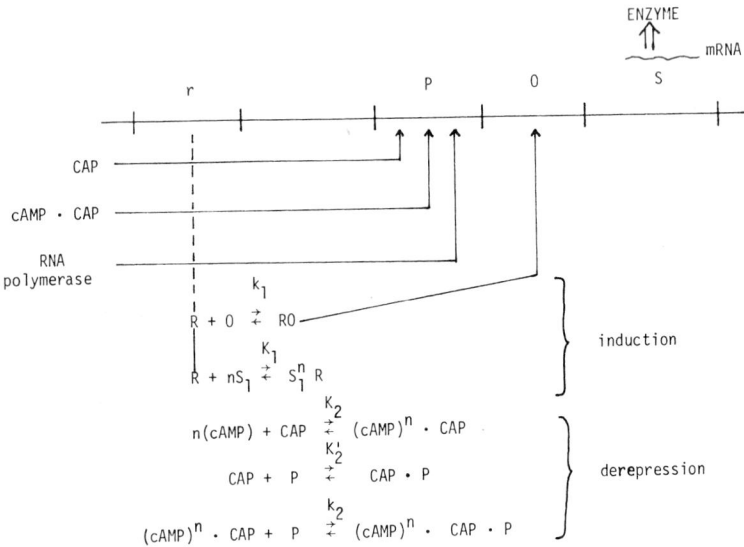

Fig. 2. Imanaka–Aiba model for a dual control mechanism (from Ref. 13). Original nomenclature used as explained in text.

$$Q_2 = \frac{1 + (K/\alpha^n) S^n}{1 + (K/\alpha^n)(1 + \eta) S^n} \tag{11}$$

where α is a proportionality constant, $K = K'_2/k_2 K_2$, and $\eta = 1/K'_2 [CAP]_t$.

Another step toward the comprehensive formulation of a structured growth model has been recently given by Fredrickson [14]. He showed that the material balance on the jth component of the biomass leads to eq. (12) for the time rate of change of the intracellular concentration of component j:

$$\frac{dc_j}{dt} = \sum_i r_{ij} - \mu c_j \tag{12}$$

In eq. (12) the term $-\mu c_j$ represents the dilution of the intracellular components caused by the growth of the biomaterial. This term has been neglected in most structured models published so far, and its possible quantitative insignificance can only be established through a reassessment of the previous models.

So far, none of the above structured models have been adequately tested with experimental data.

PARAMETERS FOR THE VAN DEDEM–MOO-YOUNG MODEL

The model for diauxic growth, which has been described in detail in a previous publication [6], is outlined in Figure 3. Here, the parameters used later are identified. It is assumed that there is a common intermediate, F, to both substrates P_1 and P_2; this common intermediate is the substrate of a branched pathway which leads to either cell mass, X, or a high-energy intermediate

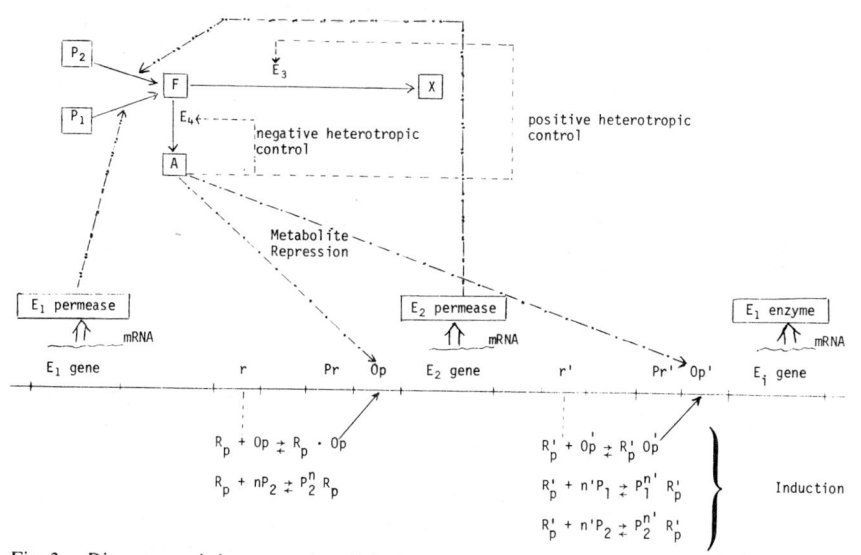

Fig. 3. Diagrammatic interpretation of the basic diauxic growth model of van Dedem and Moo-Young [6,11]. Symbols defined in Nomenclature section and text.

product, A, such as ATP. The rate of change of concentration of the common intermediate is given by

$$\frac{dF}{dt} = r_1(P_1, E_1) + r_2(P_2, E_2) - k_{15}r_3(F,A) - k_{16}r_4(F,A) \quad (13)$$

The terms r_1 and r_2 are the rates of the specific permeases for substrates P_1 and P_2, respectively. Assuming that the permeases follow Michaelis–Menten kinetics we have

$$r_1(P_1, E_1) = \frac{k_{17}(E_1)(P_1)}{K_{P_1} + (P_1)} \quad (14)$$

$$r_2(P_2, E_2) = \frac{k_{18}(E_2)(P_2)}{K_{P_2} + (P_2)} \quad (15)$$

It is further assumed that the permease E_1 for the favored carbon source is produced constitutively while the permease E_2 for the second carbon source is subjected to induction by this substrate and to metabolite repression. Enzymes E_3 and E_4 control the branched pathway and are assumed to be under heterotropic control by the high-energy compound A. High levels of A will inhibit enzyme E_4 through a negative heterotropic mechanism and stimulate E_4 activity through a positive heterotropic action. Monod et al. [15] obtained the following relationships for positive [eq. (16)] and negative [eq. (17)] heterotropic reactions, respectively:

$$r_3(F, A) = \frac{\phi(1+\phi)^{m-1}}{[L_1/(1+\alpha)^m] + (1+\phi)^m} \quad (16)$$

$$r_4(F, A) = \frac{\phi(1+\phi)^{m-1}}{L_2(1+\alpha)^m + (1+\phi)^m} \quad (17)$$

where $\phi = F/k_f$ and $\alpha = A/k_A$.

The rate of change of messenger RNA (mRNA) specific for the inducible permease, E_2, is given by

$$\frac{d(R)}{dt} = k_{20} \mu Q_1 - k_{21}(R) \quad (18)$$

where R is the concentration of mRNA and Q_1 is the fraction of free operators on the operon specific for E_2 and is given by

$$Q_1 = \frac{1 + k_{22}[r_2(P_2, E_2)]^2}{1 + k_{22}[r_2(P_2, E_2)]^2 + k_{23}} \quad (19)$$

The rate of formation of the permease E_2 is determined by the metabolite repression (assumed to occur at the translational level) according to eq. (20):

$$\frac{d(E_2)}{dt} = k_{24} Q_2(R) \quad (20)$$

where Q_2 is an arbitrary inhibition function dependent on the concentration of A and given by

$$Q_2 = [1 + k_{25}(A)]^{-1} \tag{21}$$

The rate of change of A is given by

$$\frac{d(A)}{dt} = k_{16} r_4 (F, A) - k_{15} r_3 (F, A) \tag{22}$$

For the concentration changes in the fermentation broth eqs. (23)–(25) are used:

$$\frac{d(X)}{dt} = k_{15} r_3 (F, A)(X) \tag{23}$$

$$\frac{d(P_1)}{dt} = -k' r_1 (P_1, E_1)(X) V_c \tag{24}$$

$$\frac{d(P_2)}{dt} = -k'' r_2 (P_2, E_2)(X) V_c \tag{25}$$

where V_c is the specific cell volume.

An estimate of the respiration rate, assuming that most of the ATP is obtained by oxidative phosphorylation, can also be made by

$$\text{respiration rate} = k_{26} r_4 (F, A)(X) \tag{26}$$

These equations are easily adapted to the case of a chemostat culture. Equations (13)–(22) remain the same since they represent intracellular concentrations, and eqs. (23)–(25) become

$$\frac{d(X)}{dt} = -D(X) + k_{15} r_3 (F, A)(X) \tag{27}$$

$$\frac{d(P_1)}{dt} = D[(P_1, 0) - (P_1)] - k' r_1 (P_1, E_1)(X) V_c \tag{28}$$

$$\frac{d(P_2)}{dt} = D[(P_2, 0) - (P_2)] - k'' r_2 (P_2, E_2)(X) V_c \tag{29}$$

where D is the dilution rate and no recycling is assumed.

The model can also be extended to the case of growth in tubular fermentors by assuming that a dispersed plug flow model is applicable [16]. As before eqs. (13)–(22) remain unchanged and eqs. (23)–(25) become

$$\text{De} \frac{d^2(X)}{dz^2} - v \frac{d(X)}{dz} + k_{15} r_3 (F, A)(X) = 0 \tag{30}$$

$$\text{De} \frac{d^2(P_1)}{dz^2} - v \frac{d(P_1)}{dz} - k' r_1 (P_1, E_1)(X) V_c = 0 \tag{31}$$

$$\text{De} \frac{d^2(P_2)}{dz^2} - v \frac{d(P_2)}{dz} - k'' r_2 (P_2, E_2)(X) V_c = 0 \tag{32}$$

MATERIALS AND METHODS

Fermentation Experiments

Microorganisms

The following strains were used in this study: *Candida lipolytica* (Diddens et Lodder), courtesy of the Unilever Research Laboratories, Vlaardingen/Duiven, The Netherlands; *Trichoderma viride* (ATCC 13631), obtained from the American Type Culture Collection, Rockville, MD. The *C. lipolytica* cultures were carried out at 25°C, pH 7.1 and the *T. viride* cultures at 30°C, pH 5.6.

Fermentation Equipment

Batch fermentations were usually performed in a 20-liter Marubishi fermentor (model MSJ). The fermentor includes a continuously variable stirrer speed control, pH and dissolved oxygen recording controllers, and a temperature-recording controller. The exhaust gas was dried and fed into a Servomex dual-channel oxygen analyzer (model OA 184).

Most of the continuous culture experiments and also some batch runs were performed in a 2-liter laboratory fermentor (model MAO 201) from Fermentation Design. The fermentor was equipped with a bottom-driven magnetic stirrer, temperature and pH controllers, dissolved oxygen probe, and automatic foam controller. The exhaust gas was passed, after drying, through the Servomex oxygen analyzer. Continuous feed of medium was achieved using the continuous metering module MM1 from Fermentation Design. A New Brunswick sterilizable metering plunger pump (model SP-5) was used for harvesting the fermentation broth. Where a liquid carbon source immiscible with water, such as oleic acid, was used in a continuous culture experiment the liquid was metered in with a Sage (model 355) syringe pump.

The tubular fermentor was designed and built at the Biochemical Engineering Laboratory of the University of Waterloo. Two 6-ft lengths of 2-in. i.d. acrylic pipe, with a 1/4-in. wall thickness, were flattened over their full length over a width of 5/8 in. Slots, 2 ft long and 1/4 in. wide, were milled in the middle of the flat side separated by 1/4-in. intersections. In the middle of the slots, $13/1000$-in. holes were drilled, 2 cm apart. Flanges were mounted on both sides of the pipe, fitting to end caps provided with inlets and outlets. Acrylic strips, 1/2 in. wide and 25 in. long, were provided with hose connectors and rubber linings. Size 44 hose clamps were used to tightly press the strips against the slots on the pipe. Opposite the slots, sampling ports were installed provided with rubber septums. The pipe was mounted horizontally with the sampling ports on top.

Fermentation Medium

The following media were used in this work.
Medium A. 2.0 g/liter KH_2PO_4; 3.0 g/liter $(NH_4)_2SO_4$; 0.3 g/liter $CaCl_2$;

0.3 g/liter MgSO$_4$·7H$_2$O; 5.0 mg/liter FeSO$_4$·7H$_2$O; 1.4 mg/liter ZnSO$_4$·7H$_2$O; 1.6 mg/liter MnSO$_4$·H$_2$O; 2.0 g/liter CoCl$_2$; 0.1 µg/liter biotin; 20.0 µg/liter Ca-panthotenate; 20.0 µg/liter niacin; 10.0 µg/liter riboflavin; 20.0 µg/liter thiamine; 20.0 µg/liter pyridoxin-HCl; 100.0 µg/liter inositol; 10.0 µg/liter p-aminobenzoic acid.

Medium B. same as medium A, but without vitamins and plus 0.3 g/liter urea.

Medium C. Same as medium A, but with 2 g/liter Casamino acids.

Inoculum Preparation

Inocula were grown overnight previous to a fermentation experiment in the case of *C. lipolytica,* or over a 2–3-day period in the case of *T. viride.* The same medium as the fermentation run was used for inoculum preparation.

Analytical Techniques

Cell Dry Weight

When the medium was homogeneous, an accurately measured amount of broth was filtered through a preweighed 0.8-µm Millipore filter. The residue on the filter was thoroughly washed with deionized water and dried overnight at 80°C and subsequently weighed. When the medium contained a nonmiscible liquid as a substrate component, such as oleic acid, 25 ml broth were placed in a 250-ml separatory funnel and acidified with 4N HCl. Nonanoic acid was added as an internal standard for gas chromatography analysis (0.004 ml/ml broth) and 5 ml chloroform were added and the mixture shaken for 1 min. The mixture was filtered through a preweighed 3-µm Millipore filter and the residue retained in the filter washed with 5 ml chloroform. The filtrate was poured back into the separatory funnel, allowed to separate, the bottom layer of chloroform removed, and the water layer extracted once more with 5 ml chloroform. The residue on the filter was washed with water and processed as outlined before for the case of a homogeneous medium.

Carbohydrate

When only one sugar was present in the sample, a reducing group assay was performed using an alkaline potassium ferricyanide solution as coloring agent [17]. If the medium contained glucose only, the immobilized glucose oxidase method was used. When a mixture of oligosaccharides was present in the medium, a gas chromatographic method was used as follows: Two ml sample were placed in a vial, frozen in liquid nitrogen, and lyophilized. Pyridine (1 ml), 0.2 ml hexamethyldisilazane (HMDS), and 0.1 ml trimethylchlorosilane (TMCS) were then added to the vial, the mixture was vigorously shaken, and heated at 70°C for 5 min [18]. The liquid was then analyzed in a (model 5750) Hewlett-Packard gas chromatograph with the following conditions: 6 ft × ¼ in. stainless-steel column, packed with 3% SE 52 on 80–100 mesh acid-washed silanized chromosorb W, carrier gas (helium) flow rate of 60 ml/min, inlet pressure 50 psi, oven temperature linearly programmed from 100 to 250°C at 10°C/min, injection port temperature 300°C, detector temperature 300°C.

Lipid Analysis

The chloroform extracts obtained during the cell dry weight procedure were combined in a wide test tube and the solvent was left to evaporate. Five μl of the oily residue was placed in a 1-dram vial and 0.2 ml bismethylsilylacetamide (BSA) was added. The mixture was shaken for 1 min and gas chromatographed in a (model 5750) Hewlett-Packard gas chromatograph with the following conditions: 24 ft × 1/8 in. stainless-steel column, packed with 3% JXR on 100–200 mesh Gas Chrom Q, carrier gas (helium) flow rate of 120 ml/min, inlet pressure 60 psi, oven temperature programmed linearly from 100 to 200°C at 10°C/min, injection port temperature 250°C, detector temperature 250°C.

Lipase Activity

Lipase activity was measured by the fluorometric technique of Guilbeault et al. [19].

Carboxymethylcellulase Activity

CMCase activity was determined by the viscometric technique [20]. The cylindrical beaker of a Brookfield viscometer was filled with 20 ml 0.05M Na-acetate buffer, pH 5.0, containing 3 g/liter CMC (Hercules type 7 HSP). Two ml of a conveniently diluted sample were added to the beaker and the beaker contents were rapidly mixed. The beaker was mounted around the rotor of the viscometer, lowered into the water bath, and the viscometer motor started at 60 rpm. Readings were taken at 30-sec intervals. The enzyme activity was expressed as viscometer readings/min ml of culture filtrate.

Filter Paper Activity

One ml of a previously desalted enzyme solution was added to each of two vials containing 1 ml 0.1M Na-acetate buffer, pH 5.1, preserved with 0.2% NaN$_3$. One of the vials contained also a 1 × 6 cm^2 strip (50 mg) Whatman No. 1 filter paper. The vials were incubated for 1 hr at 50°C and the enzymes were then inactivated by heating the vials for 10 min at 100°C. Reducing sugars were then determined by the potassium ferricyanide test described before. The filter paper activity was expressed as the mg reducing sugars [as glucose equivalent] released during the incubation period, taking the necessary dilution steps into account.

ATP Assay

ATP levels were measured by the firefly extract method based on estimating the quantity of light released by the reaction with the counting equipment described previously [21]. After sampling the fermenter, 1 ml broth was immediately transferred to a test tube containing 1 ml dimethylsulfoxide (DMSO) and kept for 15 min, after which the mixture was frozen with liquid nitrogen and kept at $-15°C$, or immediately assayed. A slurry of firefly extract (Sigma Chemicals Co., catalog No. FLE-500) was prepared and kept at room temperature for about 1 hr. The slurry was then filtered through a 3-μm Millipore filter (No. SSWP 04700) and the filtrate refrigerated for at least 20 hr. In a 12×75 mm^2 test tube, 2.5 ml firefly extract were used to derive the background reading. Then, 0.5 ml sample was added to the tube, a stopwatch was started at the time of mixing, and readings taken at periodic time intervals.

Computer Simulations

For the model equations, the following transformations to dimensionless parameters are made:

$$\phi = F/k_F, \quad \alpha = A/k_A, \quad \kappa_{17} = k_{17}/k_F, \quad \kappa_{18} = k_{18}/k_F$$
$$\kappa_{15} = k_{15}/k_F(E_1), \quad \kappa_{16} = k_{16}/k_F(E_1), \quad \theta = (E_1)t$$
$$\kappa_6 = k_{16}/k_A(E_1), \quad \kappa_5 = k_{15}/k_A(E_1), \quad \rho = (R)/k_{20}, \quad \mu' = \mu/(E_1)$$
$$\kappa_{21} = k_{21}/(E_1), \quad \eta_2 = (E_2)/(E_1), \quad \kappa_{24} = k_{24}k_{20}/(E_1)^2, \quad \rho_i = (R_i)/k_{28}$$
$$\psi_1 = (P_1)/(P_1,0), \quad \kappa_{P_1} = K_{P_1}/(P_1,0), \quad \kappa_1 = k_{18}/k_{17}, \quad \kappa_2 = (P_2,0)/(P_1,0)$$
$$\kappa_{P_2} = K_{P_2}/\kappa_2(P_1,0), \quad \psi_2 = (P_2)/\kappa_2(P_1,0), \quad \kappa_{26} = k_2/(E_1)$$
$$x = k_{17}(X)V_c/(P_1,0), \quad \kappa'_{15} = k_{15}/(E_1)$$
$$\eta_i = k_{17}(E_i)/k_{28}k_3(P_1,0), \quad \kappa_4 = k_4/(E_1)$$

where (E_i) is the extracellular enzyme activity, and (R_i) is the concentration of mRNA specific for extracellular enzyme.

The model equations in dimensionless form can now be obtained as

$$\frac{d\phi}{d\theta} = \frac{\kappa_{17}\psi_1}{\kappa_{P_1} + \psi_1} + \frac{\kappa_{18}\eta_2\psi_2}{\kappa_{P_2} + \psi_2} - \kappa_{15}r_3 - \kappa_{16}r_4 \tag{33}$$

$$\frac{d\alpha}{d\theta} = \kappa_6 r_4 - \kappa_5 r_3 \tag{34}$$

$$\frac{d\rho}{d\theta} = \mu' Q_1 - \kappa_{21}\rho \tag{35}$$

$$\frac{d\psi_1}{d\theta} = -\frac{\psi_1 x}{\kappa_{P_1} + \psi_1} \tag{36}$$

$$\frac{d\psi_2}{d\theta} = -\frac{\eta_2 \psi_2 x}{\kappa_{P_2} + \psi_2}\frac{\kappa_1}{\kappa_2} \tag{37}$$

$$\frac{dx}{d\theta} = \kappa'_{15} r_3 x \tag{38}$$

$$\frac{d\eta_i}{d\theta} = \rho_i Q_i' x - \kappa_4 \eta_i \tag{39}$$

$$\frac{d\rho_i}{d\theta} = \mu' Q_i - \kappa_{26}\rho_i \tag{40}$$

$$\frac{d\eta_2}{d\theta} = \kappa_{24} Q_2 \rho_2 \tag{41}$$

where Q_i and Q_i' are expressions for the extent of induction and metabolite repression, respectively.

For continuous culture, the equations for the bulk concentration in the fermentation broth become

$$\frac{dx}{d\theta} = -D'x + \kappa'_{15} r_3 x \tag{42}$$

where $D' = D/(E_1)$.

$$\frac{d\psi_1}{d\theta} = D'(1 - \psi_1) - \frac{\psi_1 x}{K_{P_1} + \psi_1} \tag{43}$$

$$\frac{d\psi_2}{d\theta} = D'(1 - \psi_2) - \frac{\eta_2 \psi_2 x}{K_{P_2} + \psi_2} \frac{\kappa_1}{\kappa_2} \tag{44}$$

$$\frac{d\eta_i}{d\theta} = -D'\eta_i + \rho_i Q_i' x - \kappa_4 \eta_i \tag{45}$$

For the dispersed plug-flow case

$$\xi = \frac{z}{L_r}, \qquad \eta_1 = \frac{(E_1) L_r}{v}, \qquad \frac{1}{\text{Pe}} = \frac{\text{De}}{v L_r}$$

$$\frac{1}{\text{Pe}} \frac{d^2 x}{d\xi^2} - \frac{dx}{d\xi} + \kappa'_{15} \eta_1 r_3 x = 0 \tag{46}$$

$$\frac{1}{\text{Pe}} \frac{d^2 \psi_1}{d\xi^2} - \frac{d\psi_1}{d\xi} + \frac{\eta_1 \psi_1 x}{K_{P_1} + \psi_1} = 0 \tag{47}$$

$$\frac{1}{\text{Pe}} \frac{d^2 \psi_2}{d\xi^2} - \frac{d\psi_2}{d\xi} + \eta_1 \frac{\kappa_1}{\kappa_2} \frac{\eta_2 \psi_2 x}{K_{P_2} + \psi_2} = 0 \tag{48}$$

$$\frac{1}{\text{Pe}} \frac{d^2 \eta_i}{d\xi^2} - \frac{d\eta_i}{d\xi} + \eta_1 \rho_i Q_i' x = 0 \tag{49}$$

The parameter values used in the simulations presented in this paper were

$\kappa_1 = 0.7$, $\quad \kappa_2 = 1.0$, $\quad K_{P_2} = 0.02$, $\quad \kappa_5 = 7 \times 10^3$, $\quad \kappa_6 = 2 \times 10^4$
$\kappa_{15} = 1.4 \times 10^4$, $\quad \kappa_{16} = 4 \times 10^4$, $\quad \kappa_{17} = 2.5 \times 10^4$, $\quad K_S = \kappa_{17}/\kappa_{18} = 4.0$
$L_1 = 10^4$, $\quad L_2 = 100.0$, $\quad \kappa_{21} = 1.0$,
$\kappa_{24} = 1.0$, $\quad \kappa_{26} = 1.6$, $\quad \kappa_4 = 0.1$, $\quad K_{P_1} = 0.01$

RESULTS AND DISCUSSION

Computer Simulations

The results of the simulation of batch growth on P_1, the substrate for which the permease is constitutive, are shown in Figure 4. P_1 also acts as the inducer for the synthesis of extracellular enzyme. It is evident (from Fig. 4) that metabolite repression is the controlling mechanism in this situation, since enzyme activity is released only when cell growth declines and ultimately ceases. As expected, the simulated respiration rate parallels the cell growth curve quite closely and drops off when growth stops.

The simulated time courses of cell mass, respiration rate, and extracellular enzyme activity for the batch growth on P_2 only are shown in Figure 5. P_2 acts as the inducer of mRNA synthesis for the extracellular enzyme. Compared with the previous simulation it can be seen that it takes considerably longer for the cell mass to start to increase, and release of metabolite repression also occurs at the later stages of cell growth. It is interesting to observe that, in this case, the peak in the respiration rate occurs before the cell mass peak.

Figure 6 presents the simulated time courses of cell mass, respiration rate, substrate, and enzyme parameters for the growth on both P_1 and P_2. It is clear

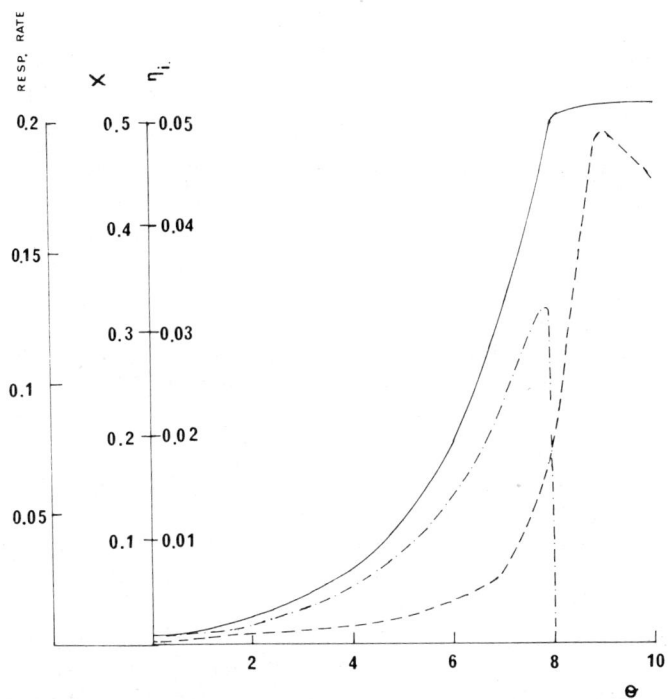

Fig. 4. Simulated time course of the cell mass, enzyme, and respiration rate parameters for growth on P_1 only, and where P_1 acts as the inducer for extracellular enzyme synthesis. (——) Cell mass; (- - -) enzyme activity; (-·-) respiration rate.

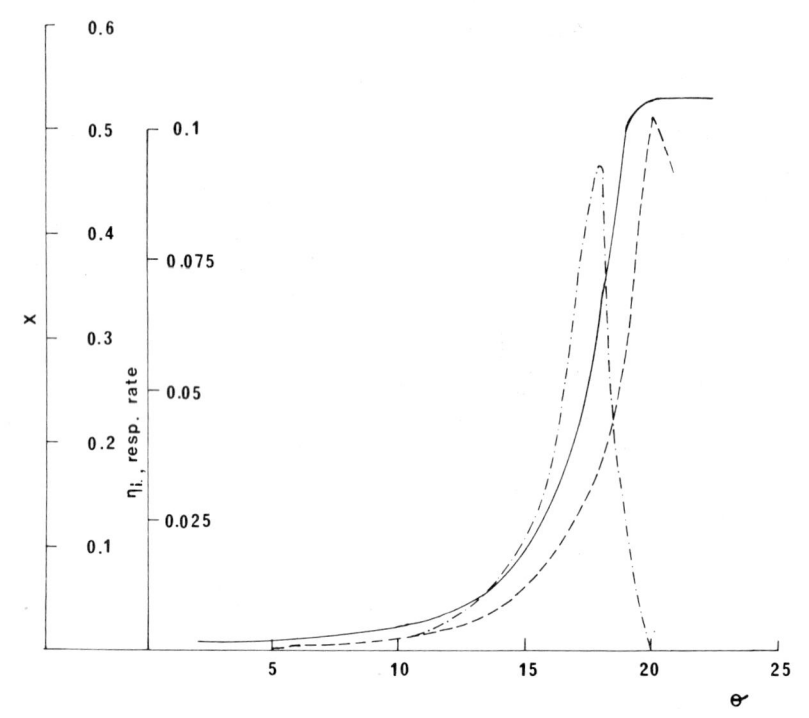

Fig. 5. Simulated time course of the cell mass, enzyme, and respiration rate parameters for growth on P_2 only, and where P_2 is the inducer for extracellular enzyme synthesis. (——) Cell mass; (- - -) enzyme activity; (-··-) respiration rate.

that the diauxic growth is predicted in this fermentation with P_2 consumption starting after P_1 is essentially depleted from the growth medium. The diauxic character is also evident from the change in the slope of the growth curve when the shift in substrate utilization occurs. (The respiration rate curve shows two peaks, one in each stage of growth, indicative also of the diauxic character of the fermentation.) Extracellular enzyme synthesis is repressed during the period of utilization of P_1 and only when the inducer P_2 starts to be metabolized does the extracellular enzyme activity increase. However, the highest value for the differential rate of enzyme synthesis only occurs late in the growth cycle when catabolite repression due to the catabolism of P_2 is released.

Computer simulations were also performed for continuous culture fermentation, utilizing a mRNA decay constant, κ_{26}, equal to 0.4 (dimensionless) corresponding to a half-life of 6.9 hr. Figure 7 shows the relationship between cell mass, enzyme, and substrate parameters and the dilution rate for the case of P_1 as carbon and energy source. It is observed that, at high dilution rates, catabolite repression is offset by the combined effect of the higher concentration of inducer and the increase in the steady-state quantity of the specific mRNA for the extracellular enzyme. This latter effect is weaker when P_2 is considered the inducer, as seen by the shape of the lower curve for enzyme activity in Figure 7.

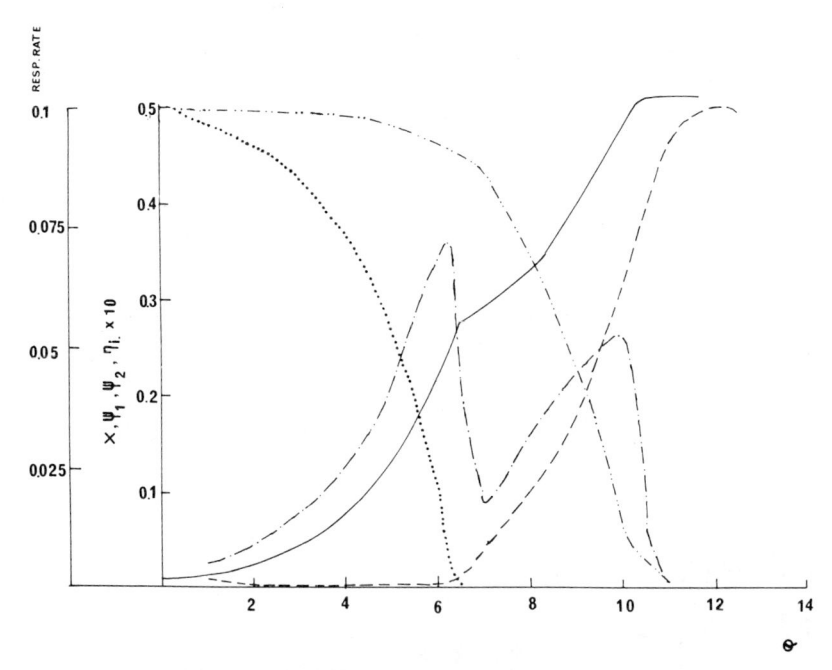

Fig. 6. Simulated time course of the cell mass, substrate, enzyme, and respiration rate parameters for growth on a mixture of P_1 and P_2 and where P_2 is the inducer. (—) Cell mass; (·····) P_1 parameter; (–··–) P_2 parameter; (- - -) enzyme activity; (-··-) respiration rate.

In Figure 8, a continuous culture simulation is presented for the case of growth on a mixture of P_1 and P_2. As expected, both substrate concentrations increase with increasing dilution rate. The enzyme activity curve remained the same for induction by either of the two substrates.

Simulation studies were also performed for the case of a tubular fermentor. A sample of the data generated is shown in Figure 9 for the case of growth on P_1 alone and for a dispersion number of 0.2. In this particular case catabolite repression was modeled by the simpler equation originally suggested by van Dedem and Moo-Young [12]:

$$Q_2 = \frac{1 + K_2(P_1)^n}{1 + K_2(P_1)^n + K_1 K_2(R)_t(P_1)^n} \quad (50)$$

Fermentation Experiments

The time courses for growth of *T. viride* on glucose only ($P_{1,0} = 5$ g/liter) are shown in Figure 10. The maximum specific growth rate was 0.21 hr^{-1} and the enzyme activity was undetectable during the growth. The fermentation time courses for an experiment using lactose as the sole carbon and energy source, at an initial concentration of 5 g/liter, are shown in Figure 11. The cell mass data in Figure 11 correspond to a maximum specific growth rate of 0.068 hr^{-1}.

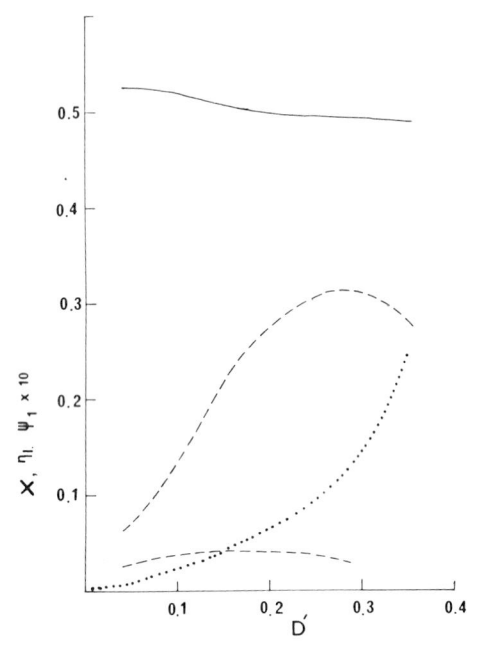

Fig. 7. Simulated dependence of the cell mass, enzyme, and substrate parameters on the dilution rate for chemostat culture with P_1 as the carbon and energy source. (—) Cell mass; (·····) P_1 parameter; (- - -) enzyme activity (upper curve: P_1 = inducer, lower curve: P_2 = inducer).

The data show that enzyme activity is released at the same time there is a decrease in the intracellular concentration of ATP. It is also interesting to note that CMCase and filter paper activities follow different patterns with CMCase activity being released much earlier in the fermentation. A much longer lag phase than the one observed during glucose growth is seen for this case.

Growth of *T. viride* on mixed substrates is shown in Figure 12. It can be observed that there is a transition in the cell mass curve, at approximately 25 hr of fermentation time, which corresponds to the point of transition from glucose to lactose consumption. The intracellular ATP level decreased during the shift from one substrate to the other and also during the period of consumption of lactose. The growth rate during the glucose-supported growth stage was 0.1 hr^{-1} and decreased to 0.06 hr^{-1} during the subsequent period of lactose metabolism. Cellulase activity was only detected after 30 hr of fermentation when the glucose repression was released and the cells started metabolizing lactose which induced cellulase biosynthesis, as postulated by the model.

The results obtained with *C. lipolytica* in batch fermentations are shown in Figures 13–15. The maximum specific growth rate for growth on glucose alone was 0.2 hr^{-1}, and for growth on oleic acid alone a maximum specific growth rate of 0.28 hr^{-1} was obtained. In Figure 13, it is seen that an increase in the enzyme production rate occurred at the same time the ATP concentration decreased. When the cells were grown on oleic acid only, as is the case shown in Figure 14,

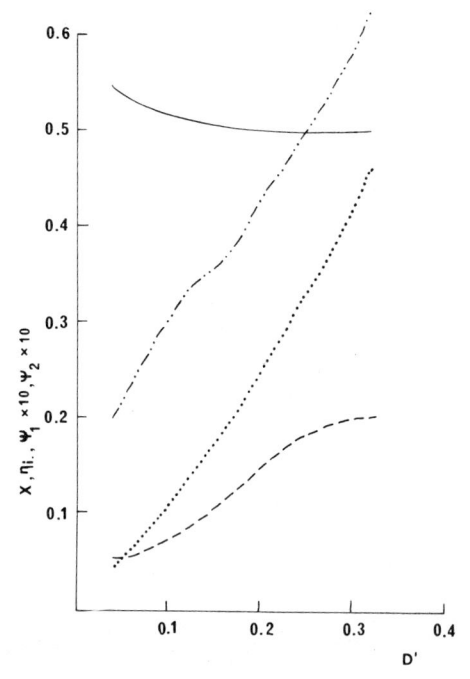

Fig. 8. Simulated dependence of the cell mass, enzyme, and substrate parameters on the dilution rate for chemostat culture with both P_1 and P_2 as carbon and energy sources (—) Cell mass; (·····) P_1 parameter; (-···-) P_2 parameter; (---) enzyme activity.

the final lipase activity level was about the same as in the glucose experiment but much higher ATP levels (about twofold) were observed. A careful observation of the data shown in Figure 14 also indicates that lipase activity increased dramatically at the final stage of the growth cycle when the ATP concentration started to decrease.

The data obtained for batch growth on a mixture of glucose and oleic acid (presented in Fig. 15) show that oleic acid was preferentially consumed although some glucose utilization did occur during the period of preferential consumption of oleic acid. The maximum specific growth rate was 0.18 hr^{-1} during the first stage of growth and decreased to 0.14 hr^{-1} during the glucose consumption period. The ATP concentration increased gradually up to 17 000 µg/g during the oleic acid consumption period and started to decrease after the shift to glucose utilization. Lipase activity was detected at high levels after the oleic acid was depleted from the growth medium, as predicted by the model.

Typical results obtained with *C. lipolytica* grown in continuous culture are shown in Table I. When glucose is used as carbon source it can be seen that at dilution rates of 0.06 and 0.1 hr^{-1} the lipase activity is considerably higher than at a dilution rate of 0.01 hr^{-1}. However, in the presence of oleic acid, lipase activity is much higher than the activity measured during the glucose experiments at the same dilution rate (data at $D = 0.03$ hr^{-1}). The data seem to in-

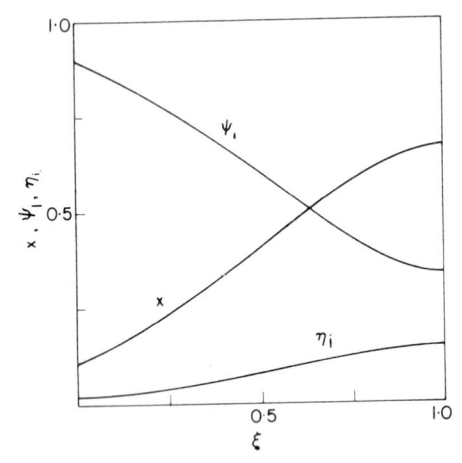

Fig. 9. Simulated profiles for cell mass, substrate, and extracellular enzyme activity in a tubular fermentor as a function of the dimensionless reactor length (dispersion number = 0.2).

TABLE I
Continuous Culture Data for *C. lipolytica* Grown on Medium A

Carbon Source	Dilution Rate (h^{-1})	Lipase Activity (U/ml)
Glucose	0.01	2.7
Glucose	0.03	7.5
Glucose	0.06	20.1
Glucose	0.095	24.0
Oleic acid	0.03	17.1
Oleic acid	0.184	78.0

dicate then that oleic acid is an inducer of lipase activity. Although the final levels of lipase activity are about the same when either glucose or oleic acid is used in batch fermentation, the results of continuous culture experiments favor the apparent induction action by oleic acid.

Some interesting considerations can be advanced regarding the significance of the ATP levels during the growth cycle. In our studies, we selected ATP as a first attempt to identify the high-energy compound represented as A in the diauxic growth model. For *T. viride* grown on glucose and/or lactose, the ATP

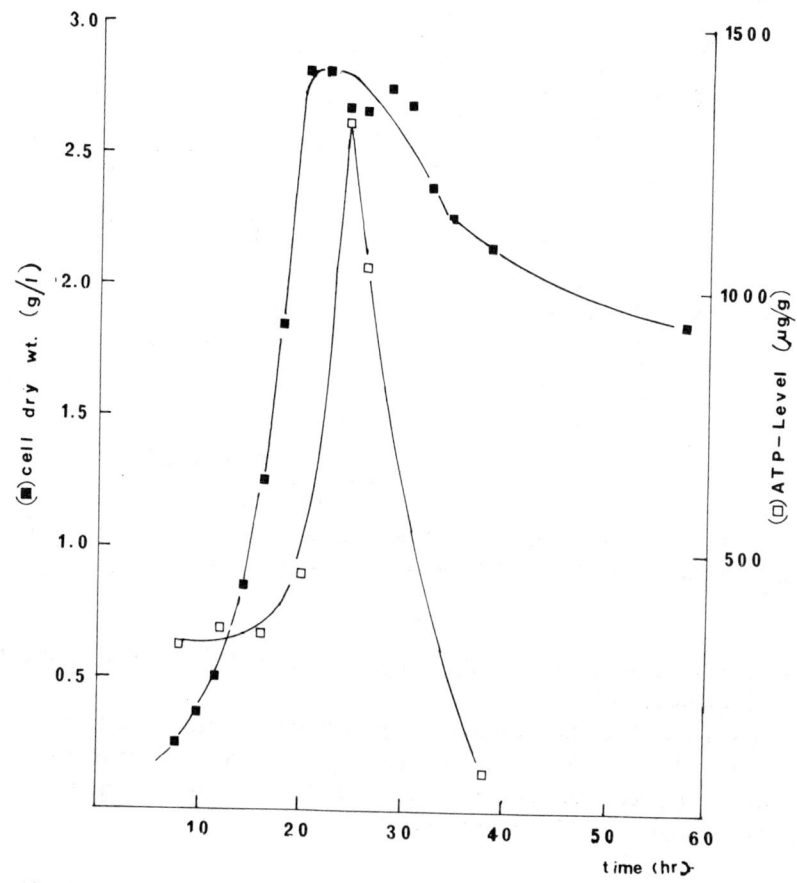

Fig. 10. Time course of cell mass and ATP levels of *T. viride* growing on a medium C with 5 g/liter glucose. (Cellulase levels undetected.)

profile does agree well with the model predictions: a high level of ATP during active growth on glucose and a decrease in ATP concentration during growth on lactose at a lower specific growth rate. Furthermore, extracellular enzyme synthesis is repressed while ATP levels are high. A more complex situation occurs for *C. lipolytica* growing on glucose and/or oleic acid. ATP levels are also higher during growth on the preferred substrate, oleic acid, which is energetically more efficient than glucose. At the same time there is evidence suggesting that oleic acid acts as the inducer for lipase biosynthesis. The high levels of ATP during this induction stage may be associated with fulfillment of the energy requirements for enzyme synthesis. This conflict between catabolite repression and energy supply to drive biochemical reactions probably points out that the absolute value of ATP concentration may not be a good and/or general way of assessing the extent of catabolite repression. Thus, the Imanaka–Aiba model has some merit. However, one difficulty in evaluating this model will be how to accurately

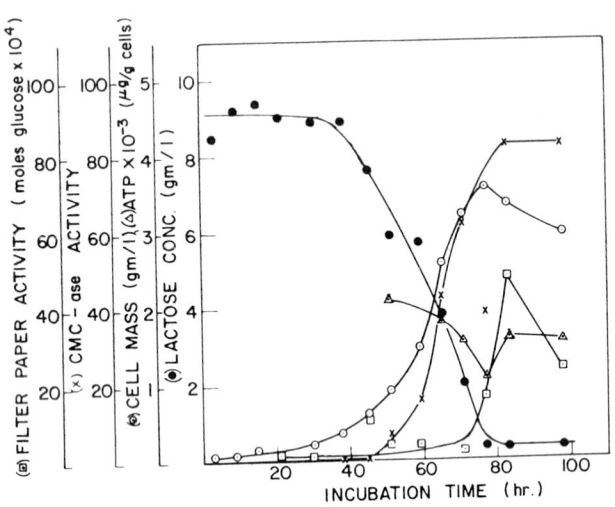

Fig. 11. Time course of lactose, cell mass, ATP, CMCase, and filter paper activity levels for *T. viride* growing on medium B with 5 g/liter lactose.

determine the concentration of the complex compounds involved in the cAMP regulatory mechanism.

Growth of *C. lipolytica* in a tubular fermentor was tested on medium C with 1 g glucose/liter. Typical cell mass, glucose, and enzyme profiles are shown in Figure 16. It is seen that the experimental results obtained follow quite closely the model predictions shown in Figure 9.

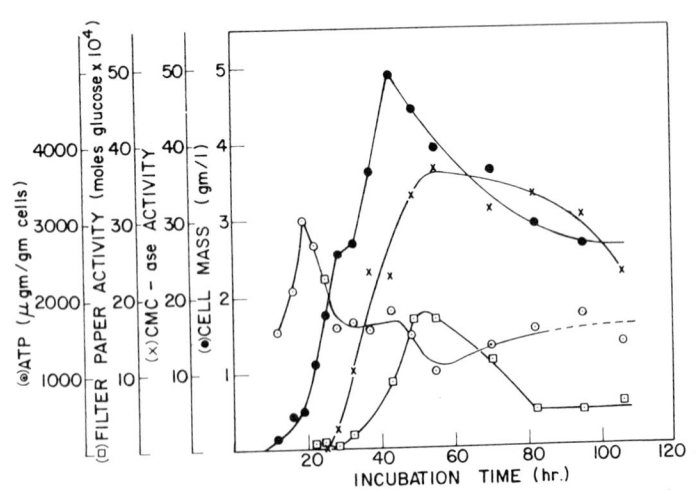

Fig. 12. Time courses of cell mass, ATP level, CMCase, and filter paper activities for *T. viride* growing on medium B with 5 g/liter glucose and 5 g/liter lactose.

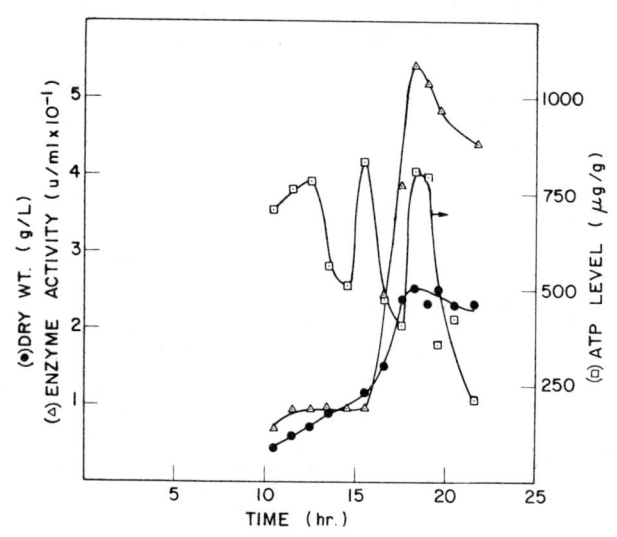

Fig. 13. Time course of cell mass, lipase activity, and ATP level for *C. lipolytica* growing on medium A with 5 g/liter glucose.

CONCLUSIONS

A diauxic growth model has been described that takes into consideration the biochemical mechanisms of enzyme induction and metabolite repression. The model attempts to describe the phenomena that occur during cell growth and

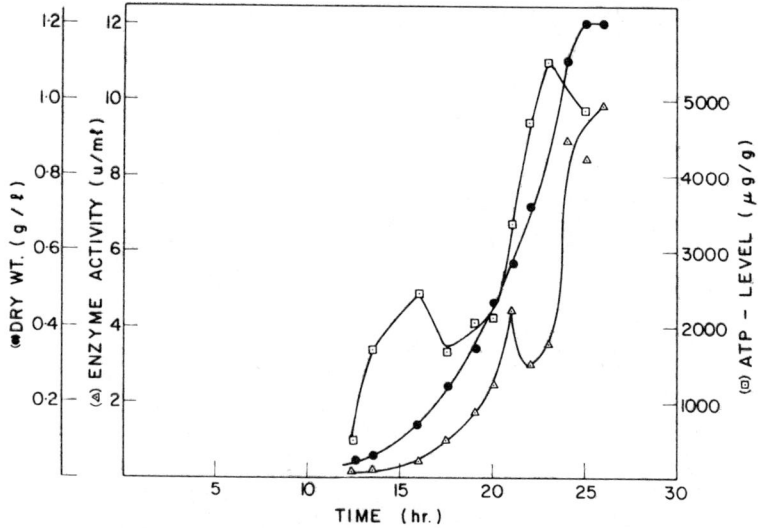

Fig. 14. Time course of cell mass, lipase activity, and ATP level for *C. lipolytica* growing on medium A with 1.79 g/liter oleic acid.

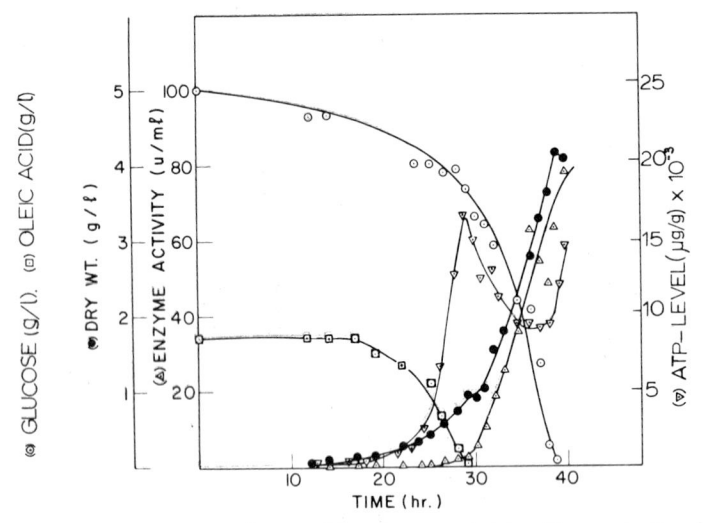

Fig. 15. Time course of cell mass, lipase activity, glucose, and oleic acid concentrations for *C. lipolytica* growing on medium A with 5 g/liter glucose and 1.79 g/liter oleic acid.

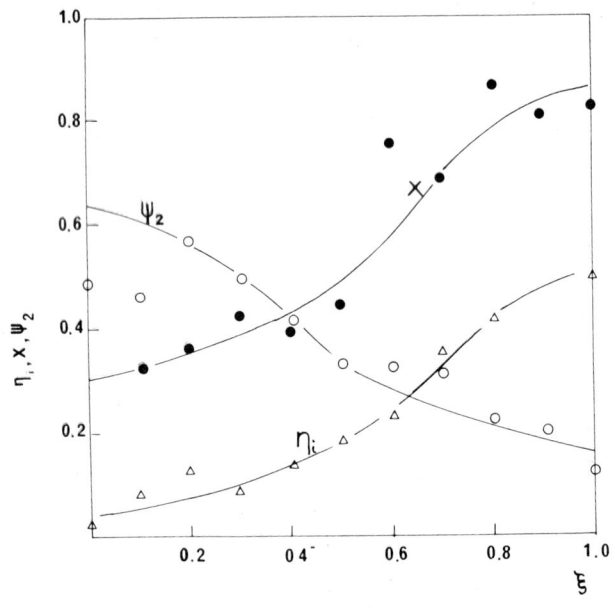

Fig. 16. Profiles of the reduced cell mass, enzyme activity, and substrate concentration with the reduced reactor length coordinate for a flow rate of 374 ml/hr. (-●-●-●-●) Dimensionless cell mass; (O-O-O-O) dimensionless glucose concentration; (-△-△-△-) dimensionless enzyme activity.

is not concerned with activities that may take place when the microbial cells enter a stationary phase. However, these effects can be easily formulated [22] and incorporated into the model.

No attempt was made during our studies to optimize the parameter values of the model; instead, reasonable estimates were made based on published information. The model is flexible enough to predict a number of the experimentally observed phenomena, such as the relative length of lag phases, the sequential uptake of carbon substrates, and diauxic character of the cell growth curves with a decrease in specific growth rate as the metabolism shifts from the favored substrate to the other. The model predictions for the case of growth in a tubular fermentor are also close to the experimentally derived results.

One of the most crucial issues in the formulation of the model is the inhibition function [eq. (21)] assumed for catabolite repression. Exact mathematical formulation of this phenomenon requires a detailed understanding of the biochemical mechanisms associated with regulation of enzyme synthesis. The concept developed by Imanaka and Aiba for cAMP action seems to be the best approach at the moment and could be incorporated into the dauxic growth model. However, more basic biochemical studies need to be done in order to elucidate the mechanism that ultimately controls the intracellular level of cAMP in response to environmental changes.

In conclusion, our model for diauxic growth is a basic one and can be the starting point for more refined formulations. The model can also be upgraded to describe more complex situations, such as growth on more than two substrates; when coupled to optimization routines it could be useful in computer control and optimization of fermentation processes.

Nomenclature

A	intracellular ATP concentration (mol/liter)
CSTR	continuous stirred-tank reactor
D	dilution rate (T^{-1})
D'	dimensionless dilution rate
De	dispersion coefficient (m^2/sec)
E_i	extracellular enzyme activity (U/ml)
E_1	intracellular concentration of constitutive permease (mol/liter)
E_2	intracellular concentration of inducible permease (mol/liter)
e	normalized enzyme activity
F	intracellular concentration of intermediate (mol/liter)
$K_1, K_2, \ldots$	Monod constants
k_A	binding constant for ATP (mol/liter)
k_F	binding constant for intermediate F (mol/liter)
$k_1, k_2, \ldots$	constants
L_r	reactor length (m)
L_1, L_2	constants in eqs. (16) and eq. (17), respectively
m	number of binding sites for enzymes, E_3 and E_4
n, n'	number of binding sites for repressor molecules
$O_P, O_{P'}$	operator genes
Pe	Peclet number

P_1	bulk medium concentration of substrate 1 (mol/liter)
P_2	bulk medium concentration of substrate 2 (mol/liter)
Pr, Pr'	promotor genes
Q_i	induction function for mRNA specific for extracellular enzyme
Q_i'	catabolite repression function for mRNA specific for extracellular enzyme
Q_1	fraction of free operators on the operon specific for permease 2
Q_2	catabolite repression function
R	intracellular concentration of mRNA specific for permease 2 (mol/liter)
R_i	intracellular concentration of mRNA specific for extracellular enzyme (mol/liter)
$R_P, R_{P'}$	repressor molecules
$r_1, r_2, \ldots$	rate expressions (mol/liter) (T^{-1})
r, r'	repressor genes
$S, S_1, S_2, \ldots$	bulk medium substrate concentration (mol/liter)
t	time (T)
V_c	specific cell volume (liter/mol)
v	linear velocity in tubular reactor (m) (T^{-1})
X	bulk medium cell mass concentration (mol/liter)
x	dimensionless cell mass concentration
z	reactor length coordinate (m)

Greek

α	dimensionless ATP concentration
η_i	$k_{17}(E_i)/k_{28}k_3(P_1,0)$
η_1	$(E_1)L_r/v$
η_2	$(E_2)/(E_1)$
θ	dimensionless time
$\kappa_1, \kappa_2, \ldots$	dimensionless constants
μ	specific growth rate (hr^{-1})
μ_m	maximum specific growth rate (hr^{-1})
ξ	dimensionless reactor length
ρ, ρ_i	dimensionless mRNA concentration
ϕ	dimensionless intermediate concentration
ψ_1, ψ_2	dimensionless substrate concentration

Our thanks are due to the National Research Council of Canada for support of this study.

References

[1] R. W. Lodge and C. N. Hinshelwood, *Trans. Faraday Soc., 40,* 571–579 (1944).
[2] G. T. Tsao and T. P. Hanson, *Biotechnol. Bioeng., 17,* 1591–1598 (1975).
[3] A. H. E. Bijkerk and R. J. Hall, *Biotechnol. Bioeng., 19,* 267–296 (1977).
[4] I. H. Lee, A. G. Fredrickson, and H. M. Tsuchiya, *Appl. Microbiol., 28,* 831–835 (1974).
[5] H. Yoon, G. Klinzing, and H. W. Blanch, *Biotechnol. Bioeng., 19,* 1193–1210 (1977).
[6] G. van Dedem and M. Moo-Young, *Biotechnol. Bioeng., 17,* 1301–1312 (1975).
[7] J. Monod, *Ann. Rev. Microbiol., 3,* 371–394 (1949).
[8] F. Jacob and J. Monod, *J. Mol. Biol., 3,* 318 (1961).
[9] G. Yagil and E. Yagil, *Biophys. J., 11,* 11–27 (1971).
[10] G. Terui, M. Okazaki, and S. Kinoshita, *J. Ferment. Technol., 45,* 497 (1967).
[11] G. van Dedem and M. Moo-Young, *Biotechnol. Bioeng., 15,* 419 (1973).
[12] K. Toda, *Biotechnol. Bioeng., 18,* 1117 (1976).
[13] T. Imanaka and S. Aiba, *Biotechnol. Bioeng., 19,* 757 (1977).
[14] A. G. Fredrickson, *Biotechnol. Bioeng., 18,* 1481 (1976).

[15] J. Monod, J. Wyman, and J. P. Changeux, *J. Mol. Biol., 12,* 88 (1965).
[16] M. Moo-Young, G. van Dedem, and A. Binder, *Biotechnol. Bioeng., 21,* 593 (1979).
[17] Technicon Auto Analyzer N method file, No. N-26, Technicon Corporation, Tarrytown, NY.
[18] C. C. Sweeley, R. Bentley, M. Makita, and W. W. Wells, *J. Am. Chem. Soc., 85,* 2497 (1963).
[19] G. G. Guilbeault, J. Hieserman, and M. H. Sacler, *Anal. Lett., 2,* 185 (1969).
[20] G. R. Weston, M.A.Sc. thesis, University of Waterloo, 1971.
[21] R. K. Ghai, Ph.D. thesis, University of Waterloo, 1973.
[22] J. V. Maxham and W. J. Maier, *Biotechnol. Bioeng., 20,* 865 (1978).

Modeling and Optimal Control of Bakers' Yeast Production in Repeated Fed-Batch Culture

P. PERINGER and H. T. BLACHERE

Station de Genie Microbiologique, INRA, 17 rue Sully, 21034 Dijon, France

INTRODUCTION

The term "fed-batch" culture [1] designates a semicontinuous culture technique in which the fresh medium is fed with a defined profile in the course of time. The more or less empirical application of this technique has been realized for more than 20 years in industrial bakers' yeast production on molasses where it avoids an intensive formation of ethyl alcohol. By an appropriate control of the substrate feed rate, this technique constitutes a means of overcoming, in general, the catabolite repression phenomenon [2] due to certain substrates or excreted products.

The behavior of a semicontinuous culture at constant feed rate was analyzed by Dunn and Mor [3], while Lim et al. [4] examined the consequences resulting from the application of exponential feed rates. Edwards et al. [5] termed "extended cultures" semicontinuous operations where the growth-limiting substrate concentration was kept constant throughout the process by an appropriate control of the feed rate. Pirt [6] defined the conditions at which a "steady state" is reached where both the substrate and the biomass concentrations remain constant.

A perspective of computer control to enhance the productivity in bakers' yeast cultivation was presented by Aiba et al. [7] using the respiratory quotient value as a parameter to control and minimize the aerobic fermentation. Wang et al. [8] applied an on-line computer-coupled system to optimally control the feed rate of sugar to the fermentor. Their method is based on material balances [9] permitting an indirect measurement of the biomass concentration and on direct monitoring of fermentation parameters for which there are established sensors. Recently Yamané et al. [10] have developed an optimal control algorithm based on Pontryagin's maximum principle to enhance the bakers' yeast productivity.

All these optimal computer control methods operate essentially by modifying the feed rate stepwise or continuously such that the respiratory quotient ranges from 1 to 1.2 throughout the process. However, this procedure, which is very efficient for establishing a purely oxidative metabolism of the yeast, does not necessarily lead to a maximum productivity.

In this paper a new approach, based on the "catabolite inhibition" idea, takes both the respiratory quotient and the oxygen uptake rate of the culture into consideration to improve the bakers' yeast productivity.

EXPONENTIAL PROFILE OF THE FEED RATE

In order to obtain maximal efficiency of the carbon substrate assimilation, the formation of incompletely oxidized products, such as glycerol or ethyl alcohol, must be avoided. These substances are produced either below a threshold value of the dissolved oxygen tension or above a critical value of the sugar concentration. Thus, in a controlled environment, one can only expect to reach a maximal productivity by working, respectively, above and below of these critical values. In fact, the results of glucose-limited aerobic continuous cultures [11] and dissolved oxygen limited continuous [12] or batch [13] cultures show in a more precise way that the maximal productivity of yeasts sensitive to the glucose effect could be obtained at concentrations of the dissolved oxygen and residual sugar very close to the critical values.

In these conditions, the optimization of the culture based upon maximizing productivity is equivalent to maintaining these concentrations around their optimal values $\overline{S}$ and $\overline{C}_L$ estimated by the critical threshold values which are determined experimentally.

Constant values for S and C_L, expressed by $dS/dt = 0$ and $dC_L/dt = 0$, result in constant values for the specific substrate and oxygen uptake rates (Q_s and Q_{O_2}) and consequently for the growth yield Y and the specific growth rate μ [14]. Then, the material balance in extended culture reduces to eqs. (1)–(4), which describe the evolution of the substrate feed rate F, the volume V, the biomass concentration X, and the volumetric oxygen transfer coefficient K_v:

$$\frac{dV}{dt} = F = \overline{\mu} V_0 k \exp(\overline{\mu} t) \tag{1}$$

$$V = V_0 [1 - k + k \exp(\overline{\mu} t)] \tag{2}$$

$$X = \frac{X_0 \exp(\overline{\mu} t)}{1 - k + k \exp(\overline{\mu} t)} \tag{3}$$

$$K_v = \frac{Q_{O_2} X_0 V_0 \exp(\overline{\mu} t)}{C_i - C_L} \tag{4}$$

with

$$k = X_0 / \overline{Y}(S_0 - \overline{S}) \tag{5}$$

These kinetics depend on the k value, i.e., if S_0 is fixed, on the inoculum concentration X_0. Equation (3) shows in particular that during the first cycle of the fed-batch culture, X increases if $k < 1$, decreases if $k > 1$, and does not change if $k = 1$. When the culture is repeated several times, the inoculum concentration of the ith cycle is the biomass obtained at the end of the $i - 1$th operation. At the end of the nth cycle, the biomass concentration would be given by the eq. (6) where T_{fi} is the time required for the ith operation.

$$X_n = \frac{X_0 \exp \bar{\mu} \left(\sum_{i=1}^{n} T_{fi} \right)}{1 - k + k \exp \left(\bar{\mu} \sum_{i=1}^{n} T_{fi} \right)} \quad (6)$$

It is clear that at the end of a few cycles the biomass concentration practically reaches its limit:

$$X_m = X_0/k = \bar{Y}(S_0 - \bar{S}) \quad (7)$$

which it will preserve during the following cycles. For $k = 1$, i.e., for an inoculum such as $X_0 = \bar{Y}(S_0 - \bar{S})$, eqs. (1)–(3) reduce to eqs. (8)–(10):

$$F = \bar{\mu} V_0 \exp(\bar{\mu} t) \quad (8)$$

$$V = V_0 \exp(\bar{\mu} t) \quad (9)$$

$$X = \bar{Y}(S_0 - \bar{S}) \quad (10)$$

Thus, optimal bakers' yeast production in extended culture requires the application of an exponential profile of the substrate feed rate expressed by eq. (8) and defined such that both biomass and residual sugar concentrations remain constant during the repeated cultivation cycles.

"CATABOLITE INHIBITION" TO MAXIMIZE THE PRODUCTIVITY

Constant values for dissolved oxygen, carbon substrate, and biomass concentration define a "steady state" in extended culture [6] which may be simulated according to Lim et al. [4] by an equivalent continuous culture. However, such results are not observed when one compares the productivities of the two systems. The productivity of an extended culture cycle is given by eq. (11), where V_0 is the initial volume of the medium and V_f is the final volume reached at the end of a time T_f:

$$P_E = \frac{V_f X_f - V_0 X_0}{V_f T_f} = \frac{(V_f - V_0) X}{V_f T_f} = \frac{(V_f - V_0) \bar{Y}(S_0 - \bar{S})}{V_f T_f} \quad (11)$$

$$T_f = \frac{1}{\bar{\mu}} \cdot \ln \frac{V_f}{V_0} \quad (12)$$

The expression for T_f in eq. (12) is obtained by making $V = V_f$ in eq. (9). Substituting T_f in eq. (11), the fed-batch culture productivity will be given by eq. (13):

$$p_E = \frac{V_f - V_0}{V_f \ln(V_f/V_0)} \bar{\mu} \bar{Y}(S_0 - \bar{S}) \quad (13)$$

where it is easily recognized that the expression of the equivalent continuous culture productivity is defined by eq. (14):

$$P_c = \bar{\mu}\bar{Y}(S_0 - \bar{S}) \tag{14}$$

The P_E/P_c ratio given by eq. (15) is, for all $V_f > V_0$, less than unity and indicates that the extended culture performance is necessarily lower than that of an equivalent continuous culture placed in identical physicochemical conditions:

$$\frac{P_E}{P_c} = \frac{V_f/V_0 - 1}{(V_f/V_0)\ln(V_f/V_0)} < 1 \tag{15}$$

This ratio becomes even lower when V_f/V_0 is great. For example when $V_f/V_0 = 2$, P_E/P_c is only 0.721. It is only when V approaches V_0 that the extended culture approaches the continuous system.

Experimental evidence shows, however, that in the presence of derepressed yeasts, one can temporarily establish a residual sugar concentration over the critical value, i.e., a specific growth rate higher than the optimal value $\bar{\mu}$, without catabolite repression and deterioration of the growth yield. Figure 1 shows the results of such an experiment for a glucose limited continuous culture of *Saccharomyces cerevisiae* at a dilution rate $D = 0.22$ hr^{-1}. At this dilution rate, the yeast metabolism is purely oxidative; the respiratory quotient is around 1, and the residual glucose concentration does not exceed 30 mg/liter. At time indicated by the arrows, a concentrated glucose solution is added to the medium. The upper graph gives the evolution of the residual glucose concentration, and the lower one shows the growth yield and the respiratory quotient (RQ). (Note that in the calculation of the RQ the response times of the paramagnetic oxygen and the infrared CO_2 analysers were taken into account.)

The lower graph shows that the growth yield remains constant during the experiment while the RQ increases rapidly a few seconds after the glucose ad-

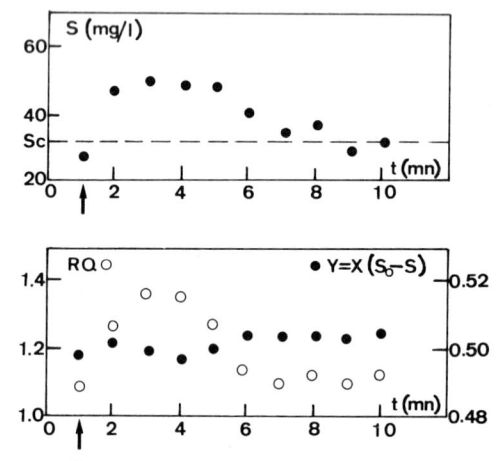

Fig. 1. Primary glucose effect. Arrows indicate addition of glucose to the continuous culture of *S. cerevisiae* operating at dilution rate $D = 0.22$ hr^{-1}. S = residual glucose concentration; S_c = threshold value; RQ = respiratory quotient; Y = growth yield.

dition reaches the values of 1.3 to 1.4. These values are characteristic of a partially repressed yeast population. The RQ drops after 4 to 6 min to its initial value of 1.1. This transitory increase in the RQ does not seem to be linked to the glucose effect since during the entire experiment the residual glucose concentration remains above the critical threshold value for catabolite repression in continuous culture. This is confirmed by the experiment described in Figure 2, where the repeated additions were made as soon as the glucose concentration fell below its critical value, S_c. The figure shows a slow but definitive increase in the RQ superimposed to the transitory increase observed at each addition due to the glucose effect. This initiates the production of ethanol and causes a significant drop in the growth yield.

The "primary glucose effect" is a phenomenon in which a sharp increase of the glucose concentration in a derepressed yeast culture provokes instantaneously a lower increase of the oxygen uptake rate than that of the CO_2 production rate—i.e., a transitory increase of the respiratory quotient. This effect is probably catabolite inhibition and not repression. The intensity of the catabolite inhibition decreases as the degree of catabolite repression increases and at a RQ > 5, the phenomenon is practically unnoticeable.

In fed-batch cultures, starting with a derepressed yeast population, when the frequency and the extent of glucose additions are conveniently chosen, the catabolite repression does not appear, although the specific growth rate is higher than that corresponding to the optimal productivity conditions in an equivalent continuous culture.

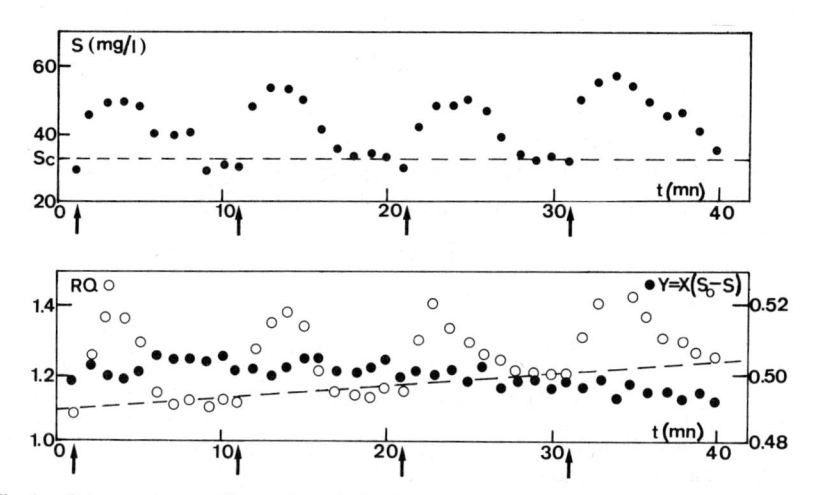

Fig. 2. Primary glucose effect and catabolite repression. Arrows indicate glucose additions to the continuous culture of *S. cerevisiae* operating at dilution rate $D = 0.22$ hr^{-1}. Dashed line in the lowest graph shows the definitive increase of RQ. S = residual substrate (glucose) concentration; S_c = threshold value; RQ = respiratory quotient; Y = growth yield.

INITIAL CONDITIONS OF THE CULTURE

When one takes catabolite inhibition into account, the P_E/P_c ratio will be given by eq. (16):

$$\frac{P_E}{P_c} = \frac{V_f/V_0 - 1}{(V_f/V_0)\ln(V_f/V_0)} \frac{\mu^* Y^*(S_0 - S^*)}{\bar{\mu}\bar{Y}(S_0 - \bar{S})} \quad (16)$$

where μ^* is the mean specific growth rate of the extended culture in which an appropriate substrate feeding strategy is applied.

Since S^* and $\bar{S}$ are negligible before S_0 and Y^* is practically equal to $\bar{Y}$, the eq. (16) becomes

$$\frac{P_E}{P_c} = \frac{V_f/V_0 - 1}{(V_f/V_0)\ln(V_f/V_0)} \frac{\mu^*}{\bar{\mu}} \quad (17)$$

Equation (17) shows that the V_f/V_0 ratio must be suitably chosen so that the productivity on the extended culture is equal to that of an equivalent continuous culture. For example, if $\mu^*/\bar{\mu} = 1.25$, (17) gives, for $P_E/P_c = 1$, a V_f/V_0 ratio value of 1.6.

When the V_f/V_0 ratio is defined by the knowledge of the $\mu^*/\bar{\mu}$ ratio, the maximization of the productivity implies maximizing the biomass concentration. This latter should be compatible with the maximum oxygen transfer capacity of the fermentor and does not therefore overtake the value given by eq. (18)

$$X_{max} = \frac{K_{max}}{V} \cdot \frac{C_i - \bar{C}_L}{\bar{Q}_{O_2}} = K_{La\ max} \frac{C_i - \bar{C}_L}{\bar{Q}_{O_2}} \quad (18)$$

X_{max} being known, eq. (19) allows the determination of the substrate concentration S_0 in the feed medium:

$$S_0 = X_{max}/\bar{Y} + \bar{S} \quad (19)$$

The initial conditions being conveniently chosen, the optimal control strategy of bakers' yeast production should be based on the on-line evaluation of the oxygen uptake rate (OUR) and of the CO_2 production rate (CPR) given by eqs. (20) and (21), whose ratio defines the respiratory quotient RQ in eq. (22):

$$\text{OUR} = Q_{O_2} x V \quad (20)$$

$$\text{CPR} = Q_{CO_2} x V \quad (21)$$

$$\text{RQ} = Q_{CO_2}/Q_{O_2} = \text{CPR}/\text{OUR} \quad (22)$$

$$Q_{CO_2} = \rho_1 Q_s + \rho_2 Q_{O_2} \quad (23)$$

The relationship that links Q_{CO_2} to Q_s and Q_{O_2} in eq. (23) was obtained from the equations of molecular balances established by means of an integrated metabolic scheme of yeasts [14]. The parameters ρ_1 and ρ_2, which depend on the culture medium composition, can be experimentally identified. A similar equation has been derived by Endo [15].

Taking eqs. (20)–(23) into account, the substrate feed rate given by eq. (24)

can be put into the form of eq. (25), where it only depends on the oxygen uptake rate and the respiratory quotient:

$$F = \mu V = Q_s YV = Q_s \frac{x}{S_0 - S} V \tag{24}$$

$$F = \frac{\text{OUR}}{\rho_1(S_0 - S)}(RQ - \rho_2) \tag{25}$$

The feed rate expressed by this equation is a "physiological" feed rate which corresponds to the real measurable state of the energy yielding metabolism of the yeasts.

Equations (26) and (27) allow the calculation of the optimal feed rate F^* theoretically observable when applying the optimal control strategy based on the catabolite inhibition idea:

$$F^* = \frac{\text{OUR}^*}{\rho_1(S_0 - \overline{S})}(RQ^* - \rho_2) \tag{26}$$

$$\text{OUR}^* = \overline{Q}_{O_2}\overline{Y}(S_0 - \overline{S})V_0 \exp(\mu^* t) \tag{27}$$

The real physical feed rate corresponding to the substrate feed rate of the dosing pump will then be defined at each moment of the process by the application of a control strategy which is able to minimize the difference between F^* and F and at the same time to avoid the catabolite repression. Thus, both

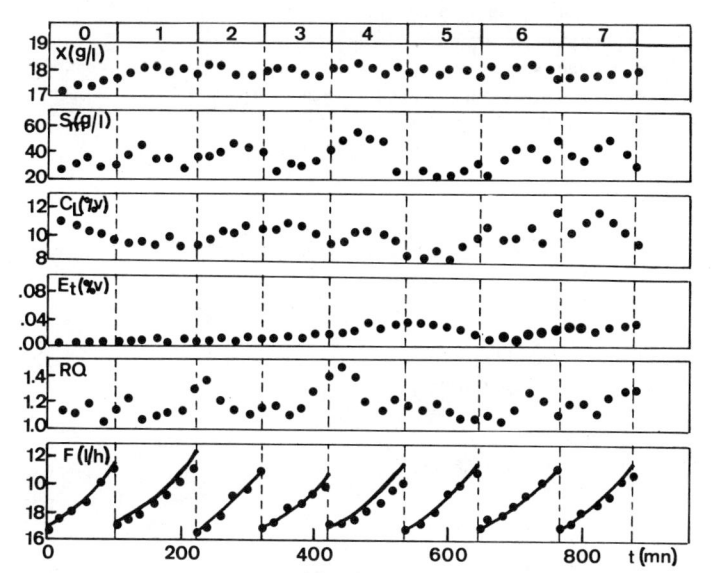

Fig. 3. Fed-batch culture of *S. cerevisiae* on synthetic glucose medium. Results of eight repeated cultivation cycles optimally controlled. Dissolved oxygen tension was regulated by means of aeration and agitation rates. X = biomass; S = residual glucose; C_L = dissolved oxygen; E_t = ethanol concentrations; RQ = respiratory quotient; F = substrate feed rate; (———) calculated optimal profile.

the respiratory quotient and the oxygen uptake rate of the culture should be used for optimal control of the production.

The strategy, using RQ as a qualitative index and the OUR as a quantitative index of the energy yielding metabolism of the yeasts, was applied in a repeated fed-batch culture of *S. cerevisiae* growing on a synthetic glucose medium. Ignoring the details of the optimal control strategy, Figure 3 shows only some of the results obtained for the initial conditions below:

$$X = X_0 = 18 \text{ g/liter}, \quad S_0 = 35.8 \text{ g/liter}, \quad V_0 = 25 \text{ liter},$$
$$V_f = 40 \text{ liter}$$

$$\rho_1 = 10.15 \text{mM/g}, \quad \rho_2 = 0.35, \quad \overline{S} = 32 \text{ mg/liter},$$
$$\overline{Q}_{O_2} = 6.5 \text{mM/g/hr}$$

$$\overline{\mu} = 0.22 \text{ hr}^{-1}, \quad \mu^* = 0.275 \text{ hr}^{-1}, \quad RQ^* = 1.2$$

The most significant result concerns the mean experimental productivity of 3.8 g/liter/hr while the productivity of an equivalent continuous culture is 4 g/liter/hr and the productivity of an extended culture exclusively based on RQ regulation is only 3.2 g/liter/hr.

Nomenclature

C_i	oxygen concentration in the bulk gas (mM)
C_L	dissolved oxygen concentration (mM)
CPR	carbon dioxide production rate (mM/hr)
EtOH	ethanol concentration (% vol)
$F, (F^*)$	physiological (optimal) substrate feed rate (1/hr)
K_{La}	oxygen transfer coefficient (hr^{-1})
K_v	volumetric oxygen transfer coefficient (1/hr)
OUR	oxygen uptake rate (mM/hr)
$P_E(P_c)$	fed-batch (continuous) culture productivity (g/liter/hr)
Q_{CO_2}	specific CO_2 production rate (mM/g/hr)
Q_{O_2}	specific oxygen uptake rate (mM/g/hr)
Q_S	specific substrate uptake rate (g/g/hr)
RQ	respiratory quotient (mM/mM)
S	residual substrate concentration (g/liter)
S_0	concentration of the substrate in fresh medium (g/liter)
t	time (hr)
$V_0(V_f)$	initial (final) volume of the culture (liter)
X	biomass concentration (g/liter)
X_0	concentration of inoculum (g/liter)
Y	growth yield (g/g)
μ	specific growth rate (hr^{-1})
ρ_1, ρ_2	model coefficients (mM/g; mM/mM)

References

[1] F. Yoshida, T. Yamané, and K. I. Nakamoto, *Biotechnol. Bioeng.*, 15, 257 (1973).
[2] B. Magasanik, *Cold Spring Harbor Symp., Quantum Biol.*, 26, 249 (1961).
[3] I. J. Dunn and J. R. Mor, *Biotechnol. Bioeng.*, 17, 1805 (1975).
[4] H. C. Lim, B. J. Chen, and C. C. Creagan, *Biotechnol. Bioeng.*, 19, 425 (1977).

[5] V. H. Edwards, M. J. Gottschalk, A. Y. Noojin, L. B. Tuthill III, and A. L. Tannahill, *Biotechnol. Bioeng.*, *12*, 975 (1970).
[6] S. J. Pirt, *J. Appl. Chem. Biotechnol.*, *24*, 415 (1974).
[7] S. Aiba, S. Nagai, and Y. Nishizawa, *Biotechnol. Bioeng.*, *18*, 1001 (1976).
[8] H. J. Wang, C. L. Cooney, and D. I. C. Wang, *Biotechnol. Bioeng.*, *19*, 69 (1977).
[9] C. L. Cooney, H. J. Wang, and D. I. C. Wang, *Biotechnol. Bioeng.*, *19*, 55 (1977).
[10] T. Yamané, E. Kume, E. Sada, and T. Takamatsu, *J. Ferment. Technol. (Jpn.)*, *55*, 587 (1977).
[11] A. Fiechter and H. K. von Meyenburg, "Regulatory properties of growing cell populations of *Saccharomyces cerevisiae* in continuous culture system," *Proceedings of the Second International Symposium on Yeast*, Bratislava (1966), p. 387.
[12] E. Oura, "The effect of aeration on the growth energetics and biochemical composition of baker's yeast," thesis, University of Helsinki, 1972.
[13] P. Peringer, H. Blachere, B. Corrieu, and A. G. Lane, *Biotechnol. Bioeng.*, *16*, 431 (1974).
[14] P. Peringer and H. Renevey, "Some new aspects in the modeling of continuous cultivation of yeasts," *Abstracts of Papers, Continuous Cultivation of Microorganisms, 7th International Symposium, Prague July 10-14, 1978*, p. 72.
[15] I. Endo and I. Inoue, "Analysis of yeast metabolisms in batch culture," *Abstracts of Papers of the 5th International Fermentation Symposium*, H. Dellwey, Ed. (Springer-Verlag, Berlin, 1976), p. 83.

Review of Alternatives and Rationale for Computer Interfacing and System Configuration

WILLIAM B. ARMIGER and DENNIS M. MORAN

BioChem Technology, Inc., Malvern, Pennsylvania 19355

1. INTRODUCTION

Advances in biochemistry and enzymology have resulted in a more complete understanding of enzyme-catalyzed reactions with respect to substrate availability and enzyme activity. These developments have enabled the biochemist to employ kinetic models to describe biochemical processes. It is the function of the biochemical engineer to interpret these basic kinetic models in light of the reactor configuration in order to transfer the process from the laboratory to production scale.

In parallel with these developments, computer science, in terms of both hardware and programming software, has advanced to the stage where it is now cost effective to utilize this technology in small-scale pilot-plant operations as well as in full-scale production plants [1-4]. Thus, by employing minicomputers and microprocessors, it is possible to expand process control beyond independent closed-loop feedback control of parameters such as temperature, pH, or flow to more sophisticated strategies of interactive control. Eventually, on-line process optimization, based upon kinetic models, will be possible by utilizing an adaptive or feed-forward, interactive control strategy. This combination of computer technology and kinetic modeling in the hands of the fermentation process development engineer is resulting in greater efficiency and higher process productivity.

In order to accomplish on-line process optimization, it is necessary to develop a system that encompasses both the hardware and software needed to generate raw data, analyze the resulting information, determine optimal solutions, and implement control decisions. A highly instrumented computer-coupled system is capable of meeting these tasks. The design alternatives and the rationale for computer interfacing and system configuration are the topics discussed in the presentation.

2. RATIONALE FOR SYSTEM DESIGN

The interfacing and configuration of computer systems for use in fermentation technology are very dependent upon the end use of the system [5-8]. For example, requirements for a pilot plant are quite different from those for a production plant. In a pilot plant, the system is utilized primarily as a research tool

(Table I); whereas, in a full-scale production plant, the computer is used for process monitoring, control, and optimization aimed at maintaining quality control and product uniformity (Table II). It is a mistake to incorporate these two separate functions into one system application. These functions should be clearly defined from the beginning in order to design a system that will perform as an integrated unit. Furthermore, with these goals in mind, it is extremely important to recognize that the successful implementation of a computer-coupled system is an evolutionary process (Table III). The first step is data acquisition, which includes on-line process monitoring, data logging, process alarming, and data analysis. The next step involves data interpretation by utilizing the information generated on-line to elaborate the system kinetics. An on-line computer system reduces the time required for determining the dynamics of a process, for testing control strategies, and for modeling the process. Finally, since minicomputer systems are capable of simultaneously performing on-line data acquisition tasks and running on-line process simulations based upon kinetic models, the process development engineer can quickly refine a mathematical model and identify the model parameters so that methods can be developed for predicting the interactive relationships between the process variables. Thus, with sufficient refinement of the kinetic model, computer-coupled fermentation systems can advance to the final stage of automatic process control, i.e., on-line optimization based upon model reference control.

3. FUNCTIONS OF A COMPUTER-COUPLED SYSTEM

There are certain functions of a computer-coupled system that are common to all applications [9-12]. If the system is designed with these functions in mind, it will be flexible enough for a variety of purposes. Table IV summarizes these functions. Data logging monitors the status of each unit operation. It keeps track of the time and initiates the scanning of the data at the proper frequency. The system should be designed for random access of each channel so that variables can be scanned at different rates rather than requiring a preprogrammed sequence of scanning. The data checking and testing function monitors key parameters during each scan and checks those values against alarm units. In the event of an alarm situation, messages are sent to the operator along with instructions for dealing with the problem. On-line data analysis performs necessary conversions of the raw input data that could include scaling, comparing with calibration curves, incorporating correction factors, or converting units. This is followed by tasks for analyzing the data with regard to the specific process that could involve solving relatively simple algebraic equations or more complex differential equations.

The system should include automatic procedures for handling power failures. A short-term power failure can disable the computer system while a long-term power failure can affect the process. In order to protect the integrity of the computer system, a power-failure restart function allows the computer system to smooth out intermittent short-term power fluctuations with no apparent loss

TABLE I
Objectives of Pilot-Plant System

Define process parameters
Develop methods for on-line data analysis
Evaluate process models
Test control and optimization strategies

TABLE II
Objectives of Plant-Scale System

Product uniformity and quality control
Process monitoring
    Raw material
    Process variables
    Alarms
    Products
Increased process reliability
Simplified plant operation
Process control

TABLE III
Evolutionary Steps in System Development

Data acquisition
    Process monitoring
    Data logging
    Process alarming
    Data analysis
Data interpretation
    Process modeling
    Process dynamics
Process control
    Interactions among variables
    Model reference control

TABLE IV
Functions of a Computer-Coupled System

Data logging
Data checking and testing
On-line data analysis
Power failure procedures
    Computer system
    Process system
Process control
    Sequencing operations
    Control of process parameters (individual loops)
    Control of entire process
    (multiple input/multiple output)
---

of service or data while at the same time maintaining logical consistency within the computer system itself and within the application tasks. In the event of a long-term power failure, the computer should initiate a sequencing procedure to place the process in a fail-safe status. When the power is restored, the computer brings the process to its working state and issues special instructions to the operator to check the proper functioning of the entire system.

There are three levels of process control that should be incorporated into the system. Each higher level involves a greater degree of programming sophistication and requires more knowledge of the dynamics of the process. The first level of control involves sequencing operations, such as manipulating valves or starting and stopping pumps associated with system startup, batch or semibatch operations, instrument recalibration, on-line maintenance, and fail-safe shutdown procedures. For most of these operations, the time base is on the order of minutes or hours so that high-speed manipulations are not critical to the process. The major advantage that a computer system offers over a traditional hardward system of timers and relays is the ease with which the sequencing logic can be changed either in its format or time base to accommodate improvements in operating procedures.

The next level of control involves manipulation of the process parameters in individual closed loops. There are two basic concepts of process parameter control: direct digital control (DDC) and digital setpoint control (DSC). Determining an economic justification of one method over the other is a complicated issue. There is no general answer because the proper solution depends upon the particular circumstances. The economic comparison for hardware costs of DSC to DDC is simply the cost of N controllers/recorders with their interfacing costs versus the cost of N signal holding units with manual backup and the portion of the computer allocated to DDC. Thus, the cost depends on the number of control loops. In most cases, the cost of a DDC system with backup controllers will be more than a DSC system. However, cost is not the only consideration in many applications. In general, DDC can provide for better control of a process since the control algorithms are mathematically stored functions rather than an electrical analog. Therefore, for a large-scale continuous process, what appear to be minor improvements in process control result in major savings because of improved plant operation. For a pilot-plant operation, small improvements in yield are generally secondary considerations because the labor and engineering costs usually are much higher by comparison. Therefore, for a pilot-plant system, DSC is usually the best concept to follow.

The third level of control is concerned with increased productivity and process optimization. It is at this level that the computer system is the most valuable since it is used to monitor many input parameters and to implement control decisions based upon the dynamics of the process. In a pilot-plant application, these control decisions are best implemented by a supervisory control system where the computer system is responsible for manipulating the setpoints on the various analog controllers. Here again, these control decisions are usually implemented at low frequencies so that the computer system is available for many other tasks other than process control. The best hardware configuration for a production plant will depend upon the specific circumstances.

4. SPECIFICATIONS FOR COMPUTER SYSTEMS

A computer system is composed of the computer hardware (processor, memory, terminal, printer, storage device, etc.) and the operating system software. The operating system, which is usually supplied by the computer manufacturer, is a group of programs written in the machine language to allow an operator to communicate with the system and to allow the various components of the computer hardware to communicate with each other. A high-level programming language (FORTRAN, BASIC, COBAL, etc.) is usually added to the operating system for ease in writing applications programs. With this kind of a basic system, an experienced programmer is required to write and implement the functions described in Sec. 3.

The engineer or technician should not be required to be a systems programmer. He or she should be able to work with the system as a "black box" to perform the desired functions. Thus, many systems include an operator-oriented software package that provides the operator with prompting messages for executing programs. There are many such packages commercially available where the operator generates a data base by "filling in the blanks." This kind of system greatly reduces the time required for on-site programming and the time required for training personnel. By utilizing the operator-oriented programming concept the system should have the capabilities summarized in Table V.

The operator should have easy access to the data base in order to examine files, edit data, format files, add off-line laboratory data, tabulate information, and transfer data between the peripheral devices. The operator should also be able to transfer data to magnetic disks or tapes for long-term and backup storage.

Since one of the primary objectives of the system is to access vast amounts of information quickly and to transfer the information into a form that can be easily interpreted, the system should automatically generate reports. Some reports are transient in nature and need only be displayed on a screen, while others will require a hard copy.

Graphical information is much easier to interpret than tables of numbers.

TABLE V
Specifications for a Computer System

Data base manipulation
Data storage
Report generation
Graphics
Process override
Operator procedures
Multiple programming
Multiple users

Therefore, a system with graphics capabilities is highly recommended. The operator should be able to create graphs from any of the data files stored on the disk. The graphics programs should support plots of multiple dependent variables from a single experiment or cross plots from several experiments. This will allow the operator to quickly compare the results of different experiments. In addition, the system should automatically generate a set of standard plots for routine reports.

The operator should be able to override the computer system in order to be able to implement operations when necessary. The entire software package is designed to assist the operator and to handle routine procedures so that there is time to interpret results rather than just react to situations. Thus, the system will respond to whatever is demanded of it and help keep track of events, but it will not be in control of the operator.

The system should store operating procedures that can function as checklists. These are particularly useful to inexperienced operators for startup, shutdown, or recalibration procedures. The procedures can include communications to the operator that indicate what options are available if he or she is not sure which command to give to the system.

Multiple programming should be included in the system to enable it to concurrently execute two or more programs residing in the memory. In addition, it should be able to schedule programs on a priority basis in response to real-time events. The advantage of this is that the system will always be ready to respond to the needs of the process, and at the same time, it will support program development and other off-line requests from the operator.

The system should accommodate multiple users so that two or more groups can access the system without interfering with each other. Thus, there would be multiple terminals on the system some of which could be in remote locations that would communicate with the computer through the telephone system.

Once the necessary functions have been determined and the specifications for the computer system have been outlined, the next step is the selection of the proper hardware, software, and a man–machine interface to meet the requirements for data base manipulation, data storage, report generation, graphical output, and program implementation. There are a number of choices that must be considered (see Table VI). These involve selecting between such alternatives as analog scanning or digital multiplexing, supervisory or direct digital control, a single processor or hierarchical configuration, multiprogramming/multitasking or simply a foreground/background operating system, a high-level machine language or an operator-oriented language. In all of these choices, it is extremely important that the hardware and software form an integrated unit. Many times mistakes are made by considering the hardware and software costs as exclusive of each other.

5. EXAMPLE OF A COMPUTER-COUPLED PILOT-PLANT SYSTEM

The computer-coupled pilot-plant system at the University of Pennsylvania is capable of generating on-line information for process scaleup and modeling,

TABLE VI
Design Choices

Analog scanning	—Digital multiplexing
Supervisory setpoint control	—Direct digital control
Single processor system	—Hierarchical configuration
Foreground–background operating system	—Multiprogramming–multitasking operating system
High-level oriented language	—Machine language

as well as interpreting on-line data for implementing strategies for process control and optimization [13]. Utilizing the signals produced by instrumentation and integrating them into a computer-controlled system involves both hardware and software development. System hardware configuration basically involves selecting electronically compatible modules. The selection and development of software for data analysis and process control are much more involved. During the configuration of an entire system, hardware and software must be structured to operate as an integrated unit. Furthermore, the design must be flexible enough to accommodate growth and expansion of both hardware and software as new products and techniques become available.

The system hardware consists of three main segments: 1) sensors and instrumentation; 2) computer interface; 3) computer system. In general, the weakest segment in any system involves the sensors. A list of the instrumentation is shown in Table VII. From this list it is obvious that the initial number of measurements is very limited. In addition, most of the sensors are measuring variables in the macroscopic environment of the process, while the key factors

TABLE VII
Monitored and Controlled Variables

Measured parameters

Inlet gas flow rate	Agitator speed
Inlet gas pressure	Power-to-shaft
Inlet gas temperature	Vessel pressure
Inlet gas carbon dioxide	Glucose concentration
Inlet gas oxygen	Culture turbidity
Cooling jacket flow rate	Culture fluorescence
Cooling jacket temperature	Exit gas carbon dioxide
Difference	Exit gas oxygen

Analog-controlled variables

Dissolved oxygen
Vessel temperature
pH
Foam
Substrate addition rate

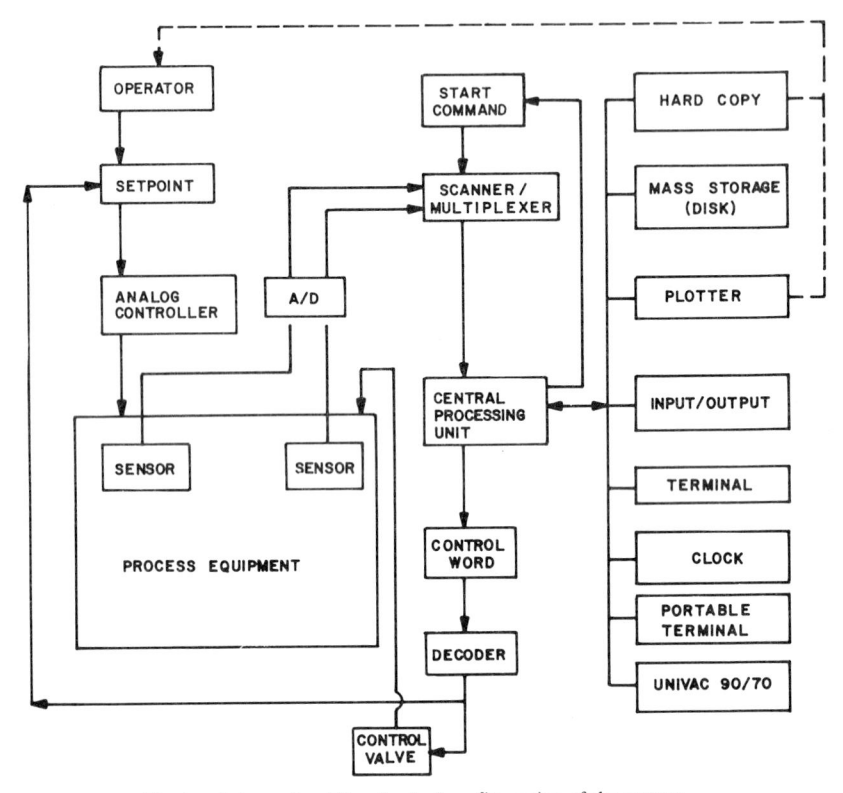

Fig. 1. Schematic of the physical configuration of the system.

(temperature, pH, and ionic strength) that control the metabolic activity of the cell involve the microscopic environment internal to the microorganism. Sensors for measuring cellular activity on a microenvironmental scale need to be developed.

The hardware for the computer interface is readily available. With sufficient attention to detail, it will function properly. However, the architecture of the interface depends upon the application. Analog scanning is the most economical method of interfacing if there are a large number (over 25) of channels involved. However, digital multiplexing, which is cost competitive on a small system (under 25 channels) offers many advantages. First, it allows each channel to function as a separate unit, and it provides the operator with a visual display of all the channels at one time. Secondly, it allows for all of the data to be logged simultaneously rather than sequentially so that calculations can be made on data from the same instant in time. For large systems, the advantages of digital multiplexing are difficult to justify because of cost. Therefore, in pilot-plant applications involving many unit operations, an analog scanning interface is the most practical design.

Computer interfacing involves a choice between flexibility in hardware or increased software requirements. For example, trade-offs occur between random

```
                    PROCESS
                       ↕
              COMPUTER INTERFACE
                       ↑
```

CENTRAL PROCESSOR (Core Memory)	SYSTEM DISK
OPERATING SYSTEM RESIDENT MONITOR	OPERATING SYSTEM PROGRAMS
OPERATOR-ORIENTED SOFTWARE PACKAGE	DATA STORAGE FILES
APPLICATIONS PROGRAMS EXECUTIVE ROUTINE	TEMPORARY BUFFER FILES
	FORTRAN IV PROGRAMS
APPLICATIONS PROGRAMS OVERLAY REGIONS	ASSEMBLY LANGUAGE PROGRAMS

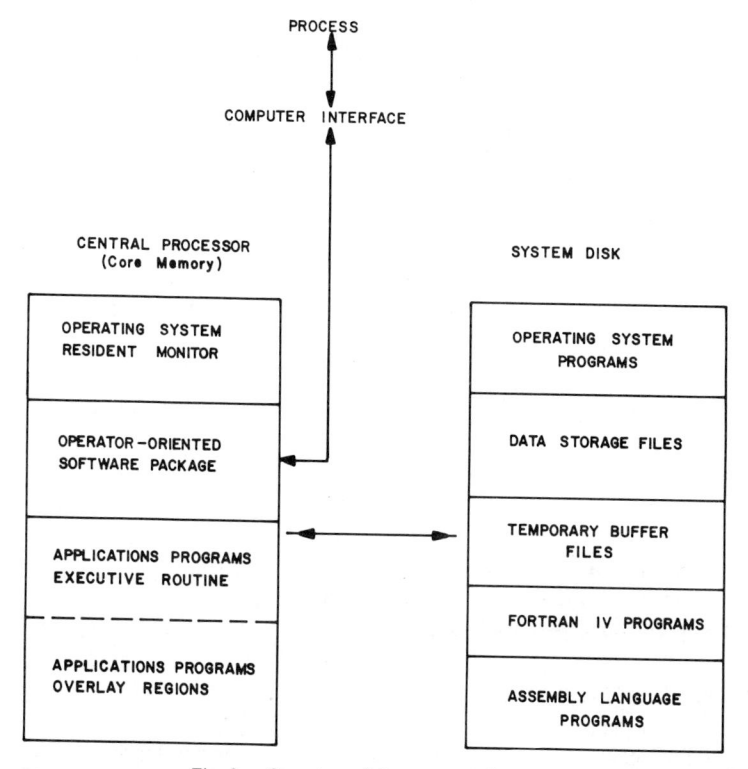

Fig. 2. Overview of the system software.

access scanning (which increases hardware costs) and less flexible sequential scanning but lower software costs. Again, the number of channels involved in the interface will determine if the cost savings are to be gained by using more sophisticated hardware or software. For a system involving several pilot-plant processes, random access scanning is recommended.

Figure 1 is a hardware schematic and signal flow diagram of the University of Pennsylvania installation [13]. The analog signals from the various sensors are fed to the interface where the analog voltage is converted to a binary coded decimal (BCD) word. The BCD output of the interface is transmitted to the computer under control of the computer software which specifies the scanning rate and sequence. Once the data reach the computer, they are stored on the system disk and analyzed by the system programs. If the results of the analysis are to be used for process control, signals from the computer are then transmitted back to the process to manipulate variables directly (DDC). The results generated can also be output on a hard copy device, stored on the disk or tapes, plotted on a graphical device, or transmitted to another computer system.

An overview of the system software organization is illustrated in Figure 2 [13]. The computer system can be conceptualized as two parts: the central processor and the system disk. All other computer equipment are really peripheral devices to support these two systems. The central processor contains the computer's instructional logic and a relatively small region of core memory that is allocated into three sections. The first part contains the operating system resident monitor, which is the software enabling the operator to issue commands from the terminal to load programs into core, execute them, and control and monitor information transmitted to or from all system devices. The second portion contains the operator-oriented software package, which is responsible for providing the man-machine interface between the operating system programs and the applications software. The third section, which is for applications programs, is partitioned into many smaller regions. The first region contains the executive routine for managing the sequence in which different program modules enter, execute their function, and leave the overlay regions.

Since the computer core is limited in minicomputers, only data and program modules required at a particular point in time reside in an overlay region. The bulk of the software remains on the disk. As time progresses and certain tasks are completed, data are transferred to a file on the disk and a new program module is read from the disk and overlayed on top of a module whose function is complete. Thus, the system disk that has no computational power is used as a storage device only. It contains vast amounts of information consisting of the operating system modules required by the resident monitor, data storage files, temporary buffer files, FORTRAN programs, and assembly language routines.

In a real-time event-driven operating system, each task has a system priority assigned to it. When a task is completed, a new task automatically begins executing. The high-priority tasks include starting the scanner, logging data on the disk, generating alarm messages, and process control. The benefit of this arrangement is that the most time-dependent functions are performed first. Clearly, it would be unacceptable if data acquisition had to be delayed because the computer was occupied by producing a report that could take as long as an hour.

Typically, the high-priority tasks can be executed very rapidly (sec) and will automatically interrupt and suspend a low-priority task if necessary. Thus, the operating system will ensure that the computer resources are allocated efficiently so that functions such as analyzing laboratory data, simulating the process, producing graphs, generating reports, or developing new programs will be run in between high-priority tasks.

6. CONCLUSIONS

There are several important points to remember when considering the design alternatives and the rationale for computer interfacing and system configuration. It is important to clearly define the objectives of the system since the requirements for a pilot plant are quite different from those of a production facility. Next, it is necessary to select the functions that are required and to specify the computer and interface that will accomplish these tasks. This will involve comparing hardware–software alternatives from both an economic and operating viewpoint. Finally, the most important consideration is that the hardware and software must function as an integrated unit. Thus, both hardware and software designs must be evaluated together in order to develop a properly designed system.

References

[1] W. B. Armiger and A. E. Humphrey, *Proceedings of the Intersectional Congress, IAMS* (IAMS, New York, 1975), Vol. 5, pp. 99-119.
[2] W. B. Armiger, D. W. Zabriskie, and A. E. Humphrey, *Proceedings of the International Fermentation Symposium* (Verlag, Berlin, 1976), Vol. 5, p. 33.
[3] D. D. Dolby and J. L. Jost, "Computer applications to fermentation operations," in *Annual Reports on Fermentation Processes,* D. Perlman, Ed. (Academic, New York, 1977), Vol. 1, p. 114.
[4] W. A. Weigand, "Computer applications to fermentation processes," *Annual Reports on Fermentation Processes,* D. Perlman, Ed. (Academic, New York, 1978), Vol 2, p. 43.
[5] R. Lundell and P. Laiho, "Engineering of fermentation plants," *Process Biochem.,* 11, 13-17 (1976).
[6] A. Meskanen, R. Lundell, and P. Laiho, "Engineering of fermentation plants. Part 3," *Process Biochem.,* 11, 31-36 (1976).
[7] J. D. Shave, "The computer in the brewhouse," *Process Biochem.,* 11, 10-12 (1976).
[8] R. Spruytenburg, A. D. P. Dang, I. J. Dunn, J. R. Mart, A. Einsele, A. Fiechter, and J. R. Bourne, "Experience with a computer-coupled bioreactor," *Chem. Eng. 310,* 447 (1976).
[9] H. Y. Wang, Ph.D. thesis, MIT, Cambridge, MA, 1976.
[10] C. L. Cooney, H. Wang, and D. I. C. Wang, "Computer-aided material balancing for prediction of fermentation processes," *Biotechnol. Bioeng., 19,* 55-69 (1977).
[11] D. W. Zabriskie, W. B. Armiger, and A. E. Humphrey, "Applications of computers to the indirect measurements of biomass concentration and growth rate by component balancing," in *Workshop on Computer Applications in Fermentation Technology 1976,* R. P. Jefferies, Ed. (GBF monograph series No. 3) (Verlag Chemie, Weinheim, 1977), pp. 59-72.
[12] L. K. Nyiri, R. P. Jefferis, and A. E. Humphrey, "Applications of computers to the analysis and control of microbiological processes," *Biotechnol. Bioeng. Symp., 4,* 613 (1974).
[13] W. B. Armiger, Ph.D. thesis, University of Pennsylvania, Philadelphia, 1977.

Microprocessor in Fermentation Control

M. CORDONNIER, J. P. KERNEVEZ, and J. M. LEBEAULT

Université de Technologie de Compiègne, BP 233, 60206 Compiegne, France

J. KRYZE

Institut de Recherche en Informatique et Automatique, Rocquencourt, 78150 Le Chesnay, France

INTRODUCTION

This paper describes the work done in order to develop a microprocessor oriented to the control of fermentation processes. Our purpose is not to describe the calculations [1-4] that are usually done to characterize cell growth, but to describe a microcomputer especially designed for the control, data logging, and data treatment of a fermentation process.

The chosen hardware structure (microcomputer and fermentor) and the software structure involved are reported here.

HARDWARE STRUCTURE

Microprocessor

A special microcomputer was built for control and data logging in a fermentation process. We preferred a problem-oriented design to a solution using some kind of existing equipment. The purpose was twofold: to reduce the cost, and to check several possibilities about the use of microprocessors for direct digital control (i.e., user's language and reliability).

From the experience accumulated with minicomputer applications, voluminous programming is needed to integrate direct digital control with data logging, supervising, and sequential procedures such as starting, sterilization, and stopping, especially if conversational interaction is demanded.

So, there is much interest in having a powerful programming language reducing both the length of code and the user's effort. Moreover, this language should be open ended to permit incorporation of special instructions needed for specific applications. Both of these requirements can be fulfilled by an interpreted language, using the microprocessor chip code as a microprogram code. This solution offers also more hardware independence, because the microprocessor chip type may be changed without need of reprogramming user's programs.

In our case the user's language is a minicomputer-type 16-bit language, with real-time facilities, such as task monitoring and scheduling for 14 levels and real-time clock and timing. Microprogrammed routines run independently of program execution and take care of all of the data acquisition tasks. There are input/output instructions manipulating items such as words, numbers, or messages as a whole (not in character terms).

There are also microprogrammed routines carrying out the usual debugging job including real-time program execution monitoring. The user has at his disposal 64K bytes of program memory and one kilobyte of data memory.

Our approach to reliability is based on noise resistance. The whole microcomputer is in high-level technology, such as canal P or MOS complementary yielding some 10 times more noise resistance than the common TTL technology.

Moreover, slow circuits were preferred to rapid ones. So, we are now using a 4004 chip, and we could in the future replace it by a MOS complementary microprocessor.

A further advantage of this approach is the very low power consumption, about 20 W for the whole system, leading to reduced heating, which improves reliability and permits a more efficient insulation against an aggressive atmosphere.

Optoelectronic isolation is applied on all peripherals. Duty pulse modulation is used for long-distance transmission of analog outputs.

The reliability of analog data acquisition is warranted by an error-detecting and error-correcting procedure implemented in the hardware and acting upon the very beginning of the data acquisition path. It eliminates virtually any drift and common mode errors and helps to maintain a precision of 20 000 points for any of the 10 available gains.

Moreover, it allows microvolt-level signals to be multiplexed by a solid-state multiplexor of up to 64 ways. It is based mainly on commutating twice the polarity of each analog input by solid-state circuitry and taking one measure of each polarity.

The microcomputer system includes different interface peripherals such as logic value inputs and outputs, smooth analog outputs, and also pulse-duty-modulated outputs, permitting direct thyristor and triac firing, as well as interface elements such as high impedence amplifiers, resistance bridges, and solid-state relays. So, it may be connected directly to measuring elements and actuators without any need for extraneous intermediate circuitry, thus achieving system homogeneity.

The microcomputer system has been operating for more than one year on an experimental installation in reduced scale, with full satisfaction.

Fermentor

As a second step, the microcomputer will be used on a sterilizable 20-liter fermentor.

This fermentor is equipped with six probes: temperature, pH, dissolved oxygen, dynamo-tachymeter for stirring-motor speed, antifoam probe coupled with a level detector used for alarm, and pressure detector used for alarm. The microcomputer is connected to the following devices for control: three pumps (acid, base, and antifoam), three heat resistances (one 2000 W for sterilization and two 200 W for fine regulation), two proportional valves (one for the cooling water input and one for the air output), and the stirring motor.

The flow rates of the pumps can be calibrated automatically using an eprouvet equipped with three level detectors connected to the microprocessor.

The dissolved oxygen is controlled either by changing the agitation speed between two preselected values or by changing aeration rate with the priority given to the speed control.

The dialog between the user and the microcomputer is realized through a teletype, but, to permit a rapid visual control, four digitalizers are connected to the microprocessor and display permanently the values of pH, temperature, dissolved-oxygen pressure, and stirring-motor speed.

It will be possible to add to this basic configuration different modules, following the user's wish, involving new probes or control devices and other programs.

SOFTWARE STRUCTURE

The software structure is composed of different tasks articulated on a central regularly activated by the internal clock.

The central task realizes the following:

i) Four analog data acquisitions, verification, comparisons to the set points, memorization and display on the digitalizers. Then, the microcomputer acts on the different control devices involved for the regulation during periods calculated from the comparisons between the acquisitions and the set points. The intensities of these actions are also memorized.

ii) The tests on the different logical inputs, and sets an alarm if necessary. If this alarm is not stopped either by the user or by the system after a while, the control devices are switched off.

iii) The process time maintenance and, therefore, the regular activation of different tasks. These activations can also be required by the user.

These tasks generally mix logical actions, regulation actions, and conversation with the user through the teletype. As an example, an initiation of the microcomputer and a start of a process is described below:

1) When the microcomputer is switched on, it asks the user the date and the hour.

2) The parameter task is activated. This task consists of a very long conversation (the number of parameters is very important) between the user and the microcomputer. When all the parameters are entered, a recapitulation is edited, for eventual modifications if mistakes have been done. At the same time, a strip of paper is perforated; this strip can be used first as archive and second to restart an identical process.

3) If the user has not given to the microcomputer the calibration values, it realizes the calibrations of the pumps or probes. For this it asks to the user different manipulations.

4) At user's request, the microcomputer starts stirring and heating for sterilization. It signals to the user the moments when to open or close the different valves for the sterilization of the annexes. During this phase, the user has the possibility, if there is something wrong, to stop the sterilization and immediately cool the broth.

5) The microcomputer resets the fermentor to the culture temperature and starts the aeration and pH regulation. When the set points are reached, it signals to the user that it is ready for inoculation.

6) At the user's signal for inoculation, it starts the process time. During the process, at moments fixed during the parameter or at user's request, it edits the average values of each parameter and the actions of regulations it has done to maintain the set-points.

These values are very useful to determine the specific growth rate. The calculation and identification of the specific growth rate will be performed by the microcomputer with another module.

CONCLUSION

Beside the economical aspect which is important, the advantage of using this type of system results from its homogeneity. This allows it to work in a research laboratory, where it gives the maximum information, as well as in production area where it can reproduce almost perfectly even very complicated processes.

References

[1] J. Kryze, M. Cordonnier, J. P. Kernevez, J. M. Lebeault, and P. Pahaut, *Design and Use of a Microprocessor in Fermentation Process* (Chemdata, Helsinki, 1977).
[2] J. P. Bravard, M. Cordonnier, J. P. Kernevez, and J. M. Lebeault, "On line identification of parameters in a fermention process," in 2nd CIC/ACS Joint Conference, Montréal, 1977.
[3] W. B. Armiger, D. Zabriskie, and A. E. Humphrey, "Design and operationnal experience with a highly instrumented computer coupled fermentation system," in 5th International Fermentation Symposium, Berlin, 1976.
[4] R. P. Jefferis, "Control structures for computer coupled fermentation," in *Proceedings of the 1st European Conference on Computer Processes in Fermentation* (INRA, Dijon, 1973).

Single-Board Microcomputer for Fermentation Control

RAYMOND P. JEFFERIS III and STUART S. KLEIN

Widener College, Chester, Pennsylvania 19013

JOHN DRAKEFORD

Intel Corporation, Santa Clara, California 95051

INTRODUCTION

When the control of a fermentation vessel by means of a dedicated microcomputer was first proposed at the Dijon conference of 1973 [1] many doubts were expressed concerning the feasibility of this notion. Could microcomputers perform the required calculations? What programming languages would be available? What operating systems could be used? How could the data be acquired? Today, five years later, we see that the microcomputer has become a fully supported member of the industrial computing hierarchy, as was predicted. The data-gathering and arithmetic capability of these devices are now higher than the minicomputers of five years ago, high-level programming languages are available, there are excellent operating systems, and process interface components of industrial quality are commonplace. Indeed, within this year several pharmaceutical manufacturers will install computer hierarchies in which microcomputers will control individual fermentation processes. This paper will illustrate some of the objectives of such systems in fermentative production and will outline the microcomputer components and programs that make such distributed control possible.

COMPUTING NEEDS FOR FERMENTATION

The computing needs that have been identified for fermentation processes [2] can be broadly separated into three categories: operator services, data processing, and communications. Other authors [3] have made a more detailed list of system requirements, which includes operator console programming, documentation of operating actions, data logging, control, the availability of high-level languages (including a fill-in-the-blanks process control language), a multipriority operating system, and failsafe shutdown with backup equipment. To this detailed list, furthermore, should be added the capability for performing batch sequential steps according to stored protocols. This is very important for the repeatability and optimization of fermentation operations. Typical batch sequencing needs of this kind, included in Figure 1, are vessel charging, sterilization, inoculation, reagent additions, broth sampling, data acquisition and

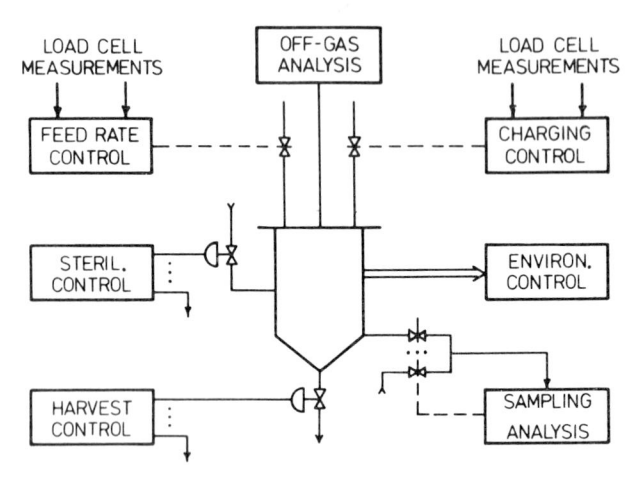

Fig. 1. Sequencing and control for fermentation processes.

control, off-gas analysis, and harvesting. Ideally a special batch control language would be available for this purpose [4]. In the proposed computing hierarchy such high-level languages would operate in a central minicomputer and would compile action commands that would be carried out by coupled remote microcomputers. These microcomputers would, in turn, monitor and control the fermentation operations according to the stored commands.

At the pilot scale it is efficient to perform the indicated functions by a single minicomputer. However, for production operations, or for pilot scale plants where computer technology is tested for eventual production applications, the higher reliability, speed, and flexibility of hierarchic control by microcomputer is preferred. In such hierocybernetic systems the functions described are assigned to individual fermentation controllers, as suggested by Figure 2. These units are completely autonomous and are coupled to the plant data-base computer by means of a data link. A typical configuration of such a controller is given in Figure 3. These controllers carry out their assigned batch sequence and control functions independently, even in the event of a failure of the plant computer. The largest unit of computer-related failure affecting the process in this case is *one fermentation controller*. This controller would revert to its backup mode upon failure, and its work load is *not more than one person can perform manually*. One could further subdivide the fermentation control functions and assign a microcomputer to each. Figures 4-6 suggest such microcomputer applications to environmental control (DDC), off-gas analysis, and substrate feed rate control, respectively. The basic structure of each of these task-oriented microcomputers is similar and includes a central processing unit (CPU), memory, communications, and data input and output devices. In the environmental DDC controller, input would typically be from an analog-to-digital (A/D) converter, and output might be through a digital-to-analog (D/A) converter. In an off-gas analysis application, however, gas sampling valves would be sequenced through the

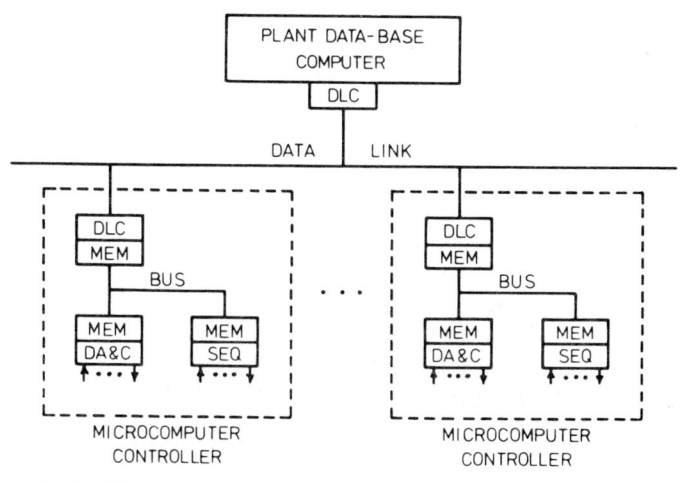

Fig. 2. Hierarchic concept for fermentation control by microcomputer.

contact closure output (CCO) interface, and the analyzer signals would be acquired by an A/D converter. A feed rate controller, while having similar (A/D) input and (CCO) output hardware, would have entirely different programming. Elaborate digital filtering techniques might be included in this device for estimation of the feed rates [5]. The subdivision of functional liability which such distributed computational units provide should yield very reliable overall plant operation and will permit simplified backup action. The trend of microcomputer applications is definitely in this direction, and further refinements are already in development.

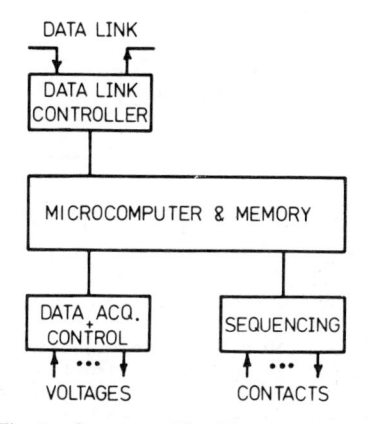

Fig. 3. Structure of local batch controller.

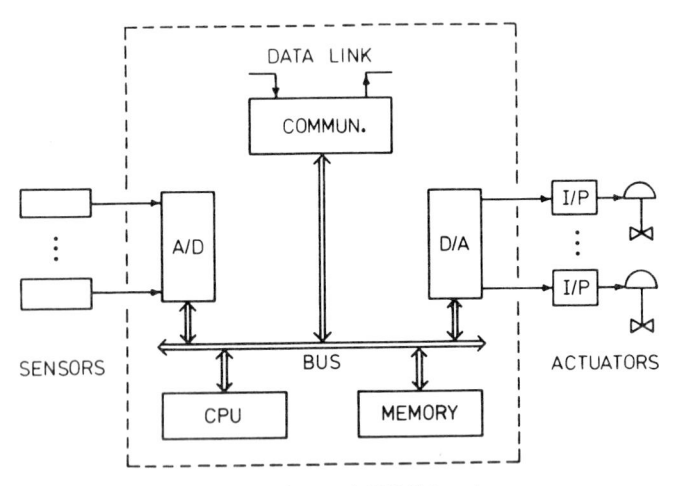

Fig. 4. Environmental control (DDC) by microcomputer.

FERMENTATION CONTROLLER ARCHITECTURE

The basic elements of a fermentation controller are a CPU, a memory, a communications channel, and process input and output (I/O) devices. Process input to the microcomputer takes the form of A/D converters or contact closure inputs (CCI). Output to the process is by means of a D/A converter or CCO's. Typically the sensor input signals over the range 4–20 mA are converted to 1–5 V by resistors at the A/D converter input. The voltages are then read by the converter, control calculations are performed, and process valves are actuated through the D/A converter channels with current-to-pressure (I/P) transformation as was shown in Figure 4. Valve sequencing usually takes place through

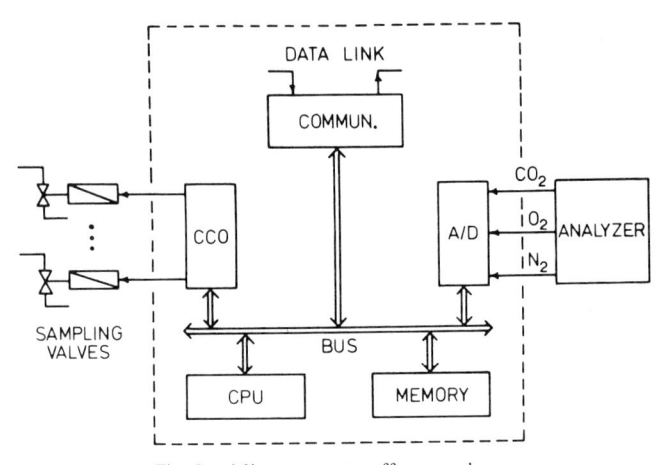

Fig. 5. Microcomputer off-gas analyzer.

SINGLE-BOARD MICROCOMPUTER 235

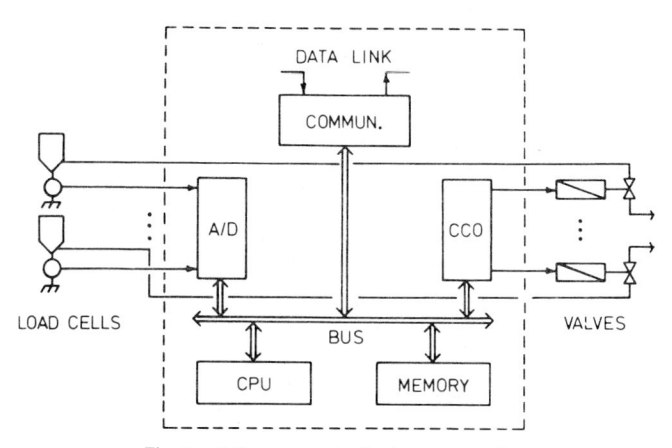

Fig. 6. Microcomputer feed rate controller.

the digital input and output (DIO) contacts, which include both CCI and CCO devices as needed. Today most of the capability for fermentation control can be obtained on two circuit boards. To prove this point the authors have assembled a fermentation controller from two Intel single-board computer (SBC) components, the SBC-80/10 CPU board with memory and the SBC-732 analog I/O board. The capabilities of this system are indicated in Figure 7. These two boards plug into a special back-plane bus, which will accept other expansion boards as well in the manner indicated in Figure 8. This bus structure yields a flexible device that can be configured, according to need, to perform any of the fermentation monitoring and control functions we have indicated, either separately or combined into a single fermentation control unit.

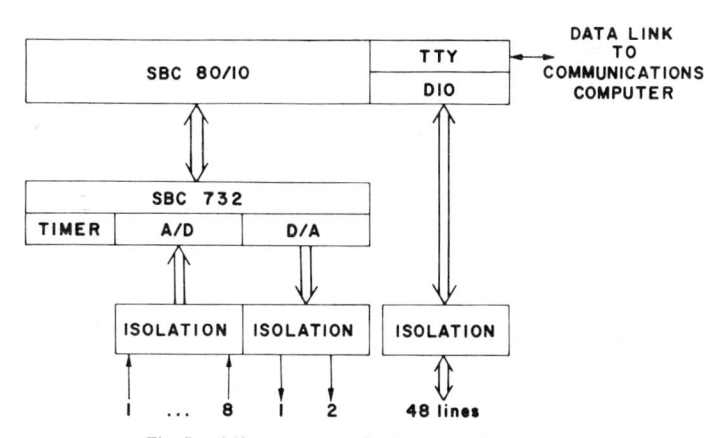

Fig. 7. Microcomputer for fermentation control.

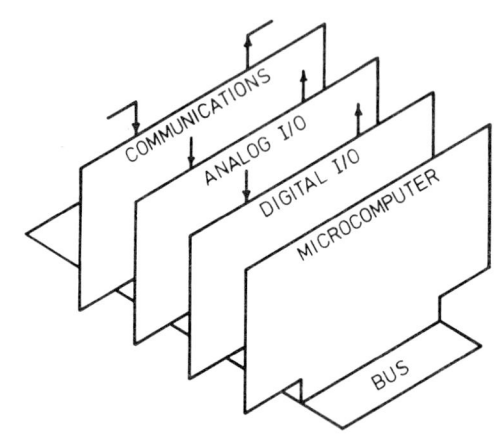

Fig. 8. Bus connection of single-board functional components.

PROGRAMMING AND PROGRAM STRUCTURE

The most general data acquisition and control needs of fermentation can be performed by the program structure illustrated in Figure 9. According to this structure a DDC program operates upon an input data base to produce an output data base which, in turn, regulates the process through the D/A converter. A sequencing program, furthermore, can access variables in the input and output data bases, compare them against preset conditions, and perform control actions on the process through the contact I/O (CIO) interface. Both input and output contact capabilities are required, since many process conditions are of a binary, rather than continuous, nature. In addition to these programs, a communications program is necessary to interpret commands from the plant computer and to report data and our process conditions for eventual interpretation and action.

The authors have programmed the two-board controller outlined above to

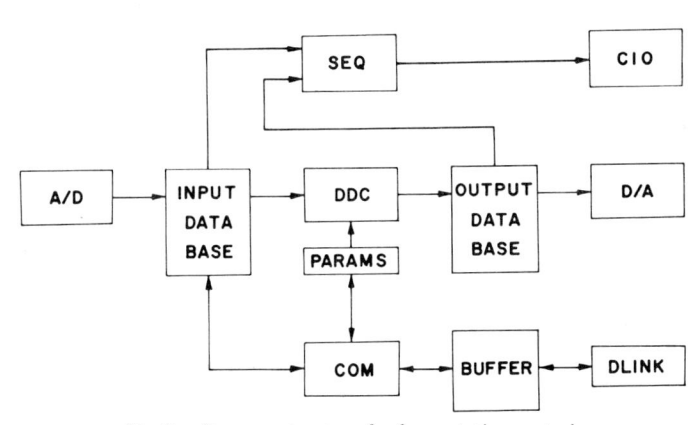

Fig. 9. Program structure for fermentation control.

perform most of the indicated tasks. Programming was implemented in the PL/M high-level language available from the equipment manufacturer [6]. In particular, a DDC program has been written which can handle 16 or more control loops, either isolated or with cascade interaction. The structure and techniques used in this program have proven effective in previous DDC installations and have already been described in the literature [7,8]. To illustrate the power of the PL/M language in such applications, Table I gives the PL/M coding used for the proportional control algorithm in this controller. This language contains very powerful data description properties such as the STRUCTURE construct illustrated in Table II, which describes the DDC data base used in the fermentation controller. In addition the language contains very powerful structures to control the program flow. Table III illustrates the use of the CASE structure to control selection of the control algorithm to be performed. This language offers the power of these data description and program control structures while preserving the efficiency and flexibility of assembly language programming to a large extent. The availability of this language is one of the contributing factors to the practicality of single-fermentor control by microcomputer.

Another equally important factor is the availability of a real-time multitasking executive program, or operating system. The authors used the RMX-80 operating system, available from the microcomputer manufacturer, which appears to be one of the most useful small operating systems available today. It can perform multiple tasks at different priorities and has a complete set of subroutines for interacting with the input/output circuit boards available from the

TABLE I

Proportional Control Algorithm Written in PL/M Language

```
ERROR=LOOP$DATA(I).SETPOINT-LOOP$DATA(I).MEASUREMENTS;
A=LOOP$PARAM(I).PROP$GAIN;
CALL MULT(ERROR,A,PRODH,PRODL);
LOOP$DATA(I).OUTPUT$DATA=PRODH;
```

TABLE II

Use of STRUCTURE Construct to Describe the DDC Data Base

```
DECLARE LOOP$DATA(16) STRUCTURE(
    SETPOINT ADDRESS,
    INPUT$DATA ADDRESS,
    MEASURE$DATA ADDRESS,
    ERROR$LAST ADDRESS,
    INPUT$BIAS ADDRESS,
    OUTPUT$BIAS ADDRESS,
    OUTPUT$DATA ADDRESS,
    RESET(2) ADDRESS) PUBLIC AT (.LOOP$DATA$POINTER);
```

TABLE III

Use of the CASE Structure to Select
the Control Algorithm

```
DO CASE ALGON
        /* CASE=0, NO CONTROL ACTION */
        DO;
             CALL NULALG;
        END;
        /* CASE=1, PROPORTIONAL CONTROL */
        DO;
             CALL PROALG;
        END;
        /* CASE=2, PID CONTROL */
        DO;
             CALL PIDALG;
        END;
END;
```

manufacturer. This greatly simplifies programming of the fermentation controller tasks. Programs under the control of this executive system send each other messages through various message exchanges, and wait for messages at these exchanges when not actually running. This technique permits programs to start other programs and to exchange data, thus providing the coordination of task execution according to priority and the continuity of data flow. The task and exchange structure used by the authors for the fermentation controller are given in Figure 10. The diagram shows the digital control program (DDC) and the sequential logic program (SEQ), which wait for messages at the user exchange (USEREX). The DDC program receives a "RUN" message from the TICKER scheduling task every second. During its calculation of all the control loop actions, the DDC program also waits at this exchange for completion of the analog input program (AINHND) or the analog output program (AOUHND), which it schedules through the RQAIEX and RQAOEX exchanges, respectively. The sequencing program, scheduled at intervals of 0.1 sec, performs in a similar manner through the contact input (CCIHND) and output (CCOHND) programs which it schedules through the RQDIEX and RQDOEX exchanges, respectively. An alarm program, also waiting for messages at the user exchange (USEREX), is scheduled every 10 sec to send messages, by means of the communications (COM) program, to the plant data-base computer. A terminal or data link handler (TRMHND) waits for these messages at the RQOUTX exchange. It then transmits the messages through the RQL7EX exchange to the plant computer. The terminal program also transfers new data and commands, arriving from the plant computer at the RQL6EX exchange, to the controller data base by means of the communications task and the RQINPX exchange. This complex diagram illustrates the power of this multitasking executive in coordinating the controller operations. It would have been difficult, indeed, to coordinate the many objectives of this fermentation controller without the availability of such an operating system.

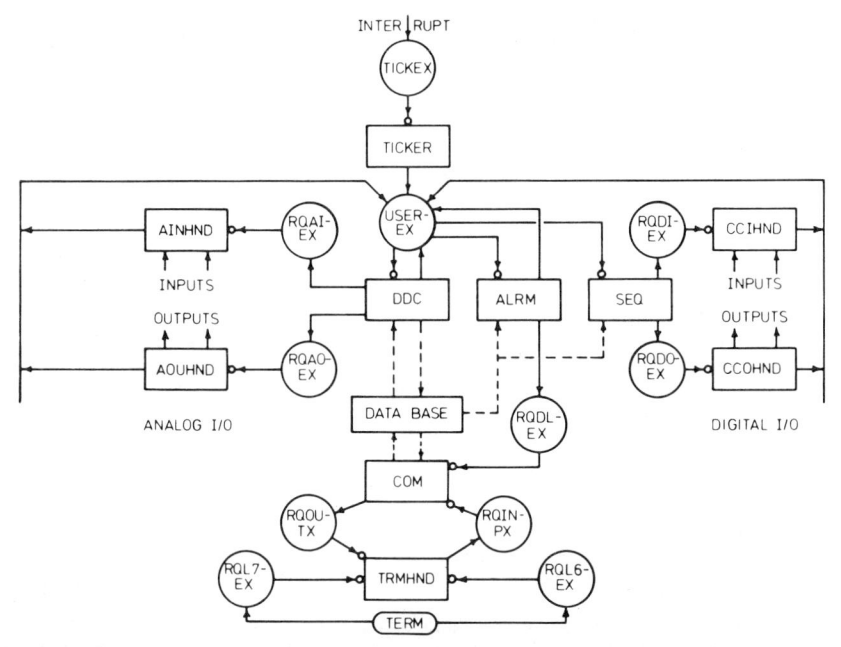

Fig. 10. Structure of tasks and message exchanges for fermentation controller operating under RMX-80.

SUMMARY

We have shown that all of the necessary electronic components and supervisory programs are available today for the application of microcomputers to monitor and control individual fermentation vessels. This capability for distributing the control over many suboperations in the plant and gathering the plant data over a communications link may provide the reliability required for computer applications to production fermentation at reasonable cost. In the next five years we should expect to see a number of pharmaceutical producers apply this technology in both pilot and production plants.

References

[1] R. P. Jefferis, *Proceedings of the 1st European Conference on Computer Process Control in Fermentation* (INRA, Dijon, France, 1973).
[2] R. P. Jefferis, in *Workshop, Computer Applications in Fermentation Technology 1976* (Verlag Chemie, Weinheim, 1977).
[3] D. D. Dobry and J. L. Jost, in *Ann. Rep. Ferment. Prog.*, 1, 95 (1977).
[4] A. E. Humphrey and R. P. Jefferis, *Proceedings of the IV GIAM Meeting* GIAM (Saõ Paulo, Brazil, 1973).
[5] R. P. Jefferis, "Parameter estimation and digital filter techniques," in *Conference on Biochemical Engineering* (New England College, Henniker, NH, 1978).
[6] D. D. Mc Cracken, *A Guide to PL/M Programming for Microcomputer Applications* (Addison-Wesley, Reading, MA, 1978).
[7] G. W. Markham, *Control Eng.*, 15(5), 87.
[8] G. V. Woodley, *Instrum. Technol.* 15(4), 57.

Fermentation Control Systems—A Time for Change

M. C. BEAVERSTOCK and G. P. TREARCHIS

The Foxboro Company, Foxboro, Massachusetts 02035

INTRODUCTION

Fermentation is an established technology for producing products of importance. Antibiotics, food products such as bakers' yeast, vinegar, beer, and wine, and waste treatment processes are only a few examples. But fermentation is also being actively considered and studied for many new and exotic applications. Here examples include single-cell protein (SCP) production from petroleum and cellulose, synthetic fuels, synthesis of complex organic molecules, alcohol from black liquor, and new waste treatment processes. These latter areas have especially brought many newcomers with various backgrounds into the fermentation field. In addition, fermentation specialists have had to learn about new technologies. One such technology is industrial process control.

Interest in control has resulted in control technologists learning about fermentation, fermentation technologists learning about industrial process control, and the computer becoming the communication link for both parties. The mutual interest in control appears to be well founded. Present indications are that the use of advanced control and optimizing techniques will provide substantial cost reductions by reducing the cost of nutrient material per unit of biomass, by improving the yield of biomass per unit time, and by decreasing the power requirements for agitation and aeration in aerobic fermentations.

To the control engineer new to fermentation, the ordinary fermentor first appears to be a relatively simple and innocuous piece of processs equipment. However, work over the past few years by many dedicated investigators has clearly indicated that this simple process involves many, if not all, of the classical control problems with which the control engineer has wrestled in applying control to chemical processes and to mechanical devices.

The role control can play in future fermentation operations is further emphasized by this conference. Here leading workers in the fermentation process development field are focusing on computer applications. This is in contrast to other process industries, where a similar meeting would primarily be attended by control engineers.

The control problems in fermentation processes are numerous. Many key measurements are unavailable on a time scale consistent with control requirements and those available are, generally, randomly noisy. The relationships among controlled variables are interactive, which poses a very complex multivariable control problem. Additionally, the processes are not totally understood, particularly where interaction exists between organisms. These individual problems are being addressed, however, as indicated by other papers at this and previous conferences.

This paper, however, will address the next level of control questions that will result when technology transfer and scale up are attempted between the pilot plant and production facilities. Three specific areas for consideration are: 1) the change in issues when considering a production facility, 2) the change in control scope from single unit to multiple operating units, and 3) the impact of a rapidly changing electronic technology on measurement and control equipment.

LABORATORIES, PILOT PLANTS, AND PLANTS

Laboratory and pilot plant installations have been using computers for data acquisition and control techniques involving respiration quotient and other inferential measurements of process variables. In production facilities, however, the computer has been used in much less imaginative ways. Generally the computer has served as a routine replacement of analog controllers, used for data acquisition, and replaced cam controllers for improved sequencing of filling, cleaning, and sterilization operations. In all, not much really new control technology has been applied or divulged in the literature. Empirical methods are still relied upon—they have simply just been automated.

A matrix summary of functions performed with computers in both laboratory and plant environments is shown in Table I. The matrix format differentiates between pilot plant (i.e., laboratory) and industrial plant environments. In each case basic existing processes (bakers' yeast, etc.) are separated from the new and more exotic ones. Among the functions listed, the "computation of non-

TABLE I
Control Function Comparison Matrix

USE / FUNCTION	PILOT PLANT — Existing Processes	PILOT PLANT — New Processes	INDUSTRIAL PLANT — Existing Processes	INDUSTRIAL PLANT — New Processes
Feedback Control			•	•
Data Acquisition	•	•	•	•
Scheduling and Sequencing			•	•
Computation of Non-measured Variables	•	•	Limited Use	Critical Areas: Work Needed
Advanced Control Techniques	•	Challenge: Work In Progress	Opportunity For Use	
Optimization	•			
Modeling, Parameter Estimation	•			

measured variables" includes such items as the calculation of biomass, while "advanced control techniques" includes optimization algorithms, modeling techniques, and the decoupling of air, nutrient, and other interactive variables. The asterisks designate where each function is a major concern and cause for study. Of interest are the areas that are blank and those with special notations.

Scheduling, sequencing, and feedback control are areas not normally considered in any detail at the laboratory level. Yet in a production environment they predominate. Economic surveys show that the major portion of the payback on automated systems comes from sequencing and scheduling. Such plant control systems in a brewery, for example, can involve 500 or more contact inputs and outputs associated with sequential control.

Simple feedback control (e.g., single variable flow, temperature, pressure, level, etc.), which represent a small part of the experimental effort, is a major control task in a plant. Two hundred or more feedback controllers and perhaps only five advanced control loops in a fermentation facility would not be unreal. In addition, any advanced control application, such as optimizing the growth rate in a fermentor, requires that the basic feedback loops be correctly operating. The correct application, specification, and maintenance of these many loops is a major plant control effort.

Other areas of interest involve the study of advanced control for the newer fermentation processes. This is an area of challenge that is currently being approached, as evidenced by the recent literature. Also, there is presently an opportunity to transfer the more advanced control techniques already tried in the pilot plant on some of the existing processes to the industrial plant. However, the acceptance of new control technology to existing plants is an extremely slow process for reasons other than technology [1].

The main concern, however, should be focused on the control techniques that will be required in the new fermentation technology plants that have yet to be designed. Control critical to these plants will inherit the same acceptance problems unless some changes begin now. The changes start by at least considering the issues involved with industrial process control back in the laboratory or pilot plant environment.

The issues an industrial control system has to take into account revolve around the characteristics of operating plants. With many control loops and thousands of pieces of mechanical equipment, a large plant is usually always working with some automatic control loop in manual or some pump or valve down and out for repair. But in spite of everything, the plant must run. Hence, safety and reliability are more critical. A computer failure that may be inconvenient in the laboratory environment could be an economic nightmare in the plant. In coping with such situations, it should be noted that in most plants no one person has direct responsibility to maintain an "on control" condition. The instrument department fixes "broken" instruments; the process engineer is responsible for the process itself; and the corporate engineer is too far away from the plant for day-to-day contact and probably would not be allowed to touch the control system anyway. Yet the plant runs.

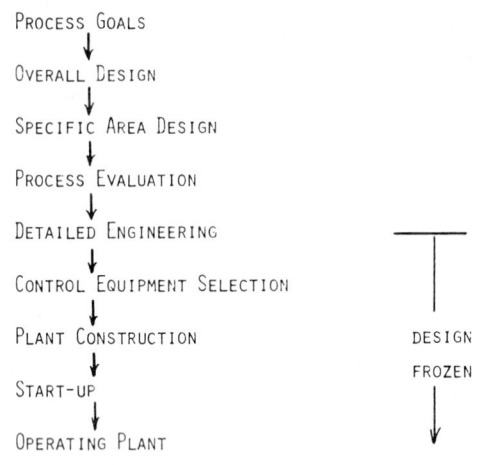

PROCESS GOALS
↓
OVERALL DESIGN
↓
SPECIFIC AREA DESIGN
↓
PROCESS EVALUATION
↓
DETAILED ENGINEERING
↓
CONTROL EQUIPMENT SELECTION
↓
PLANT CONSTRUCTION   DESIGN
↓   FROZEN
START-UP
↓
OPERATING PLANT

Fig. 1. Design approach–typical.

Finally, it should be recognized that control equipment maintenance and operator training will always be perceived as being critical even if a completely automated, reliable plant control system were available. Such a system is just not considered to exist, especially when conceived and used only in a laboratory environment. In spite of this perception, control will always be present.

When you stop and think about it, almost all the operating procedures in a plant are based on some aspect of the control system design. The steps an operator goes through to make a change depend on the layout and design of the control panel or computer interface. The control loop configurations dictate allowable control variables, while incorrect valve sizing, measurement selections, and sensor placement can lead to processing and stability problems. Additionally, interactive variables not correctly perceived can cause problems. As a result, the person who ends up specifying the control system design almost singlehandedly specifies the operating procedures and thereby influences the economics of the plant.

PROCESS GOALS ← PROCESS OPERATION
↓   ↑
OVERALL DESIGN   CONTROL DESIGN
        ↘   ↗
    SPECIFIC AREA DESIGN
            ↓
        EVALUATION
            ↓
    DETAILED ENGINEERING
            ↓   DESIGN
    PLANT CONSTRUCTION   FROZEN
            ↓
        START-UP
            ↓
    OPERATING PLANT

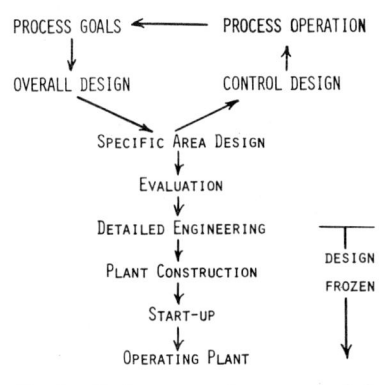

Fig. 2. Design approach–recommended.

In order to design an acceptable control system, a procedure is needed that addresses both the technical and the operational issues.

SYSTEM DESIGN TECHNIQUES

One would expect that such an important function as control system design would be of prime importance in any plant design, but it is not. Traditionally and across industry lines, a cost-effective plant design is frozen as much as possible once detailed engineering is ready for field construction. Unfortunately, control system decisions are normally carried out at this time as shown in Figure 1, and consequently, the impact of the decisions has little chance to feed back to the process design and plant operations issues with which they usually interact [2]. All the average control engineer can do at this point is design conventional control schemes, usually based on the individual plant unit operations (i.e., filtration, drying, heat transfer, etc.). These are then hurriedly combined into some form of process control system to meet plant construction and start-up schedules. In most cases, these time schedules do not allow any tailoring or innovation to better meet fermentation control needs.

Historically, this worked out reasonably well because only conventional analog hardware systems were available and most higher level control functions were carried out by the operating personnel. Hardware set the structure and complexity of the control system. Such functions as anticipating disturbances, interacting control, and changes in operating conditions were done by plant operators. Operators are also required to manually intervene in many sequencing functions.

A recommended plant design technique is shown in Figure 2 which allows for an integration of process, plant operation, and control goals early in the design stage when a free exchange of concepts among vendors and users can be accomplished. Within this structure, questions concerning process requirements, plant operation, and process economic sensitivity as well as the technical areas of measurement, control, communications, and degree of control system distribution can be considered. These and other factors to be considered during this

TABLE II
Factors Considered during System Design

PLANT GOALS	SAFETY CONSIDERATIONS
PLANT OPERATIONAL/MANAGEMENT PHILOSOPHY	STANDARD OPERATING PROCEDURES
	PROCESS ECONOMIC SENSITIVITY
STAFFING REQUIREMENTS	CONTROL STRATEGY
MAN/MACHINE FUNCTION DESCRIPTION	LEVEL OF CONTROL DISTRIBUTION
CONTROL ROOM PROCEDURES	PROCESS EQUIPMENT REVIEW
COMMUNICATIONS NEEDS	MEASUREMENT NEEDS/MEANS
AREAS FOR CONTROL ENFORCEMENT	CONTROL SYSTEM DESIGN RESPONSIBILITIES

design period are listed in Table II. Thus the process goals and plant-operating goals are considered together by the design and control engineers in an iterative fashion. However, a methodology is required in order for all parties to carry out a meaningful discussion about a complex situation.

A basic approach to this methodology involves segmenting the entire plant by level and components into a workable structure based on carefully defined purposes, goals, functional requirements, and product mix. Such segmentation results in smaller subsystems that are easier to analyze.

One of the biggest problems in examining a plant design by the technique described above is the lack of formal methods for carrying out a detailed study of a process system. Bernard [3] suggests a method that involves a horizontal breakdown based on the concept of "operating units" instead of the standard unit operations. The result is a structure with many nodes of minimum interaction. Carefully defined purposes, goals, and broad functional requirements are basic to the structure. The key to a good analysis is the willingness to write down purposes for different parts of the system and functions that seem obvious and simple. A good measure of the quality of the structuring is how little the whole system must be changed to accommodate changes in any of its parts.

An example of how an antibiotic plant [4] can be studied in this way is shown in Figure 3. This figure gives a starting point for the plant study but does not begin to contain the detail that is required. The plant breakdown should conform to control board layout and responsibilities of various organizations. In the example, the rotary filter operation is separated from the fermentation section by both control board location and operator assignment.

Each plant area shown is broken down into its equipment, measurements, valves, and other operating parameters. The detail continues to the point of individual input/output assignments for control devices. In addition to serving as a basis for plant control system decisions and discussions, the method provides a documentation of the plant that is functional in nature. Consequently, information concerning a pH controller on a fermentor is found in the fermentation section of the plant rather than as part of a long list of controllers used throughout the plant.

FUTURE DIRECTION FOR INDUSTRIAL CONTROL SYSTEMS

The methodology outlined in the previous section forms the basis for discussing how control strategies interact with plant design and operation. Such a structure is required to consider the control system options that will be available when the process we are discussing become realities. These future systems can be considered in three categories.

Non-Computer-Based Systems

To many a "non-computer-based" system may seem an archaic term, however, the vast majority of control systems today do not have computers associated with them. In fact, pneumatic instrumentation remains a major factor in plant control

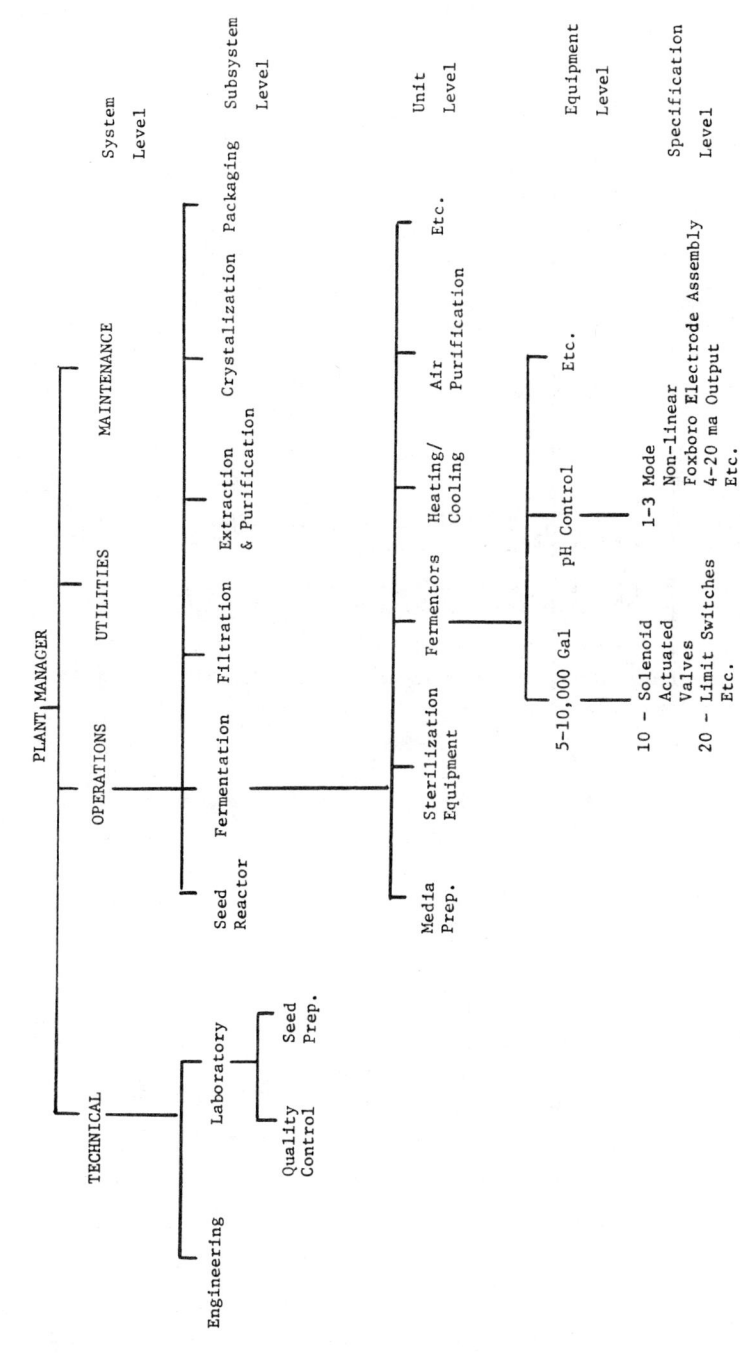

Fig. 3. Antibiotic plant functional flowsheet. Note: for clarity, details on various levels are shown for selected examples only.

Fig. 4. Single-loop system examples.

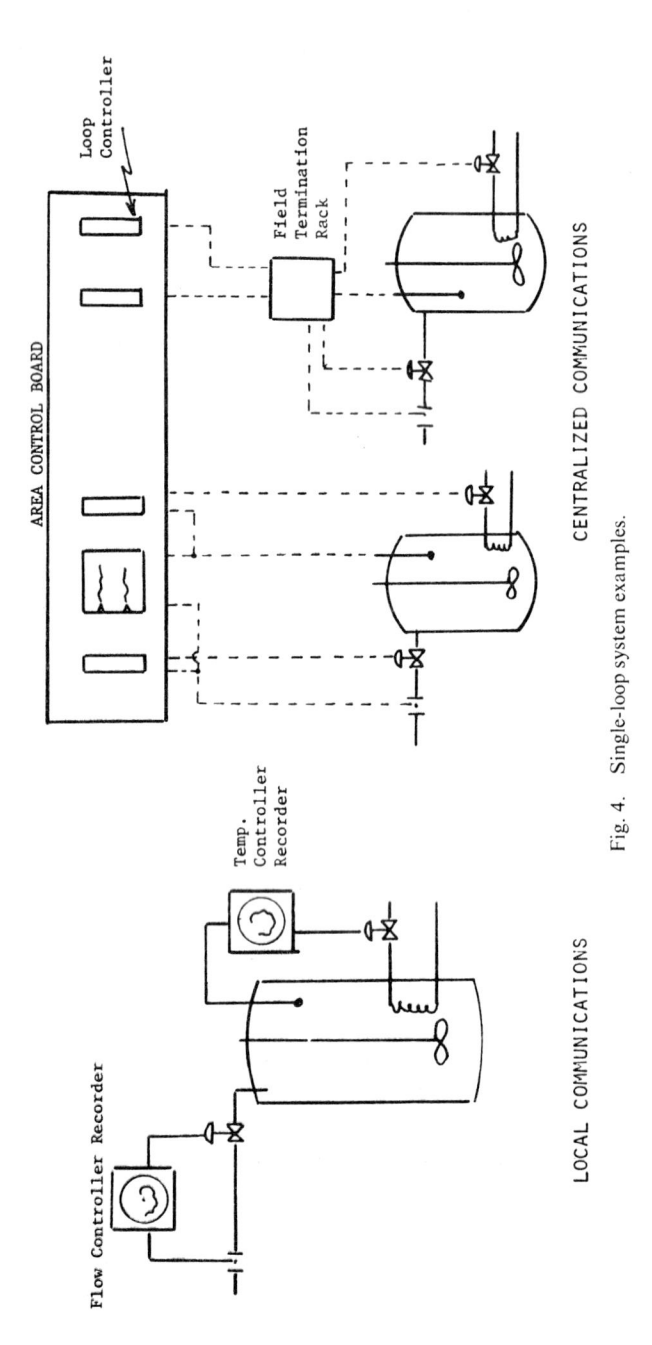

equipment. Pneumatics will continue to survive the challenge of electronics partly because no reliable replacement is seen for the pneumatic valve in the near future. Advances in pneumatics continue as evidenced by the recent development of an all pneumatic, on-line process chromatograph [5].

Designing control configurations without computers involves selection of controlled variables, specification of each individual loop (i.e., measurement-controller-valve), and algorithm and tuning parameter selection. Measurement devices and valves are directly wired to the individual controllers, and data concerning the loop are stored as meter readings or dial settings.

A major feature of such a system is the reliability level. From an equipment viewpoint, a device failure in this configuration only affects the single loop. However, this single loop concept is also part of the plant operating philosophy. Plant operation is thought of as consisting of many independent control loops as shown in Figure 4. Equipment in process areas is either left on the individual pieces of equipment (e.g., filters, dryers, etc.) or collected on local control panels. These latter decisions are usually made by plant production personnel and reflect staffing requirements.

Although conceptually simple, the non-computer-based systems continue to form the basis for many advanced control structures. Dryer control systems using an inferential moisture measurement technique and energy conservation systems are two examples [6,7]. A distributed-hardware architecture that separates field connections, controllers, and display devices is also available.

The non-computer-based systems described here represent a philosophy more than specific equipment. For example, the use of microprocessors to produce a *single*-loop controller would not change the single-loop orientation of the system even though computer technology is used. These points will become clearer as we look at the remaining categories.

Computer-Based Systems

This category represents the present state of control systems that use a digital computer as an important component. It is also representative of most computer-coupled fermentation work presently being conducted in research facilities. The possible control structures are shown in Figure 5.

The computer is the central component in the structure and carries out control or supervisory tasks for many loops [8].

From a design standpoint, several issues can be raised. Decisions have to be made concerning the use of DDC or supervisory control (see Fig. 5) and whether analog back-up is required [9]. The scope of the computer control is also important. Consideration has to be given to the areas of the plant to be involved. One computer may serve the fermentation area while other sections of the plant utilize non-computer-based systems.

Data are centrally located in the computer, which means that provisions must be made for its security. Access must be defined to prevent illegal or unauthorized use of the system (e.g., an operator changing tuning constants, which is an in-

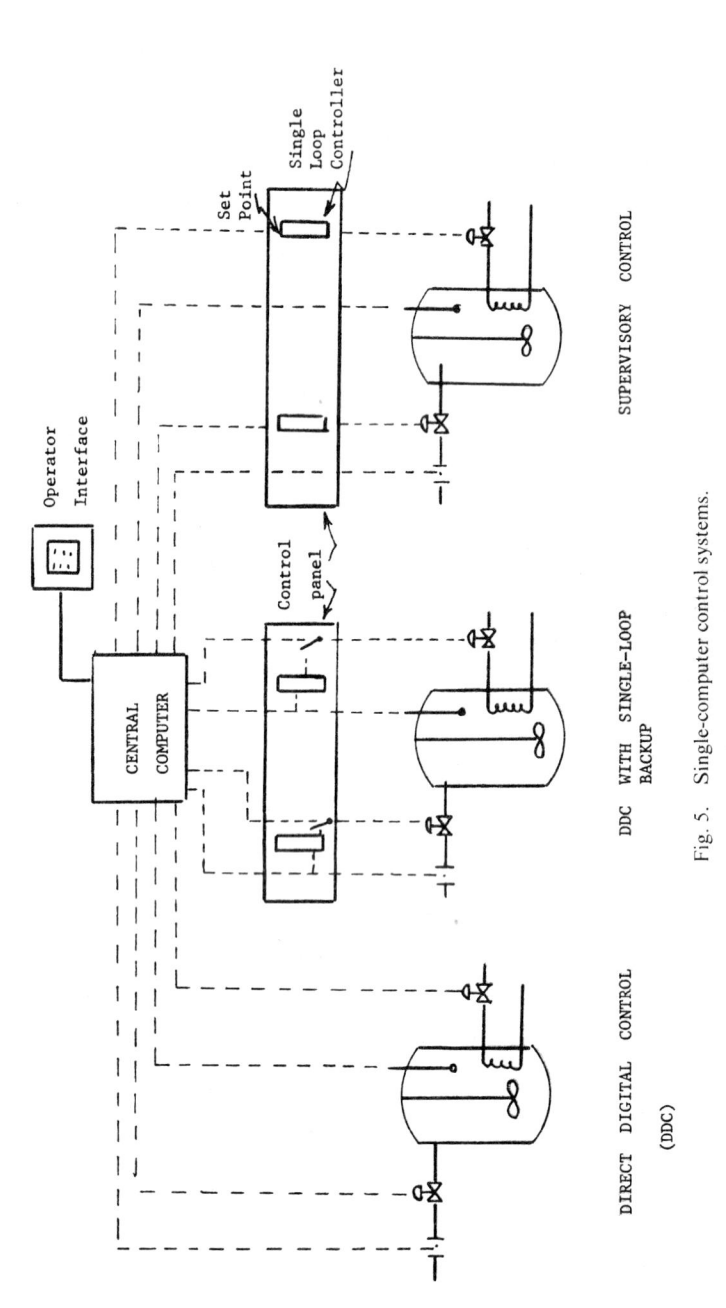

Fig. 5. Single-computer control systems.

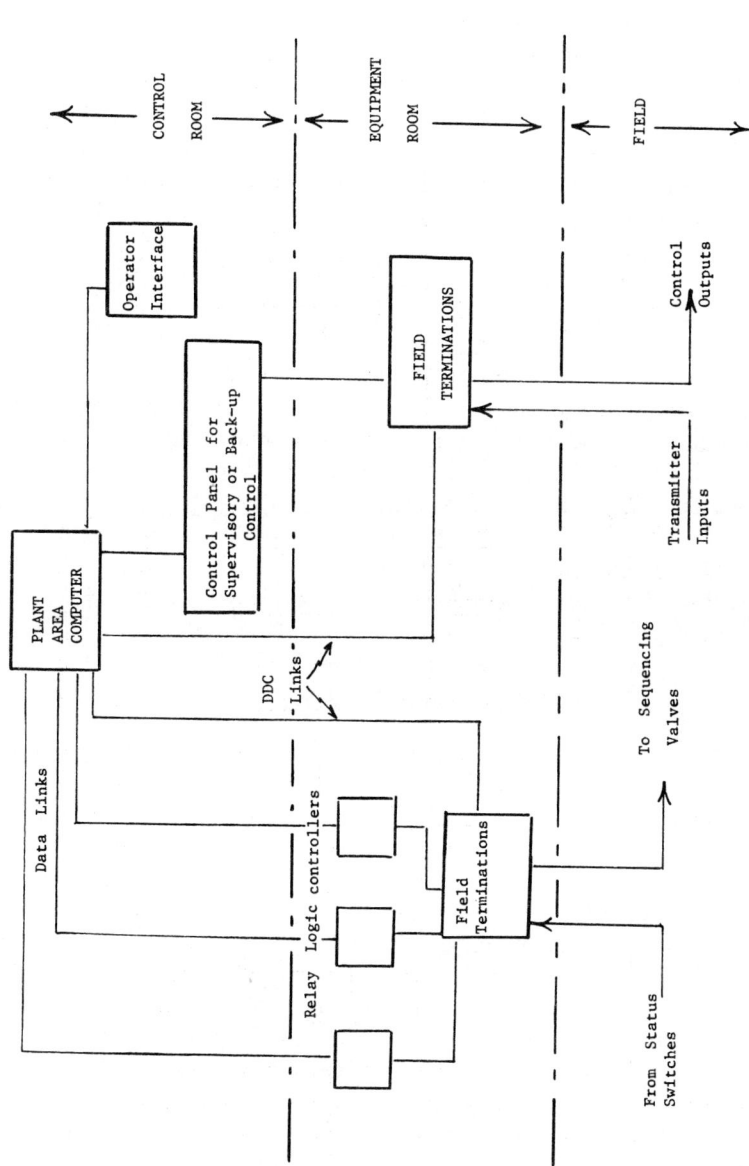

Fig. 6. Single-computer hierarchical structure.

strument man's job). Formats also must be defined so that the information stored as bits can be read as meaningful information by the operators. Process input and output, which now appears as pieces of data, must be addressed so they can be located and verified for correctness. All of this requires a new level of documentation to be created and maintained.

Computer-based systems provide for a degree of hierarchical structure. Supervisory control is one example when the lower level control work is carried out using individual controllers, while more complex calculations (models, inferential measurements, etc.) are performed by the computer. Another example is the use of a computer to drive logic controllers involved in batch fermentation, cleaning, sterilization, and other sequential activities, as shown in Figure 6.

While this system provides flexibility through the power of the computer, reliability could remain an issue since many functions are shared in the one computer. Increased reliability is possible through redundancy and analog back-up but additional techniques are required to assure that the stand-by equipment is ready to take over at the appropriate moment.

Because a section of the plant controls is focused in one piece of equipment, the operational philosophy of the plant tends to be one of individual operating units. In other words, the control systems look at the fermentation area as a unit rather than as individual loops. However, communication between areas is still provided by individuals (i.e., foremen, engineers collecting reports, etc.).

Where a computer is used in an operating unit, the computer functions must be related to plant operation and therefore production, process, and control expertise are required.

Multicomputer-Based Systems

Microprocessor technology is rapidly making the use of multiple computers economically possible, resulting in new control structures, with the terms "distributed" and "hierarchical" replacing "DDC" and "supervisory". The technical reasons for such systems are reliability and flexibility (the best features of the first two system categories mentioned above). Reliability and flexibility are intimately related. In fact, most of the proposed strategies for using distributed systems to achieve reliability depend on the exercise of flexibility. A possible distributed structure for an antibiotic plant is shown in Figure 7.

Distributed systems require the greatest interaction between process, production, and control personnel because the general philosophy is to connect all control elements together and consider an integrated plantwide control network. Communication between plant sections is via a "data link" allowing problems in the broth recovery section to be automatically transmitted back to the fermentation and seed preparation areas where appropriate action can be taken.

The data transmitted throughout the plant must have an easily understood standard format in each computer for messages to be meaningful. Data storage files must also have a common format. This is not an easy task since the variables describing a crystallizer are significantly different from those describing a

Fig. 7. Multicomputer control system.

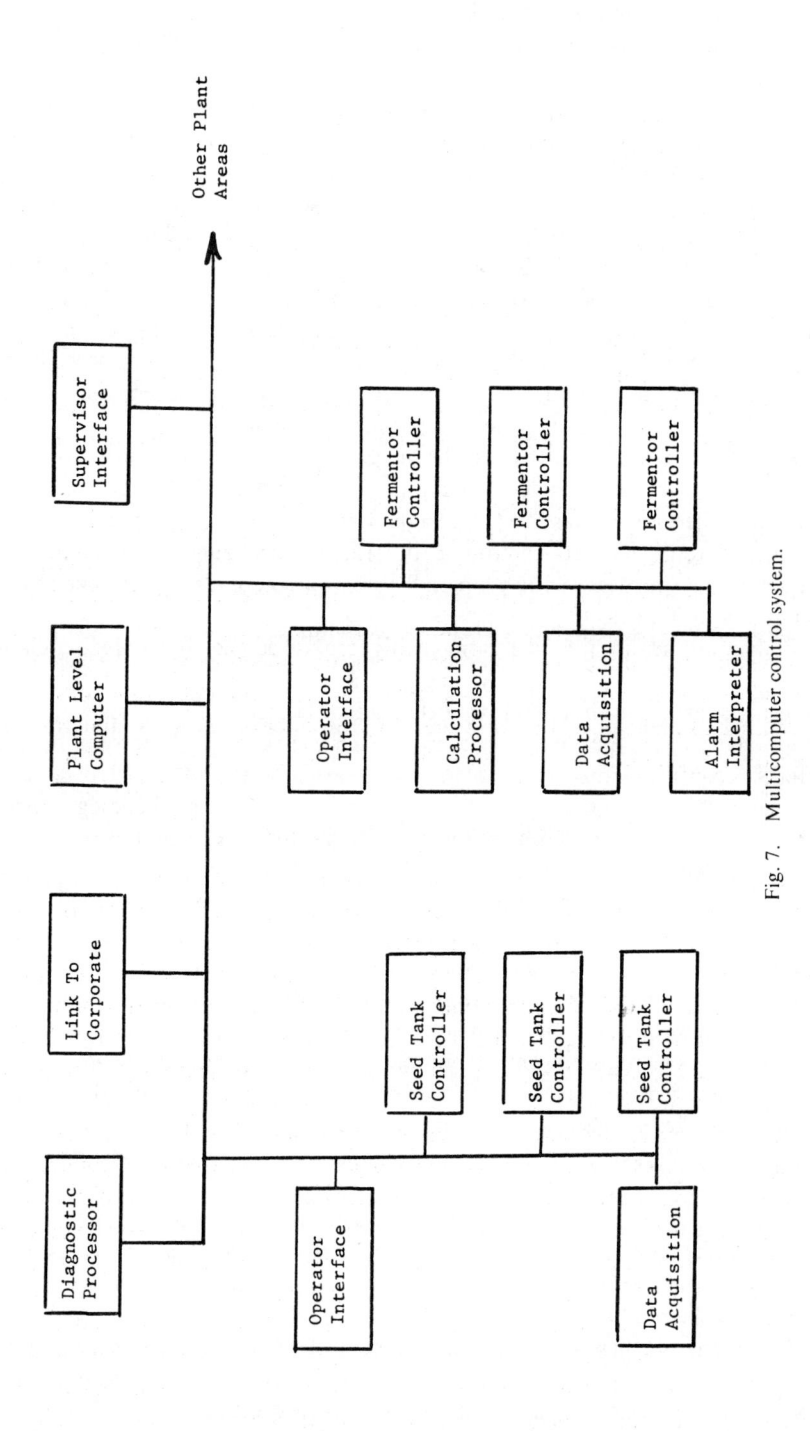

controller, extractor, or seed tank. Decisions also have to be made concerning where data will be located in the network for prompt, easy accessability by those concerned.

Measured variables such as broth temperature, dissolved oxygen, and pH may be stored in both the individual fermentor processor and a supervisory processor collecting data for the entire fermentation section. Models for control and calculated variables (oxygen uptake, respiratory quotient, and cell biomass concentration) may be of sufficient size that they only reside in a higher-level processor. Information concerning other controlled operations such as inoculation, cleaning, and sterilization procedures may also be contained in various sections of the plant. Consequently, the data network requires sufficient design effort to ensure that all records are updated correctly and error recovery procedures defined. With the data and hardware appropriately distributed, the control network will match the operational structure of the plant.

Ideally, distributed systems can be more reliable than a monolithic system. The effects of failures are localized and easily corrected. When compared to centralized organizations, a well-designed network or hierarchy can allow a reduced system-wide data communication rate by localizing function and data requirements (most presently proposed systems are actually highly centralized about a data highway).

By having the distributed system visibly structured (i.e., by analyzing plant operation with the methodology described above) failures can be recognized in terms of errors of function rather than abstract errors of computation. This functionally based design approach minimizes the necessity to account for the effects of confusing interactions between the various parts of the control system. Additionally the functional structure corresponds to traditional analog control structure and familiar human organization patterns. In practice then, the hardware elements are more thoroughly under the control of the operating people with analogous roles in plant operations, making them more acceptable.

Each processor in the system will perform a specific, well-defined task for the plant operation. Each element will be "self-aware," which is to say that the element is designed to function well in a community of other elements [10]. In short, it will be autonomous from other boxes, disallowing direct access to raw internal data or programs, and communicating only in functional terms. In this example the processor charged with operating a single fermentor would not allow a change to the biomass value stored in its data base except by a message from the processor carrying out the calculation and at a time when it expected such information.

There is a pervading interest in multicomputer-based distributed/hierarchical process control systems. This interest for use in new fermentation processes is justified by the requirements for reliability, flexibility, and computing capability to operate the plants.

A summary of these three categories of control systems just described is given in Table III. The matrix format identifies the characteristics of various plant control system design issues as they are handled by each category.

TABLE III
Control System–Characteristic Matrix

CONTROL SYSTEM \ CHARACTERISTIC	DATA STORAGE	CONTROL FUNCTIONS	COMMUNICATIONS	DESIGN PROCEDURE	RELIABILITY	FLEXIBILITY	OPERATING PHILOSOPHY
SINGLE-LOOP	At Local Unit	Built In Local Units	Human	Last Stage	Individual Loop Level	Hardware Limited: Wired Connections	Single Loop
SINGLE-COMPUTER	Collected In Central File	Centralized	Human Between Units	Pre-planning Needed	Reliable As Central Unit	Changeable: Processor Dependent	Unit Operation
MULTI-COMPUTER	Distributed Multiple Copies	Shared And Distributed	Data Link Across Units	Pre-planning Required	Shared Function	Redesignable And Expandable	Plant Operation

CONCLUSIONS

The conclusions to be drawn from this paper and how they affect industrial fermentation control system design are as follows:

1) The control system requirements and considerations for industrial plants are different from those in laboratory or pilot plant environments.

2) The present interest in computer applications for fermentation processes and microprocessor technology indicate that single-computer-based and probably multicomputer-based control systems will be a critical part of future plants.

3) Computer-based control systems require extensive consideration of plant operation in the control system design.

4) Those involved in the control system design, process design, and plant operation do not usually converse early enough in the plant design cycle. This has resulted in slow acceptance of needed control techniques.

5) The time is *now* for those with fermentation process background and those with industrial control background to work together for the successful application of control systems in future fermentation process plants.

References

[1] M. C. Beaverstock, and J. W. Bernard, "Advanced control: Ready, able, accepted?," in *5th IFAC/IFIP International Conference on Digital Applications to Process Control* (IFAC/IFIP, The Hague, The Netherlands, 1977).

[2] M. C. Beaverstock, "3 R's of control system design," in *Third Annual Advanced Control Conference* (Purdue University, Control Engineering—Sponsors), (Purdue University, Control Engineering—Sponsors) (Purdue University, Lafayette, IN, 1976).

[3] J. W. Bernard and G. M. Howard, "Organizing multi-level process control systems;" in *IEEE System Science and Cybernetics Conference* (IEEE, New York, 1968).

[4] P. E. Dlouhy and D. A. Dahlstron, "Continuous filtration in pharmaceutical production;" *Chem. Eng. Prog.* 64(4), 45 (1968).

[5] A. Waters, "A pneumatic compositions transmitter and its applications," *Chem. Eng. Prog.*, 74, 52 (1978).

[6] F. G. Shinskey and T. J. Myron, "Product moisture control for steam-tube and direct fired dryers"; *MBAA Tech. Quart.* 12(4), 235 (1975).

[7] C. A. Fauth and F. G. Shinskey, "Advanced control of distillation columns," *Chem. Eng. Prog.*, 71(6), 49 (1975).

[8] H. Paschold and N. D. L. Veerstoep, "Computer based control of fermentation processes," in *Symposium on Process Control and Instrumentation in Fermentation* (Society of Chemical Industry, London, 1977).

[9] A. Meskanen, R. Lundell, and P. Laiho, "Engineering of fermentation plants," *Process Biochem.*, 6, 31 (1976).

[10] E. H. Bristol, "The organizational requirements of the process control distributed system," in *Joint Automatic Control Conference* (1978).

Development of a Computerized Fermentation System Having Complete Feedback Capabilities for Use in a Research Environment

R. D. MOHLER, P. J. HENNIGAN, H. C. LIM, G. T. TSAO, and W. A. WEIGAND

School of Chemical Engineering, Purdue University, West Lafayette, Indiana 47907

INTRODUCTION

The use of computers for data acquisition, optimization, and control of fermentors has increased significantly in recent years. Dobry and Jost [1] reviewed modeling, control, and optimization of computer-coupled fermentors from the viewpoint of the use of the computer to assist in the operation of industrial fermentation processes. More recently Weigand [2] reviewed computer-coupled fermentation systems for research purposes. Several references to existing computer-coupled systems can be found in these reviews. There is considerable activity both in industry and academic institutions in the use of computers for fermentation optimization and control.

In a research environment the most desirable features for a computer-coupled fermentor, among others, are the flexibility to handle various modes of fermentor operation for different organisms, the ability to monitor and control the experimental conditions tightly and accurately, and the ease of operation. This paper gives a description of the computer-coupled fermentor system developed at Purdue designed to meet these needs.

DESIGN PHILOSOPHY AND SYSTEM CONFIGURATION

The goal was to develop a computerized fermentation system that would have maximum flexibility for use in a research environment. The system was to be designed to accommodate different biological systems, as well as several bioreactor modes of operation such as batch, continuous, fed-batch, or some variation thereof.

The concept of "gateway sensors," as coined by Humphrey [3], was felt to be an important consideration in the selection of measurement loops and those biologically significant parameters that can be extracted from those measurements. In addition to data acquisition, it was desired to have the ability to feedback upon the basic acquired measurements or those parameters that can be calculated from the basic measurements. It was desired to study the production of single-cell protein from a dynamic viewpoint, which entailed not only acquiring data in real time, but also perturbing the system to determine its dynamic response with the goal of eventually controlling the system to obtain the desired response. The design philosophy was to measure as many physical process variables that are within the reach of current technology, and allow for the flexibility of feeding back upon those measurements or parameters obtained from those measurements using the physically available manipulated variables. A hierarchical control structure was adopted in order to provide maximum flexibility.

It was desired to provide maximum flexibility by allowing the fermentor's operation to be defined by computer software. It is the computer-software–fermentor-hardware interface that defines the degree of flexibility obtainable. It was apparent that the coupling of measured process variables and manipulated variables must be a software association rather than a hardware link to provide the above-stated flexibility.

In view of the above requirements, it was decided to use hardware setpoint control for only the lowest-level control elements of the hierarchical control structure. The other alternative considered was to use DDC (direct digital control) for the lowest-level control loops. Lowest-level control loops are designated by those loops that directly feedback upon one of the basic process measurements. For instance, the manipulated variable of heat input or dissipation is used to control the process measurement of temperature. It was felt that the only way in which temperature would ever be controlled was by heat exchange through a transfer medium, thus there would be no loss in flexibility in coupling the temperature measurement to a single manipulated variable. Hardware setpoint control was used so that the fermentor system would provide good control of basic process measurements even when the system was not operated under computer control. Figure 1 shows the fermentation system with its associated basic process measurements and coupled manipulated variables. Table I indicates which of the measurements are set but not necessarily controlled (meaning fed-back upon). One will note that no loss of flexibility is apparent in any of the hardware coupled loops since in all cases there is only one possible association that can be made.

Direct digital control of the lowest-level control loops was considered, but its advantages and disadvantages had to be considered in evaluating overall system performance. Direct digital control is more flexible in view of the fact that different or even "so-called optimal" control algorithms can be used on each loop. The clear disadvantage of DDC without analog backup is that the system cannot operate at all without the use of a digital computer. It was felt that the lowest-level control loops were so basic that once a control scheme was employed that performed satisfactorily, that control scheme would probably never change. This combined with the desirability of having good low-level control when the computer was not operating formed the basis for using hardware setpoint control on this particular system. It is not implied that direct digital control of low-level loops should not be considered under other circumstances. In large-scale oper-

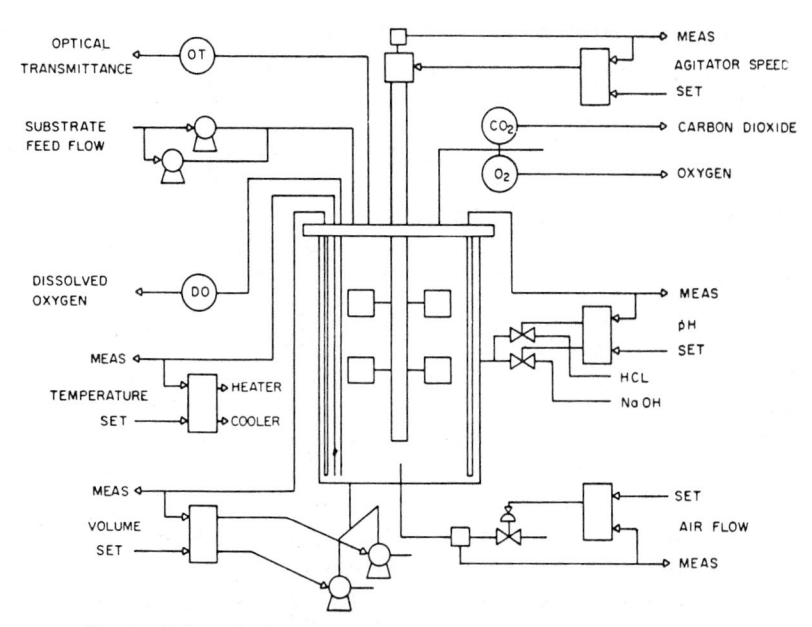

Fig. 1. Schematic diagram of fermentor and associated instrumentation.

ations there is clearly a crossover point, dependent on the number of loops, where the cost per loop by using DDC is economically justified. Direct digital control of low-level loops is attractive in that it substitutes software for hard wired logic.

TABLE I
Measured and Controlled Fermentation Variables

Measured Variables

Carbon Dioxide in the fermentor off-gas	Oxygen in the fermentor off-gas
Dissolved Oxygen	Optical transmittance

Hardware Controlled and Measured Variables	Digitally Cascaded Controlled Variables
Temperature	Dissolved Oxygen
pH	Optical density
Agitator Speed	Carbon Dioxide in Off Gas
Liquid Volume	Oxygen in Off Gas
Airflow Rate	Calculated Parameters such as R.Q., $k_L a$, etc.

Manipulated Variable

Substrate Feed Rate

COMPUTER–FERMENTOR INTERFACE

Selection and performance of basic measurement instrumentation is important in the development of a computer-coupled fermentation system. Selection of instrumentation was based primarily upon accuracy, stability, linearity, and compatibility with the overall system. Commercially available instruments were purchased for the basic measurements of optical transmittance (related to cell density), carbon dioxide in the off-gas, oxygen in the off-gas, pH, liquid volume in the reactor, and air-flow rate (measurement and control). Ideally it is best to have the measurement instrument just slightly overranged to obtain the greatest sensitivity. Since a digital readout device at the fermentor site is incorporated into the system, the output of each instrument was linearized so that the local meter would display the parameter in meaningful units.

Only a portion of the basic measurement instrumentation was purchased commercially; all other measurements listed in Table I and all of the associated analog hardware controllers were designed into the fermentation system. The computer-system–fermentor-inferface determined the fermentor operating flexibility. Before discussing the interface techniques involved, we should point out that the computer system that was used has an extensive analog-to-digital and digital-to-analog conversion system at the computer site, which was remote from the fermentor site. The remote analog–digital conversion system required that analog signals rather than digital signals be transmitted between the fermentor and the computer. The analog signal transmission worked well since hardware setpoint control of low-level loops was used. It was also decided to use totally electronic techniques employing the latest state-of-the-art components in the computer-system–fermentor interface instead of electromechanical components such as relays, servopots, and digital stepping motors.

Conceptually, all measurement loops are treated alike as are all measurement loops coupled with their associated hardware controller. Figure 2 is a block diagram of a typical measurement loop and a typical measurement loop with hardware control. All measurement loops employ some means of converting a physical measurement to an electrical signal, which is then conditioned to proper levels for signal transmission. The measurement signal is converted into a differential signal for analog transmission, which takes advantage of the noise rejection capability of the differential input amplifiers on the analog–digital converter. Each measurement is also input to a multiplexed digital display which indicates the measurement in proper engineering units at the fermentor site. The measurement part of a measurement loop with hardware control is identical to that described above. In addition, the measurement loop with hardware control has a switch selectable setpoint signal coming either from the computer site via a digital-to-analog converter or from the local fermentor site. The setpoint signal is combined with the measurement signal forming an error signal that is used to drive the controller. The setpoint is also an input to the digital display for local readout. The above-described configuration allows measurements to be displayed locally and transmitted to the computer, while setpoints are independent of each

MEASUREMENT LOOP

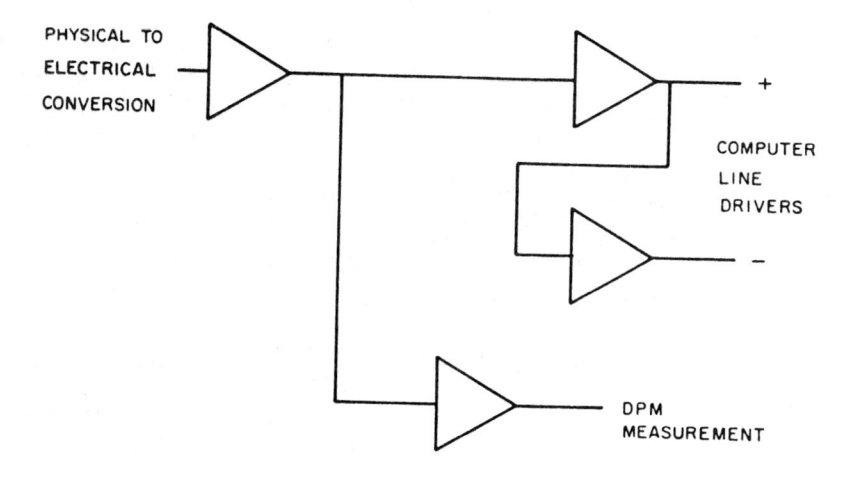

MEASUREMENT LOOP WITH HARDWARE CONTROL

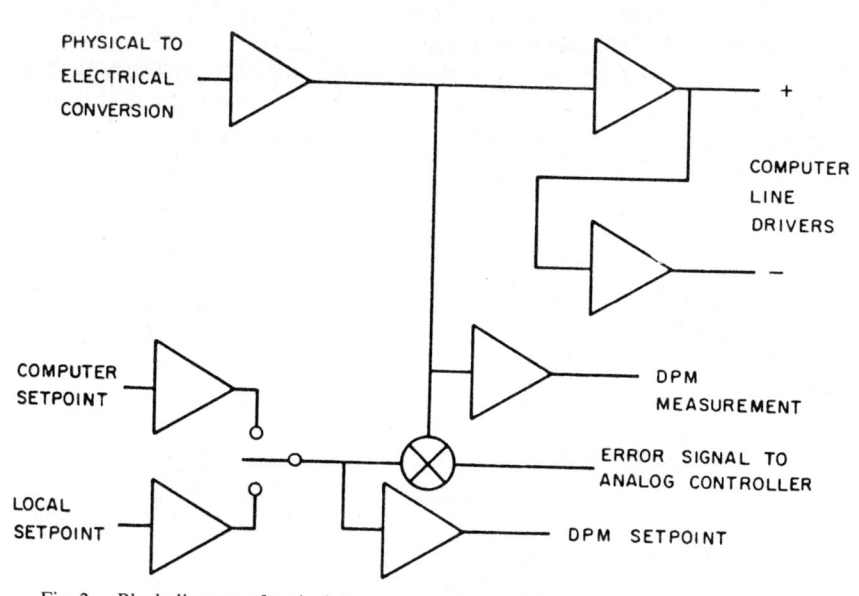

Fig. 2. Block diagram of typical measurement loops with and without hardware control.

other and set locally or programmed by computer. Another attractive feature of the local control capabilities of the system is that the user can set setpoints locally through a digital display and then walk away being assured that the measurement will attain the setpoint value. The original Microferm fermentor did not have this capability since the user acted as the controller in the sense that the potentiometers had to be manually varied until the measurements were visually determined to be that which are desired.

OPERATIONAL CHARACTERISTICS OF MEASUREMENT AND CONTROL LOOPS

It seems appropriate to briefly consider the operational and performance characteristics of each measurement and associated control loop which was designed into the system, not including those purchased commercially. Dissolved oxygen was measured using the silver–lead galvanic-type electrode. The millivolt signal was amplified, and calibration adjustments for span and zero were provided to the user to account for probe condition and age. Temperature compensation of the DO probe was handled by software when computer temperature profiling was employed. A hardware temperature compensation scheme using the fermentor temperature input has been designed and will soon be employed.

The approach to temperature and pH control is thought to be unique; therefore they will be considered together since they are very similar in many ways. The temperature measurement was obtained by using of a thermilinear which converts a temperature change into a linear resistance change. pH was measured by a commercially (Corning model 125) bought meter but had to be electrically isolated from the controller circuits since the reference probe of a pH meter holds it at the potential of the solution that it is measuring. Figure 3 is a block diagram representation of both the temperature and pH, measurement and control loops. The temperature and pH loops both use an on–off type of control where temperature is controlled by a resistance heater and a cooling water solenoid, whereas the pH loop employs both acid and base additional solenoids. Two-sided control was achieved by utilizing the bipolar output of the computer digital-to-analog

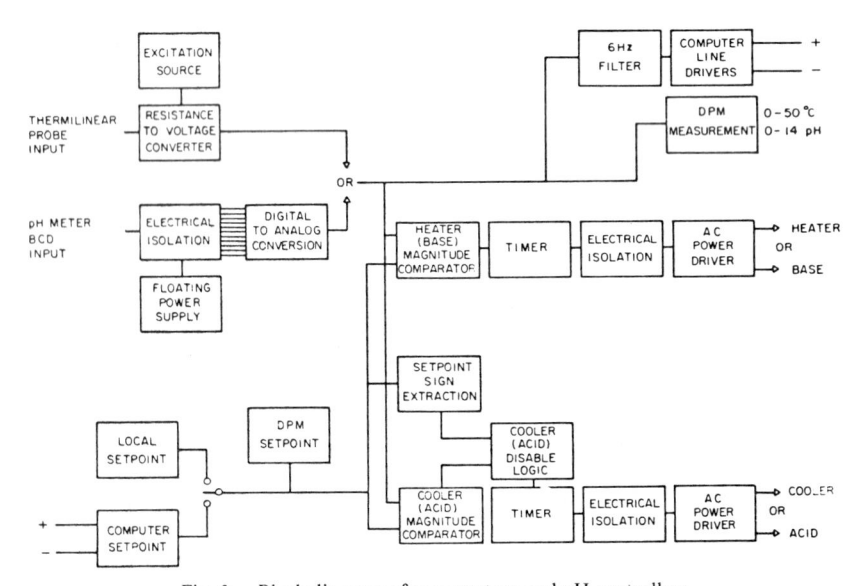

Fig. 3. Block diagram of temperature and pH controllers.

converters. The user programs a positive set-point value to indicate heat input for the temperature loop and base addition for the pH control loop. A negative setpoint of equal magnitude will cause the controller to seek the same value, but cooling or acid addition will be employed. This scheme allows the user maximum flexibility in determining whether to use single-sided control or dual-sided control with a programmable bandwidth. Caution must be exercised in setting the bandwidth on dual-sided control since the system can be made to oscillate. Figure 4 shows a typical batch fermentation where temperature was programmed to increase linearly by computer and pH to be maintained constant at 7.00. The steps in the pH profile are equivalent to 0.01 pH units and are due to the accuracy of the analog-to-digital converter and a small amount of overshoot during base addition. One can also note an increase in the base addition frequency, as it should, with the increase in the cell density.

Agitator speed control was interesting because hardware feedback linearized that which was provided with the original Microferm fermentor, while also providing a reliable agitator speed independent of liquid level or culture viscosity. As shown in Figure 5, the agitator speed is sensed by a slotted-disk optical tachometer assembly which outputs a pulse train. This is then converted to a voltage proportional to agitator speed which drives the computer lines and along with the agitator speed setpoint, forms the error signal input to the speed controller. The controller output drives an isolation circuit, which then controls the SCR-controlled DC motor. The feedback control of agitator speed transformed a nonlinear system into a linear system which made the loop more ideally suited for cascade control.

Level control, which is required for continuous fermentor operation, was implemented by a continuously modulated peristaltic pump on the reactor outlet which feeds back upon the level measurement as sensed by a capacitance-type

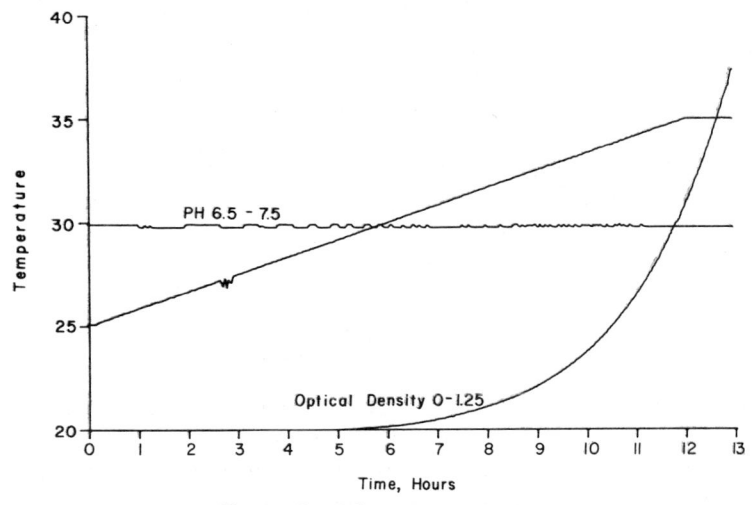

Fig. 4. Batch fermentation profile.

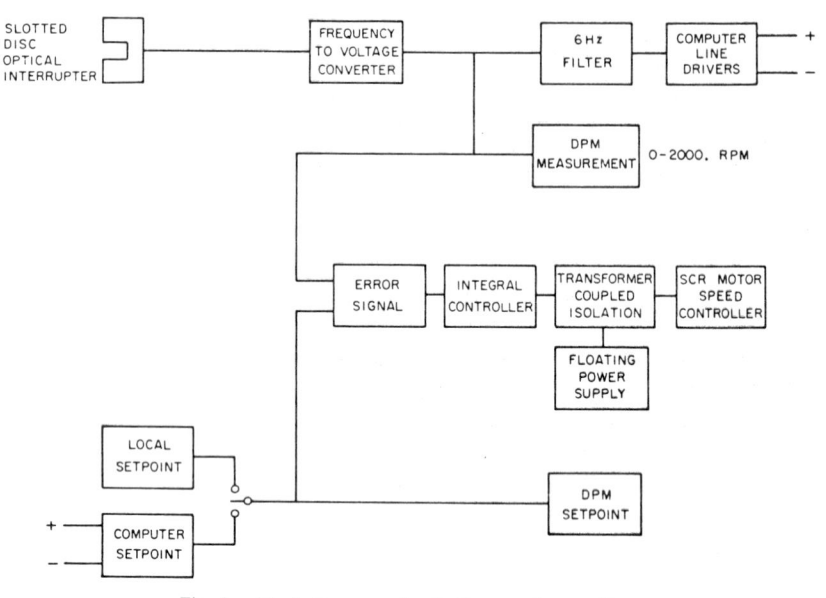

Fig. 5. Block diagram of agitation speed controller.

probe. In parallel with the modulated exit pump is a high-speed peristaltic pump which can empty the reactor quickly as required under one variation of the feed-batch mode of operation. Again the modulated pump controls in reference to a positive setpoint, whereas the high-speed pump is turned on by a negative level setpoint.

A digital readout of all fermentation process measurements and setpoints, in engineering units at the fermentor site, proved to be most convenient in operating the system. The initial fermentor system specification provided for nine basic process measurements and six setpoints driving hardware controllers. A digital display utilizing pushbutton selection of any measurement or setpoint was incorporated into the system. The display interface electronically multiplexed all measurements and setpoints, scaled all values to the correct engineering units, and selected the correct placement of the decimal point. The value of the local digital display was realized by the user programming the fermentor operation since program errors can be detected readily. The display is useful in troubleshooting both the hardware operation of the fermentor system and the software operation of the user's program.

HIERARCHICAL CONTROL STRUCTURES

The capabilities of the computer-coupled fermentation system can be truly realized with the next higher level of control within the hierarchical structure, namely, that of cascaded digital control. The previously mentioned low-level analog loops only maintain basic measurements at constant values as defined by their programmable setpoints and associated hardware controllers. Cascaded digital control allows the user to control other low-level measurement loops that were not associated with a hardwared controller or to control biologically significant parameters that are calculated from the basic measurements.

It was desired from the start of this work to have accurate dissolved oxygen control during any mode of fermentation operation. It was decided not to hardware link dissolved oxygen control to the manipulated variables of air-flow rate, agitator speed, or some combination thereof, since it is conceivable that dissolved oxygen could be controlled by inlet substrate feed rate, or some other method. Figure 6 shows the configuration currently being used for dissolved oxygen control where a digital dissolved oxygen controller is cascaded upon both the hardware setpoint controllers of air flow and agitator speed. The unique property of the dissolved oxygen control, as developed by Hennigan [4], is that the dissolved oxygen controller is driving either or both the air flow and agitator speed controller using a cascaded control scheme which greatly speeds up the response and overall quality of the dissolved oxygen control. Figure 7 is a dissolved oxygen, agitator speed, and air-flow profile during a batch fermentation where the oxygen demand is continually changing. Note that the agitator speed and air-flow rates are kept at their minimum present values until the point where the culture oxygen demand has increased enough to cause the dissolved oxygen controller to take positive action. As can be seen from Figure 7, the culture medium was saturated with oxygen at the beginning of the batch. Then, as cells grew the dissolved oxygen dropped until it reached the desired minimum value of 70%, at about the sixth hour into the batch. At this point, the agitator speed increased to maintain the 70% value. Note the quality of control during this period. At about 12.7 hr the agitator speed reached its upper limit of 1000 rpm,

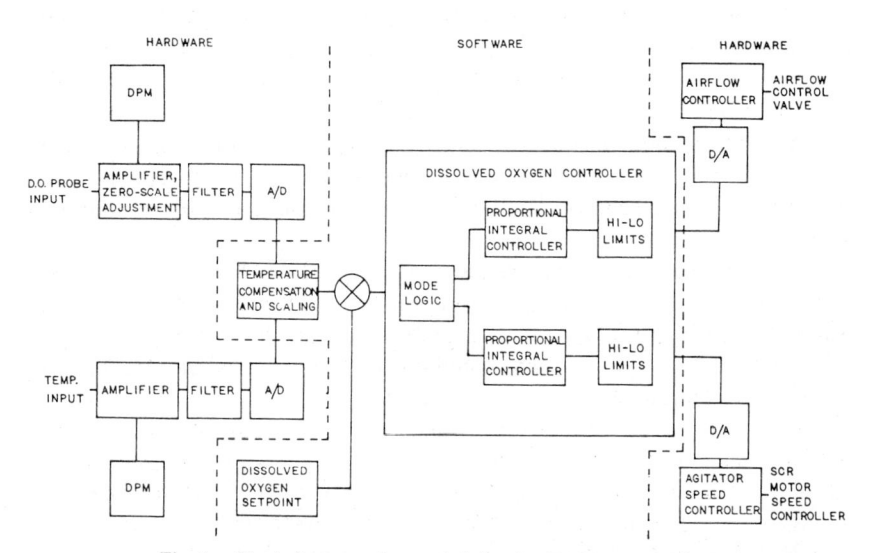

Fig. 6. Block diagram of cascaded dissolved oxygen controller.

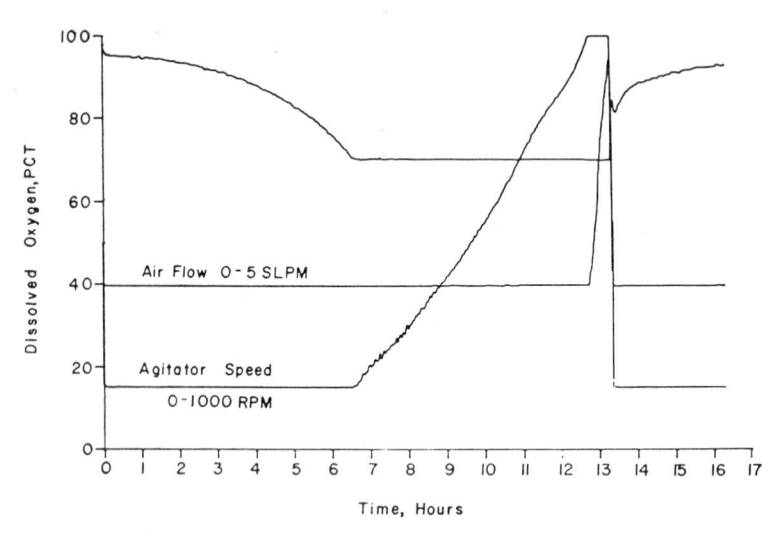

Fig. 7. Dissolved oxygen control in batch fermentation.

and then the air-flow rate quickly increased to maintain the dissolved oxygen (DO) of 70%. Note that the DO is essentially unperturbed when this occurs. Then, at about 13.3 hr the substrate was completely utilized, and the agitator speed and air flow very rapidly drop to their initial nominal values of 150 rpm and 2 slpm. Others [5] have also used both air flow and agitation speed to control DO with a computer-coupled fermentor. In this case, a conventional feedback digital control loop was used which does not appear to cascade the air flow and agitation speed. In addition, digital output pulses and a velocity output mode to stepping motors to drive the final control elements was used. The Purdue system which produced the results shown in Figure 7 uses a proportional-integral algorithm with position, i.e., full value output (not the velocity algorithm) and setpoint control for the analog controllers which manipulate the final control elements of agitator speed and air flow. This method naturally leads to the cascade mode of control which produces the quick response of air flow and agitator speed shown in Figure 7. A more detailed discussion of these and other conventional digital control methods have been presented by Williams et al.[6].

The difference in the controlled DO by the method used here is seen by comparing Figure 7 of this paper with Figures 7 and 8 of Nyiri et al.[5]. Upon examination of these figures, it is evident that a much slower response of the manipulated air flow and a correspondingly greater "ripple" in the controlled DO levels of reference [5].

Using the same type of cascaded control, cell density during the fed part of fed-batch operation has been controlled to a constant value by cascading the optical transmittance measurement upon the inlet substrate feed flow. The major point to be made is that the design of cascaded control strategies is dependent upon the needs of the experimental procedure being employed. A calculated parameter such as respiratory quotient (RQ) can be just as easily cascaded upon substrate feed flow, for instance, as dissolved oxygen was cascaded upon air-flow rate and agitator speed. The system as designed allows the user to control any measurement or parameter derived from a combination of basic measurements by directly actuating or cascading upon any manipulated variable or combination of manipulated variables.

The supervisory level of the adopted hierarchical control scheme accomplishes the objectives of the mode of operation under which the fermentor is being operated. The supervisory programs determine the setpoint inputs to the cascaded controllers and uncascaded basic hardware controllers. The supervisory programming monitors sequencing of events and performs timing control, which is considerable in the fed-batch mode of operation. It is within the supervisory program where the user defines the objectives of his/her experiment. Generally, advanced control strategies, on-line optimizations, and higher-level on-line data analysis would be handled within the supervisor programs. The supervisory program monitors and inputs to all lower levels of control. Figure 8 shows various profiles of a reactor run under repeatedly fed-batch mode of operation. On the same plot, one can see the lower-level cascaded dissolved oxygen profile and cell density profile where the cell density was maintained constant during the fed portion of fed-batch operation.

CONCLUSION

A computer-coupled fermentation system has been described which allows a researcher maximum flexibility with respect to data acquisition and control of the biological system under study. The experiment and mode of fermentor

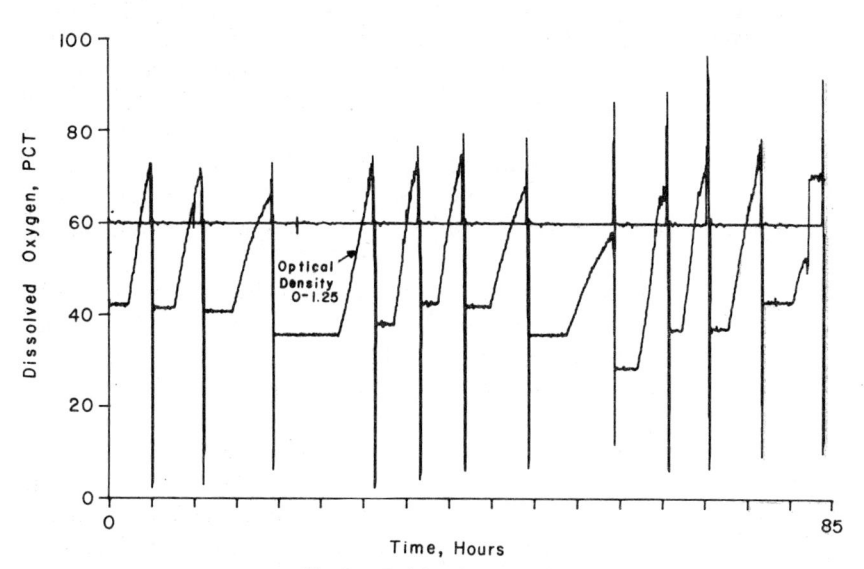

Fig. 8. Fed-batch operation.

operation are defined through a high-level FORTRAN language. The supervisory level of control defines the fermentor operation, monitors and controls sequencing of events, and calculates setpoints for lower-level cascaded and hardware-defined control loops. Cascaded digital control algorithms provide the system with its inherent flexibility by allowing the user to software define the pairing of control variables and manipulated variables. Although use of a computer is required for supervisory and cascaded digital control, the hardware setpoint control of basic fermentation process parameters enables the system to be generally useful even when not under computer control.

Added features such as the extensive digital display system are evidence of the design philosophy of flexibility and ease of operation by the user. The system has been operated under batch, continuous, and fed-batch modes where the controllability and acquired data are found to be much better than that which can be accomplished manually. The limitations of the system are inherent in the basic process instrumentation and there is a continued need for better and more sophisticated measuring devices.

This work was supported in part by the National Science Foundation Grant Nos. ENG 75-17796 and ENG 76-22309.

References

[1] D. Dobry, and J. Jost, in *Annual Reports on Fermentation Processes,* D. Perlman, Ed. (Academic, New York, 1977), Vol. 1.
[2] W. A. Weigand, in *Annual Reports on Fermentation Processes,* D. Perlman, Ed. (Academic Press, New York, 1978), Vol. 2.
[3] A. E. Humphrey, *Chem. Eng., 81,* 98 (1974).
[4] P. J. Hennigan, M.S. thesis, Purdue University, 1979.
[5] L. K. Nyiri, R. P. Jefferies, and A. E. Humphrey, *Biotechnol. Bioeng. Symp., 4,* 613 (1974).
[6] T. J. Williams, F. J. Mowle, W. A. Weigand, G. V. Reklaitis, R. E. Goodson, and H. C. Lim, *Digital Computer Applications to Process Control* (short course notes) (Purdue Laboratory for Applied Industrial Control, Lafayette, IN, 1973; revised 1976), Vols. 1 and 2.

Review of Process Control and Optimization in Fermentation

SHUICHI AIBA

Department of Fermentation Technology, Faculty of Engineering, Osaka University, Yamada-kami, Shuita-shi, Osaka, Japan

INTRODUCTION

As early as 1962, Yamashita and coworkers examined DDC (direct digital control) in glutamic acid fermentation. A pilot-scale fermentor was equipped with controllers for: i) temperature, ii) vessel pressure, iii) pH, iv) air flow rate, and v) foam. Analyses were made for i) outlet gas composition, ii) glutamate concentration in the broth, iii) dissolved oxygen concentration, and iv) microbial population density. Their approach to optimization was either by an early detection of a change in the value of a principal parameter of the kinetic model of glutamic acid fermentation in each run or by comparing the performance of each fermentation with the standard run extracted from past experience. The latter approach is heuristic and requires no mathematical model.

Regardless of the two different approaches to optimization of fermentation process, the importance of adequately instrumenting a fermentor coupled with a properly designed DDC and computer need not be exaggerated. The principal feature of "on-line" process control and optimization was clearly pointed out more than a decade ago by Yamashita et al. [1], who recognized the substantial difference of fermentation from the chemical industry in terms of control engineering.

COMPUTER APPLICATIONS

Nyiri [2] reviewed relevant contributions that had been published by many workers prior to March, 1971. According to this review most of the computer applications are concerned with "off-line" use, i.e., for construction of a microbial kinetic model, experimental data analysis, and parameter estimation. It is urged that "on-line" computer application can evaluate a process by accumulating data throughout the course of a fermentation and by simultaneously analyzing environmental as well as intracellular activities of microbes.

Using a highly instrumented fermentor to cultivate *Hansenula polymorpha* DL-1 from methanol as the sole carbon source, Swartz and Cooney presented recently another review on the use of a computer [3]. Particularly, this work referred to gateway sensors of CO_2 and O_2 in the ferment exit gas, emphasizing the difference in sensitivity of each sensor to gas flow. So far as the computer is concerned, it is for data acquisition and processing.

Model Identification and Parameter Estimation

When specific rates of limiting substrate consumption, cell mass propagation, and product accumulation are formulated, nonlinear and Monod-type formulas are usually employed in the growth-associated pattern of product formation. In particular, this latter modeling is sometimes augmented by the first-order rate of product decomposition and/or by-product accumulation. [4]

Constantinides identified a model for cell growth rate (sigmoid or quadratic equation of cell concentration) and penicillin synthesis rate (non-growth-associated minus decomposition rate of product) which fit the experimentally observed data. In addition, effect of the temperature on parameters estimated by the Gauss–Newton nonlinear regression method was presented to calculate from the maximum principle an optimal temperature profile during the batch cultivation to maximize the potency of penicillin at a specific point in time [5]. Blanch et al., on the other hand, introduced a new concept of specific rate of maturation in a model identification of gramicidin S biosynthesis. Assuming a continuous culture, rate equations of newly born cells, matured cells, and limiting substrate were presented [6]. Recently, Fishman et al. introduced the concept of "cumulative age" of multicellular microbes to model the secondary metabolite (penicillin) production rate [7].

In parallel with sophistication of a kinetic model, the number of parameters increases. Herein lies the significance of parameter estimation in nonlinear phenomena such as those of cell propagation and metabolism. As far as the parameter estimation by the use of digital computer is concerned, several algorithms are available in the library and no substantial difficulty around this estimation is expected.

Optimization (Analytical Approach—the Maximum Principle of Pontryagin)

Assuming that the temperature and pH of a culture broth are kept constant, system equations of a fed-batch culture are

$$\frac{d(V \cdot X)}{dt} = \mu'(V \cdot X) \tag{1}$$

$$\frac{d(V \cdot S)}{dt} = F \cdot S_R - \frac{1}{Y_{X/S}} \frac{d(V \cdot X)}{dt} \tag{2}$$

$$\mu = g(S) = \mu_{max} \cdot [S/(K_s + S)] \tag{3}$$

$$S = g^{-1}(\mu) \tag{4}$$

$$Q_p = \frac{1}{V \cdot X} \cdot \frac{d(V \cdot P)}{dt} = f(\mu) \tag{5}$$

$$\frac{d(V \cdot P)}{dt} = f(\mu) \cdot (V \cdot X) \tag{6}$$

$$\frac{dV}{dt} = F \tag{7}$$

The initial conditions are

$$\begin{aligned} X &= X_0 \\ V &= V_0 \end{aligned} \tag{8}$$

Constraint:

$$V \cdot X \le (V \cdot X)_{\max} \tag{9}$$

Objective function:

$$J = (V \cdot P)_f = \int_0^{t_f} f(\mu) \cdot (V \cdot X) dt \to \max \tag{10}$$

Taking X, S, and V as state variables and F as control variable, Yamane et al. discussed from the maximum principle the answer to this problem [8]. By the same token, an optimal feeding in the singular control region was presented by Fishman et al. [7]. In connection with the model identification of gramicidin S biosynthesis, Blanch et al. [9] proceeded to calculate, via the maximum principle, the profit of this antibiotic production in continuous vessels in series. Their calculations demonstrated the following:

i) Among three decision variables (retention time, pH, and temperature) the retention time (vessel volume) was most influential to the profitability. With an increase in the antibiotic price, the longer retention time was advisable in the first vessel of two-stage continuous culture, whereas the retention time of the second vessel was independent of the change in price of this product.

ii) When the price of this antibiotic is fixed, the number of vessels recommended was $n = 4$. If $n \ge 4$, the profit expected of that extension was negligible.

Different from the above approach (the maximum principle), Ohno et al. [10] applied Green's theorem to find out the feeding rate that maximizes an average production rate in a fed-batch culture. Their discussion deals with the trajectory of an objective function on a V, S phase plane. An optimal trajectory for lysine fermentation was demonstrated. Q_p in eq. (5) was taken as a quadratic function of μ, while μ was linearized with respect to S [cf. eq. (3)] and $Y_{X/S}$ in eq. (2) was taken as 0.135 throughout.

Apparently the foregoing discussion tacitly assumes that parameters involved in kinetic equations remain unchanged. However, it is urged from practical experience that microbial behaviors are frequently subjected to changes from run to run even if operating conditions (including the specific strain used) are kept exactly the same. In most cases, the reason for the behavioral change is difficult to detect. Accordingly, the analysis with this mathematical means remains to be examined from another angle of experimentation. It should be remembered that the maximum (or minimum) principle of Pontryagin does not present a solution that satisfies the necessary and sufficient conditions required for optimizing a problem in question; the solution pertains only to the necessary condition.

Optimization (Direct Approach)

If one considers the practice of producing a specific material from a fermentation process, another problem of optimization emerges. This problem involves not only the fermentation, but also unit operations such as filtration, extraction, drying, solvent recovery, waste treatment, etc., which follow the cultivation process. In this category of optimization, the search for a number of variables that minimize the cost of annual operation for a given output of final product is usually beyond the scope of the previous analytical approach.

Although system equations in each process extending from fermentation to the final operation of product recovery must be made available, kinetic equations are not always needed. For example, product concentration P, as a polynomial function of fermentation time, t, would suffice.

In general, the relationship between input and output for the ith process (ith subprocess or subsystem) is

$$\mathbf{Y}_i = \mathbf{f}_i(\mathbf{X}_i, \mathbf{u}_i) \tag{11}$$

If output is partially recycled,

$$\mathbf{X}_i = \sum_{j=1}^{n} C_{ij} \mathbf{Y}_i, \qquad C_{ii} = 0, \tag{12}$$

Inequality constraint among X_i, Y_i, and u_i:

$$g_i(\mathbf{X}_i, \mathbf{u}_i, \mathbf{Y}_i) > 0 \tag{13}$$

provided:

$$\mathbf{X}_i \in X_i, \qquad \mathbf{Y}_i \in Y_i, \qquad \mathbf{u}_i \in u_i \tag{14}$$

Objective function:

$$J = \sum_{i=1}^{n} J_i(\mathbf{X}_i, \mathbf{u}_i, \mathbf{Y}_i) \tag{15}$$

The system equation is shown by eq. (11). An optimization is attempted here to optimize the value of J in the ith system by manipulating the decision variable, $\mathbf{u}_i$, once the output, $\mathbf{Y}_i$ is given. Supposing that the annual production capacity of a metabolite is prefixed and that the production rate of the metabolite be shown as a function of fermentation time, an optimization to minimize the expenditure of a process composed of fermentation to product recovery was thus attempted by Okabe et al. [11]. In the fermentation subsystem, fermentor and boiler capacity were the decision variables. In this filtration subsystem, the amount of filter aid added per unit amount of broth and the rotation speed of a filter drum were the decision variables, whereas, in the extraction subsystem, efficiencies of the first and second extractions were the decision variables.

If the output of the fermentation subsystem in terms of V(broth volume) and product output, $P_f \cdot V$ are given, the objective function could be optimized by the modified complex method (two decision variables), provided system equations for evaluating the objective function are given [11]. The output of filtration was also assumed and the optimization of filtration subsystem could be made. Designating the optimized (suboptimized) objective function to J_i^*, each set of variables that define the input and output to and/or from each subsystem could determine the value of $\sum_i^n J_i^*$, where n = number of subsystems.

Since the latter set of variables are termed those of coordination in the second level, the global optimization to define

$$J = \sum_i^n J_i^* \to \max, \min$$

could be made by comparing directly the sum of each suboptimized value of J_i^* (the first level) when the range of coordination variables was searched by the feasible decomposition method. Okabe et al. demonstrated the global optimization of an antibiotic plant [11].

Because of the lack of a mathematical background in this approach to optimization, it is difficult to conclude the converged value of J simply as optimal. Consequently, a laborious check on this optimality remains to be made by changing the initial values of search in the modified complex method and/or by referring to the practice in relevant fermentations. Unwieldy as it may sound, the direct search for optimization of a fermentation plant as a whole is still useful from the viewpoint of process design and/or scale-up [12,13].

Computer-Aided Fermentor

Wang et al. [14] used a 14-liter fermentor with the aid of "on-line" computer to estimate the biomass of bakers' yeast in a fed-batch culture. This idea comes from a mass balance using stoichiometric equations for the culture [15]. Ammonia and molasses were fed during the run separately without evidence of the feeding-rate being controlled by the value of the respiratory quotient, although they pointed out the possibility of controlling bakers' yeast cultivation by this criterion. [15].

Humphrey urged the rationale for a computer-aided fermentor and demonstrated a means to maximize the productivity of secondary metabolite by an intermittent feeding of the limiting substrate, i.e., [16],

$$\int Q_p(t) X(t) dt \to \max$$

Recently, Zabriskie and Humphrey [17] used a computer-aided fermentor to indirectly measure the biomass of aerobic cultivation of *Thermoactinomyces* sp., *Streptomyces* sp., and *Saccharomyces cerevisiae* by a continuous measurement of oxygen uptake (OUR) and by assuming the linear correlation between OUR and growth rate,

$$\text{OUR} = m_O \cdot X + \frac{1}{Y_{GO}} \cdot \frac{dX}{dt} \tag{16}$$

This work claims that continuous measurement of OUR by the gas analysis and an instantaneous integration of eq. (16) (m_O and Y_{GO} values given separately) gave a satisfactory agreement between observation and computation of X values with respect to *Thermoactinomyces* sp. and *Streptomyces* sp. In the case of *S. cerevisiae* the change of the metabolic pathways of glucose, due to the ever-changing concentration of glucose in culture medium from a fed-batch operation, was taken into account by multiplying OUR on the left-hand side of eq. (16) with β, assuming implicitly another linearity between Y_{GO} and β.

Dynamic Analysis

A systems engineering analysis of a microbial process presented rather recently by Young and Bungay [18] will be reviewed first. The point of this presentation is to designate substrate uptake and microbial growth as input and output, respectively, leaving the microbial metabolism inside a black box. Taking deviations from steady state, a block diagram was constructed from mass balance equations to manifest transfer functions that could correlate output (growth) with input (substrate consumption). A negative feedback mechanism functioning in the black box was revealed. In the experimental check (*S. cerevisiae*) on this diagram, the Monod equation was confirmed not to be applicable to describe the dynamic behavior of microbes.

More recently, Endo et al. [19] published a functional analysis of bakers' yeast cultivation in batch culture. Differentiating the microbial system from that of substrate transport, the following functional relationship between output and input was presented:

$$\begin{bmatrix} \mu \\ Q_{CO_2} \end{bmatrix} = \begin{bmatrix} C_{11} & C_{12} \\ C_{21} & C_{22} \end{bmatrix} \begin{bmatrix} \nu \\ Q_{O_2} \end{bmatrix} \quad (17)$$

$$Q_p = Y_{P/S} \cdot \nu$$

With respect to the logarithmic growth phase, the coefficients of C_{ij} in eq. (17) were found to change in absolute values (some of C_{ij} were negative, while the others were positive), depending on the initial concentration of glucose.

COMPUTER APPLICATION (A PERSPECTIVE)

A promising application of an "on-line" computer to the control of fermentation is in the cultivation of bakers' yeast. Indeed, the metabolism of bakers' yeast has been studied by many workers, and now the information on how to control the aerobic culture, minimizing the production of ethanol is established. The first paper suggesting the possibility of controlling the cultivation by observing respiratory quotient (RQ) values (Q_{CO_2}/Q_{O_2}) with the use of exit-gas analyzers coupled with a computer and connected with a pump to control the feed rate of glucose medium was presented by Nyiri et al. [20].

Fed-Batch Culture of Bakers' Yeast [21]

Supposing that the concentration of glucose (S_R) in a fresh medium becomes instantly exhausted when the medium is fed to a well-mixed fermentor at a rate of F for Δt in the fed-batch culture of *S. cerevisiae,* the balance equation of cell material yields

$$F = \frac{\Delta V}{\Delta t} = \frac{\mu' x}{Y_{X/S} \cdot S_R} \tag{18}$$

Although F should be expressed by F_i depending on the time interval, Δt_i ($i = 1-n$ in interval numbers subdividing the whole period of fed-batch run), the subscript i is omitted for simplicity.

Supposing also that the feed rate is manipulated such that the deviation of RQ from unity is small, i.e., $1.0 \le RQ \le 1.2$ originally (practically, RQ turns out to be less than unity, if the cells consume a small amount of ethanol generated necessarily when RQ values exceed unity), the difference between Q_{CO_2} and Q_{O_2} observed during this period of fluctuation in RQ values is commensurate with the specific rate of ethanol production. This obviously comes from the stoichiometric bases of complete oxidation of glucose and the fermentation, correcting for CO_2 evolution due to respiration.

Consequently,

$$Y_{P/S} = (Q_{CO_2} - Q_{O_2})/v'$$

$$= \frac{Q_{O_2}[(Q_{CO_2}/Q_{O_2}) - 1]}{\mu'/Y_{X/S}}$$

$$= \frac{I_{O_2}(RQ - 1)}{\mu' x/Y_{X/S}} \tag{19}$$

Cancelling out $\mu' x/Y_{X/S}$ from eqs. (18) and (19),

$$F = I_{O_2}(RQ - 1)/(Y_{P/S} \cdot S_R) \tag{20}$$

Taking the experimental fact for granted that the value of $Y_{P/S}$ could be linearized against $(RQ - 1)$ within a small range of RQ ($1.0 \le RQ \le 1.2$) [21],

$$Y_{P/S} = K(RQ - 1) \tag{21}$$

From eqs. (20) and (21),

$$F = I_{O_2}/(K \cdot S_R) \tag{22}$$

Apparently, the feed rate is implicitly dependent on RQ values and explicitly determined by I_{O_2} values which could be measured continuously with an oxygen gas analyzer at the exit gas line of a fermentor.

Actually, in view of the fact that RQ values are kept controlled nearly around unity, I_{O_2} values could be determined without grave error by the following equation:

$$I_{O_2} = F_{air} \cdot p_{O_2} - F_{air} \cdot p_{O_2}$$
$$\quad\quad\quad \text{IN} \quad\quad \text{IN} \quad\quad \text{OUT} \quad \text{OUT}$$

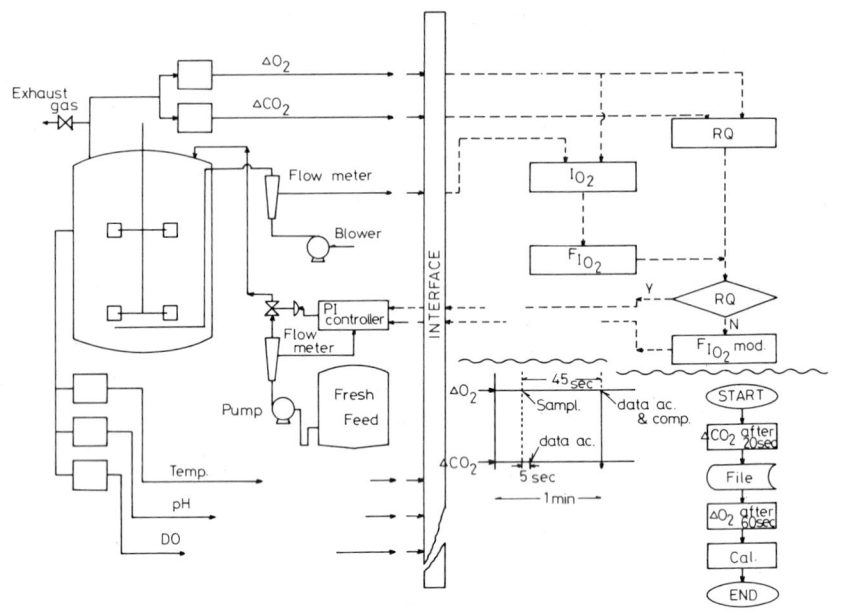

Fig. 1. Schematic diagram of "on-line" computer control of bakers' yeast cultivation.

$$\doteq F_{\text{air}} \underset{\text{IN}}{(p_{O_2}} - \underset{\text{OUT}}{p_{O_2})} \quad (23)$$
<small>IN</small>

Once the medium composition (fresh medium for feeding) is fixed [K value in eq. (22) given], oxygen and carbon dioxide gas analyzers determine the value

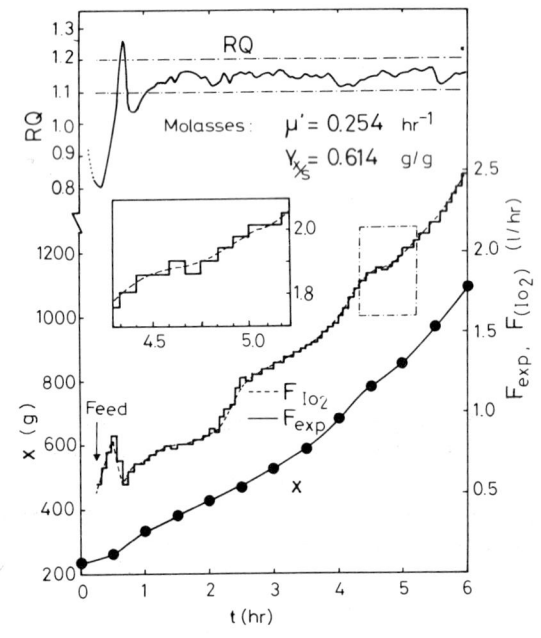

Fig. 2. Feedback control of fed-batch culture of *S. cerevisiae*.

of RQ and at the same time, I_{O_2} values; the latter values generate a pulse to control the feed rate of fresh medium, referring incessantly with the help of a computer to the prefixed range of RQ values. Figure 1 shows schematically the "on-line" computer control of bakers' yeast cultivation, while Figure 2 is an example of fed-batch culture of *S. cerevisiae* using I_{O_2} (and RQ) feedback control.

Citrate Production from *n*-Alkane by *Candida lipolytica* [22]

This example is not intended to demonstrate a good example of "on-line" computer control of the production, but to exemplify the extent to which specific rates of microbial activities observed externally can give access to a better understanding of the metabolic pathways functioning under a particular condition.

Experimental data on a NH_4^+-limited chemostat culture of *C. lipolytica* using *n*-alkane as the sole carbon source are shown in Figure 3. Q_{p_1} and Q_{p_2} are specific rates of citrate and isocitrate productions, respectively, while v and v_N are those of *n*-alkane and ammonia uptake. In this example, an optimal value of the dilution rate to maximize the specific rate of citrate production seems to be located around $D = 0.04$ hr.$^{-1}$

In order to elucidate this particular phenomenon, the metabolic pathway most probably functioning in this microorganism was constructed as shown in Figure

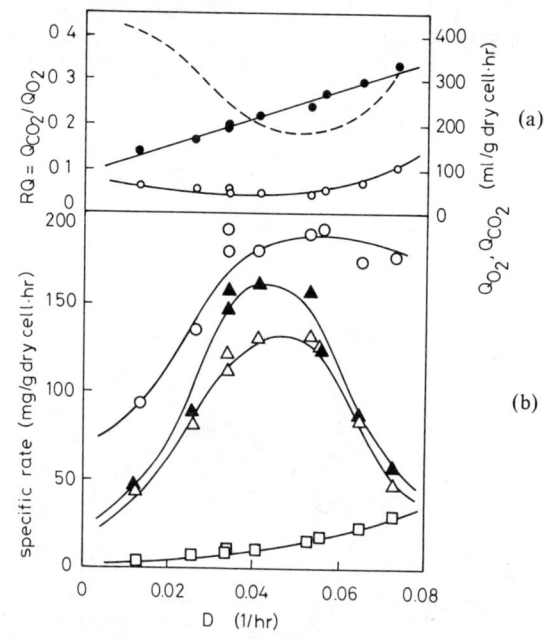

Fig. 3. NH_4^+-limited chemostat culture of *C. lipolytica* (citrate production from *n*-alkane). (a) (□)v_N; (△)Q_{P_2}; (▲)Q_{P_1}; (○)v. (b) (- - -)RQ; (●)Q_{O_2}; (○)Q_{CO_2}.

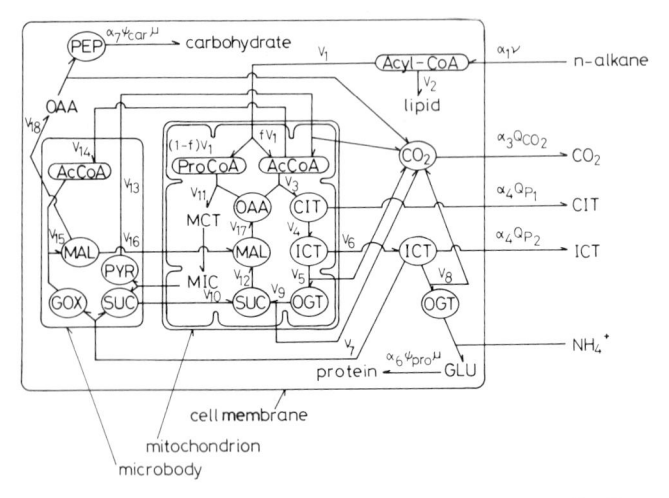

Fig. 4. Metabolic pathways of *C. lipolytica* (compartmentalized).

4, referring to the wealth of biochemical information made available, where AcCoA: acetyl-CoA; ProCoA: propionyl-CoA; CIT: citrate; GLU: glutamate; ICT: isocitrate; PEP: phosphoenolpyruvate; OGT: 2-oxoglutarate; GOX: glyoxylate; SUC: succinate; MAL: malate; OAA: oxaloacetate.

The point of the following examination is to assess the steady state flux of carbon (v_i in Figure 4, $i = 1$–18) from the external observations of v, Q_{CO_2}, Q_{O_2}, Q_{p_1}, Q_{p_2}, etc., all of which could be monitored. Mass balance of carbon per unit cell material per unit time around intermediates circumscribed by solid lines in the figure yields

$$\alpha_1 v = v_1 + v_2 \tag{24}$$

$$fv_1 + \tfrac{2}{3} v_{13} = \tfrac{1}{3} V_3 + v_{14} \tag{25}$$

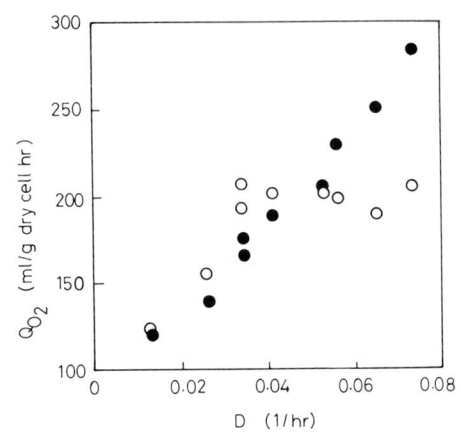

Fig. 5. Values of Q_{O_2} against D. (○) Measured; (●) assessed.

$$v_3 = v_4 + \alpha_4 Q_{p1} \tag{26}$$

$$v_4 = v_5 + v_6 \tag{27}$$

$$v_6 = v_7 + v_8 + \alpha_4 Q_{p2} \tag{28}$$

$$\tfrac{5}{6} v_8 = \alpha_6 \psi_{pro} \mu \tag{29}$$

$$\tfrac{5}{6} v_5 = v_9 \tag{30}$$

$$\tfrac{2}{3} v_7 + \tfrac{4}{7} v_{11} = v_{10} \tag{31}$$

$$\tfrac{4}{5} v_9 + v_{10} = v_{12} \tag{32}$$

$$(1 - f)v_1 = \tfrac{3}{7} v_{11} \tag{33}$$

$$\tfrac{3}{7} v_{11} = v_{13} \tag{34}$$

$$v_{14} = \tfrac{1}{2} v_{15} \tag{35}$$

$$\tfrac{1}{3} v_7 = \tfrac{1}{2} v_{15} \tag{36}$$

$$v_{15} = v_{16} + v_{18} \tag{37}$$

$$v_{12} + v_{16} = v_{17} \tag{38}$$

$$v_{17} = \tfrac{4}{7} v_{11} + \tfrac{2}{3} v_3 \tag{39}$$

$$\tfrac{3}{4} v_{18} = \alpha_7 \psi_{car} \mu \tag{40}$$

$$\tfrac{1}{6} v_5 + \tfrac{1}{6} v_8 + \tfrac{1}{5} v_9 + \tfrac{1}{3} v_{13} + \tfrac{1}{4} v_{18} = \alpha_3 Q_{CO_2} \tag{41}$$

It is possible to assess each carbon flux from the above simultaneous and algebraic equations. Once the flux is known, it is possible, for example, to estimate Q_{O_2} values immediately; by the comparison between observation and calculation (Fig. 5), it may be possible to have the insight into what is most probably occurring in the metabolic activity and/or specific activity of the relevant enzyme *in vivo*.

It must be mentioned that Q_{O_2} assessment from carbon fluxes exemplified in Figure 5 is not based on any stoichiometric and overall picture of the metabolism, but solely on carbon balance, which is examined more accurately from outside the cell. The application of more detailed metabolic activities and/or mechanisms from continuously monitoring events outside the cell is considered to be another important potential use of an "on-line" computer in fermentation.

Nomenclature

C_{ij}	matrix coefficients
D	dilution rate (hr^{-1})
F	feed rate (liter/hr)
F_{air}	air flow rate (mol air/hr)
f	function; fraction of *n*-alkane converted to acetyl-CoA

$\mathbf{f}_i$	(vector) function for ith subsystem
g	function
$\mathbf{g}_i$	(vector) function for ith subsystem
I_{O_2}	total consumption rate of oxygen (mol O_2/hr)
J	objective function
K	proportionality constant (empirical (mol EtOH mol O_2/mol glucose mol CO_2)
K_s	saturation constant (mol substrate/liter)
m_O	maintenance coefficient based on oxygen (mol O_2/g cell hr)
n	number of subsystems; number of vessels; number of intervals
OUR	oxygen uptake rate (mol O_2/liter hr)
P	product concentration in culture broth (mol/liter)
p_{O_2}	partial pressure of oxygen (atm; %)
Q_{O_2}	specific rate of O_2 uptake (ml O_2/g cell hr; mol O_2/g cell hr; g O_2/g cell hr)
Q_{CO_2}	specific rate of CO_2 evolution (ml CO_2/g cell hr; mol CO_2/g cell hr; g CO_2/g cell hr)
Q_p	specific rate of product formation (mg product/g cell hr)
Q_{p_1}, Q_{p_2}	specific rate of citrate and isocitrate production, respectively (mg/g cell hr)
RQ	respiratory quotient (mol CO_2/mol O_2)
S	limiting substrate concentration in culture medium (mol substrate/liter)
S_R	limiting substrate concentration in fresh medium (mol substrate/liter)
t	time (hr)
U_i	constraint range of decision variable for ith subsystem
$\mathbf{u}_i$	decision vector for ith subsystem
V	working volume of fermentor; broth volume (liter)
v_i	carbon flux for ith pathway (mg carbon/g cell hr)
X	concentration of cell mass in culture medium (g/liter)
X_i	constraint range of input variable for ith subsystem
$\mathbf{X}_i$	input vector for ith subsystem
x	total cell mass in fermentor (g)
Y_{GO}	yield constant based on oxygen (g cell/mol O_2)
Y_i	constraint range of output variable for ith subsystem
$\mathbf{Y}_i$	output vector for ith subsystem
$Y_{P/S}$	yield of product (EtOH) formation based on limiting substrate (mol EtOH/mol glucose)
$Y_{X/S}$	yield of cell growth based on limiting substrate (g cell/mol substrate; g cell/g substrate)

Greek

$\alpha_1, \alpha_3, \alpha_4, \alpha_6, \alpha_7$	stoichiometric constants (α_1 = 0.849 g carbon/g n-alkane used [22]; α_3 = 0.536 g carbon/liter CO_2; α_4 = 0.375 g carbon/g citrate or isocitrate; α_6 = 0.465 g carbon/g protein [22]; α_7 = 0.444 g carbon/g carbohydrate [22])0
β	correction factor
μ	specific growth rate (hr^{-1})
μ_{max}	maximum value of specific growth rate (hr^{-1})
μ'	specific growth rate based on total cell mass (hr^{-1})
ν, ν'	specific rate of substrate uptake (mol substrate/g cell hr; mg/g cell hr; mol glucose/g cell hr)
ν_N	specific rate of ammonia uptake (mg/g cell hr)
ψ_{car}	carbohydrate fraction in cells

ψ_{pro}	protein fraction in cells
Subscripts	
f	time at the end of batch culture
i	ith subsystem; ith path; interval number
0	at $t = 0$
IN, OUT	inlet and outlet of fermentor, respectively
I_{O_2}	calculated from I_{O_2}
exp	experimental
Superscript	
*	(sub)optimized

References

[1] S. Yamashita, H. Hoshi, and T. Inagaki, in *Fermentation Advances,* D. Perlman, Ed. (Academic, New York, 1969), p. 441.
[2] L. K. Nyiri, in *Advances in Biochemical Engineering,* T. K. Ghose, A. Fiechter, and N. Blakebrough, Eds. (Springer-Verlag, Berlin, 1972), Vol. 2, p. 49.
[3] J. R. Swartz and C. L. Cooney, *Process Biochem., 13,* 3 (1978).
[4] S. Aiba and M. Okabe, Progress Report No. 93, Institute of Applied Microbiology, University of Tokyo, 1976.
[5] A. Constantinides, J. L. Spencer, and E. L. Gaden, Jr., *Biotechnol. Bioeng., 12,* 803, 1081 (1970).
[6] H. W. Blanch and P. L. Rogers, *Biotechnol. Bioeng., 13,* 843 (1971).
[7] V. M. Fishman and V. V. Biryukov, *Biotechnol. Bioeng. Symp., 4,* 647 (1974).
[8] T. Yamane, T. Kume, E. Sada, and T. Takamatsu, *J. Ferment. Technol., 55,* 587 (1977).
[9] H. W. Blanch and P. L. Rogers, *Biotechnol. Bioeng., 14,* 151 (1972).
[10] H. Ohno, E. Nakanishi, and T. Takamatsu, *Biotechnol. Bioeng., 18,* 847 (1976).
[11] M. Okabe and S. Aiba, *J. Ferment. Technol., 53,* 730 (1975).
[12] S. Aiba and M. Okabe, in *Advances in Biochemical Engineering,* T. K. Ghose, A. Fiechter, and N. Blakebrough, Eds. (Springer-Verlag, Berlin, 1977), Vol. 7, p. 111.
[13] H. T. Blachère, P. Peringer, and G. V. Corrieu, in *Workshop on Computer Applications in Fermentation Technology,* R. P. Jefferis III, Ed. (Verlag Chemie, Weinheim, 1977), p. 1.
[14] H. Y. Wang, C. L. Cooney, and D. I. C. Wang, *Biotechnol. Bioeng., 19,* 69 (1977).
[15] C. L. Cooney, H. Y. Wang, and D. I. C. Wang, *Biotechnol. Bioeng., 19,* 55 (1977).
[16] A. E. Humphrey, in *The 1st European Conference on Computer Process Control in Fermentation* (Institut National de la Recherche Agronomique, Dijon, France, 1973).
[17] D. W. Zabriskie and A. E. Humphrey, *AIChE J., 24,* 138 (1978).
[18] T. B. Young, III and H. R. Bungay, III, *Biotechnol. Bioeng., 15,* 377 (1973).
[19] I. Endo, K. Ohtaguchi, T. Nagamune, and I. Inoue, *Kagaku-Kogaku-Ronbun-shu (Jpn.), 3,* 543 (1977).
[20] L. K. Nyiri, G. M. Toth, and M. Charles, *Biotechnol. Bioeng., 17,* 1663 (1975).
[21] S. Aiba, S. Nagai, and Y. Nishizawa, *Biotechnol. Bioeng., 18,* 1001 (1976).
[22] S. Aiba and M. Matsuoka, *Eur. J. Appl. Microbiol., 5,* 247 (1978).

Application of Modern Control Theories to a Fermentation Process

T. TAKAMATSU, S. SHIOYA, M. SHIOTA, AND T. KITABATA

Department of Chemical Engineering, Kyoto University, Kyoto, 606 Japan

INTRODUCTION

It is a well-known fact that a microorganism acts as a reactor by internally control of very complex enzyme reactions. It may be very difficult to erect an exact mathematical model of a fermentor from the phenomenological viewpoint. However, some approximate mathematical models are usually necessary to design and control a fermentation plant. From this viewpoint, several mathematical models of cell growth have been reported, especially based on the transient state [1–3]. The authors have reported a mathematical model introducing "activity" as a dummy state variable to describe the behavior of the lag phase in batch culture and the transient phase in continuous culture [4]. In that paper, it was shown that "activity" strongly correlated with RNA content and that the rate equation of "activity" could be determined empirically. In this paper, first the rate equation of "activity" is experimentally discussed and skillfully explained from the viewpoint of the number of cells in a floc. Secondly, the effects of several control systems for a fermentor are discussed by using the mathematical model mentioned above; i.e. some single-variable control policies by a classical PI controller and multivariable control by a noninteracting controller are discussed by computer simulation. Also, an introduction to state observer based on the mathematical model is considered.

MODEL WITH A DUMMY STATE VARIABLE

The model we have proposed in the batch phase [4] is as follows:

$$\frac{dX}{dt} \triangleq \mu_X = a\mu_M X = a\frac{\mu_m S}{K_S + S} X \tag{1}$$

$$\frac{dS}{dt} \triangleq \mu_S = -\frac{1}{Y} a\mu_M X - \frac{V_d S}{K_d + S} X - A_k S \tag{2}$$

$$\frac{da}{dt} \triangleq \mu_a = c_1 \exp\left(\frac{-ba}{1-a}\right) \frac{S}{K_R + S} - a(a+d)\mu_m \left(\frac{S}{k_S + S} - \frac{S}{K_R + S}\right) \tag{3}$$

where X, S, and a are biomass concentration, substrate concentration, and the dummy state variable termed as "activity," respectively. μ_m, K_S, K_d, K_R, Y, b, c_1, and d are constant parameters. From eq. (1), the "activity" a will be understood as the correcting factor for the Monod-type equation. It has also been shown in previous work [4] that "activity" a had a linear relation to RNA content. In eq. (2), the rate of endogenous respiration was introduced. The evaporating term of ethanol, $A_k S$, is not neglected here because of experimental techniques, but because it is not essential [4]. Equation (3) was deduced from a concept of the active site, but mainly was estimated from empirical data. If the biochemical information of the inside of the cell could be summarized into the information of "activity" a, the quantitative representation based on "activity" a may be enough to use for designing and controlling a practical fermentor. But the question of what the physical meaning of eq. (3) is still remains. Therefore it may be desirable if a phenomenological explanation were possible for eq. (3). This can be done by introducing a number of cells in a floc as shown later. This model could explain the experimental results of batch, continuous, steady state, and transient operations.

EXPERIMENT

The microorganism used in this study is *Pichia mogi* IFO-062. The yeast assimilates ethanol as a sole carbon source. The composition of medium, experimental apparatus, and method of analysis have been reported in a previous paper [4]. In addition to the previous experiments, the number of cells in a floc and the number of flocs, are counted by using a microscope and a counting chamber. Cell concentration (dry cell g/liter), ethanol concentration [vol. %], average number of cells in a floc (given by definition stated later), and RNA content (in unit dry cell) [optical density (OD) based on 650 nm/mg dry cell] are measured during the batch fermentation.

It is often said that the length of the lag phase in a batch culture is strongly dependent on the inoculum phase. That is, when the culture is seeded with inoculum in the stationary phase, the length of lag phase becomes long. On the contrary, the lag phase is shorter when the inoculum is transferred during the log phase. After the cell growth rate became zero (i.e., reached the stationary phase), the cell was maintained for 53 hr at the same experimental conditions to the seeding except for having a substrate condition of the normal batch culture. Then, after addition of ethanol and other inorganic media, the batch culture was restarted. The results are shown in Figure 1. Of course, the reappearance of the result could be guaranteed experimentally. The common knowledge about the length of the lag phase in batch culture does not fit our experimental results, as seen in Figure 1.

It seems that one of the factors governing the lag phase is the coagulating state of the cell. The following facts could be recognized from microscopic observation. Cells always gather in a relatively large floc in the lag phase of a batch culture, e.g., a floc is composed of about 10 cells. As microorganisms grow, the floc be-

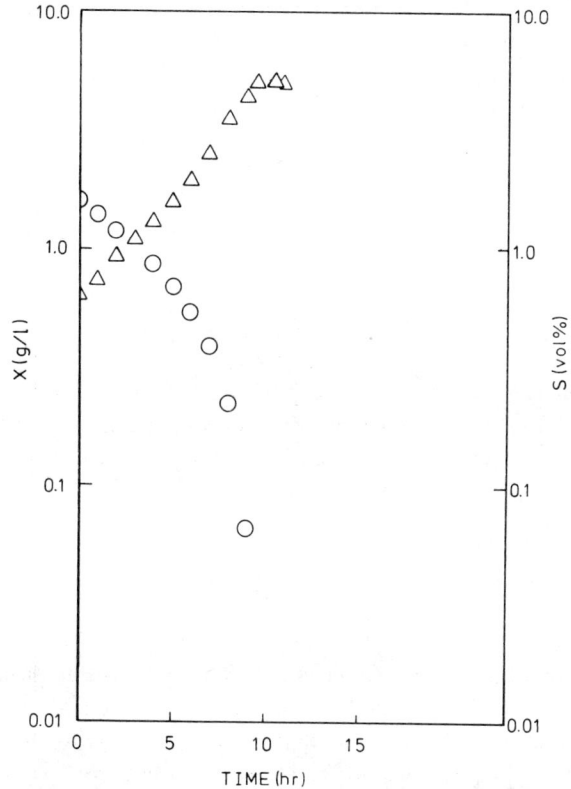

Fig. 1. Time course of restarted batch culture. Run 10: (△)X; (○) S.

comes smaller. In the exponential growth phase, the number of cells in a floc is about two or three. After consuming substrate ethanol in a batch operation, the number of cells in a floc does not change. The number of cells in a floc has a strong correlation with the growth rate. The growth rate in an early period of batch operation when ethanol is not so limited is plotted against the inverse value of the average number of cells in a floc defined here as $\bar{a}$ in Figure 2. This figure shows that there is a linear relationship between μ and $\bar{a}$. This relationship shows that the above explanation concerning the length of lag phase may be reasonable.

Another example of a batch culture that has no lag phase is shown in Figure 3.

CELL POPULATION BALANCE AND MATHEMATICAL MODEL

A cell mass balance is introduced based on the fact that yeasts grow by budding. The balance equations are deduced from the following assumptions: i) when the average number of bud scars per cell reaches l, the cell loses its ability

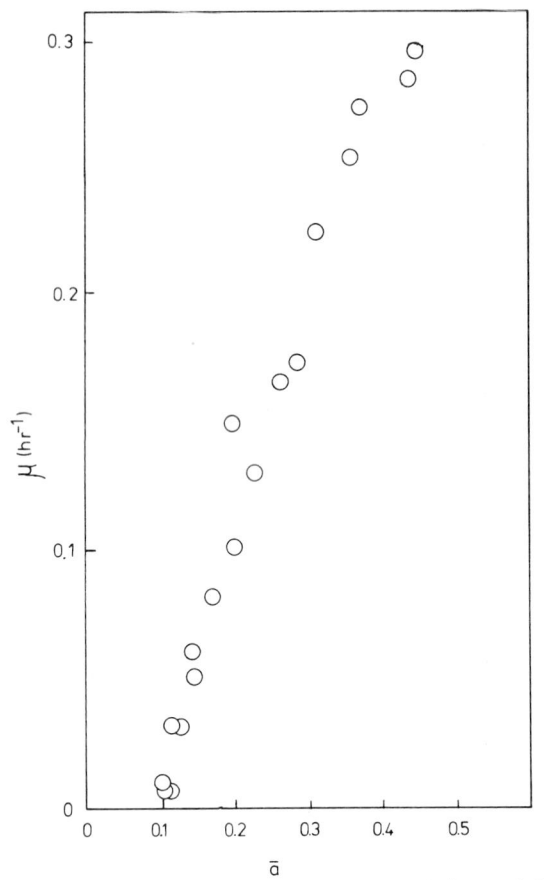

Fig. 2. Relationship between (average number of cells in a floc)$^{-1}$, $\bar{a}$, and the specific growth rate.

to bud; ii) the number of cells in a floc has the maximum value k; iii) the daughter cells never construct a floc by themselves. This means that a floc consists of nonseparated forms of mother and daughter cells.

The following notations are used: $f(n,m)$ is the number of flocs where the total number of bud scars is m and the total number of cells is n; ν_n^m is the rate of budding in the case of newly budded daughter cells being together with a mother cell or a floc; γ_n^m is the rate of budding in the case of newly budded daughter cell being separated from a mother cell or a floc; ξ_n^m is the death rate. Balance equations are as follows.

In the case of a single cell:

$$\frac{df(1,0)}{dt} = \sum_{n=1}^{k} \sum_{m=n-1}^{nl-1} \gamma_n^m f(n,m) - \gamma_1^0 f(1,0) - \nu_1^0 f(1,0) - \xi_1^0 f(1,0)$$

$$\frac{df(1,1)}{dt} = \gamma_1^0 f(1,0) - \nu_1^1 f(1,1) - \gamma_1^1 f(1,1) - \xi_1^1 f(1,1)$$

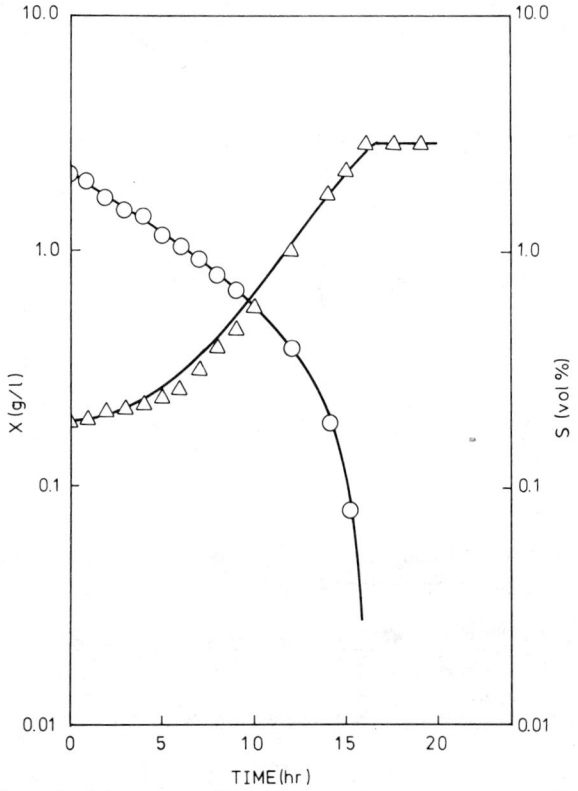

Fig. 3. Example of time course of batch culture with lag-phase. Run 8: (△) X; (○) S.

$$\vdots \tag{4}$$

$$\frac{df(1,l)}{dt} = \gamma_1^{l-1} f(1,l-1) - \xi_1^l f(1,l)$$

In the general case of $f(n,m)$,

$$\frac{df(n,m)}{dt} = \nu_{n-1}^{m-1} f(n-1, m-1) + \gamma_n^{m-1} f(n,m-1)$$

$$- \nu_n^m f(n,m) - \gamma_n^m f(n,m) - \xi_n^m f(n,m) \tag{5}$$

$$n = 2,3,\cdots,k, \quad m = 1,2,\cdots,kl, \; n-1 \leq m \leq nl$$

where,

$$\gamma_n^{n-2} = 0$$

$$\nu_n^i = 0, \; \gamma_n^i = 0 \quad \text{for} \quad i \geq nl$$

$$\nu_k^j = 0 \quad \text{for} \quad j \geq k-1$$

Introducing the following definition and assumptions,

$$\sum_{m=0}^{l} f(1,m) \triangleq x$$

$$\sum_{n=2}^{k}\sum_{m=n-1}^{nl} f(n,m) \triangleq y, \qquad \sum_{n=2}^{k}\sum_{m=n-1}^{nl} nf(n,m) \triangleq z \qquad (6)$$

$$\nu_n^m = \nu, \qquad \gamma_n^m = \gamma, \qquad \xi_n^m = \xi \quad \text{for all} \quad n,m \qquad (7)$$

x, y, and z represent the number of single cells, the number of flocs except for single cells, and the total number of cells in flocs without single cells, respectively. By using the assumption in eq. (7), natural approximation, and summation of eq. (5), the following simple equations can be obtained:

$$\dot{x} = (\gamma - \nu - \xi)x + \gamma y \qquad (8)$$

$$\dot{y} = \nu x - \xi y \qquad (9)$$

$$\dot{z} = 2\nu x + \nu y - \xi z \qquad (10)$$

Furthermore, the following linear relationship between the microorganism concentration and total number of cells,

$$X = \alpha(x + z) \qquad (11)$$

is determined experimentally where α is constant. From eq. (11), the specific growth rate μ in a batch operation is rewritten

$$\mu = \frac{1}{X}\frac{dX}{dt} = \frac{\dot{x}+\dot{z}}{x+z} = \left(\bar{a} - \frac{\xi}{\gamma+\nu}\right)(\gamma+\nu) \qquad (12)$$

where $\bar{a}$ is the inverse value of the average number in a floc defined as

$$\bar{a} \triangleq (x+y)/(x+z) \qquad (13)$$

which is already used in Figure 2. The dynamics of the state variable $\bar{a}$ can be written as follows by using eqs. (8)–(10):

$$\frac{d\bar{a}}{dt} = -(\gamma+\nu)(\bar{a})^2 + \gamma\bar{a} \qquad (14)$$

The equation of substrate ethanol concentration is assumed to be the same as eq. (2):

$$\frac{dS}{dt} = -\frac{1}{Y}\mu X - \frac{V_d S}{K_d + S}X - A_k S \qquad (15)$$

In these equations, γ, ν, and ξ are assumed to be Monod-type as follows:

$$\gamma + \nu = \beta\mu_m S/(K_s + S) \qquad (16)$$

$$\gamma = \mu_\gamma S/(K_\gamma + S) \qquad (17)$$

$$\xi = \mu_\xi S/(K_\xi + S) \qquad (18)$$

The set of eqs. (12) and (14)–(18) is an expression of the system from the viewpoint of a cell population balance. Based on eqs. (8)–(10), (12), (14), and (15), the parameter values in these equations are estimated from the experimental data. The result is shown in Table I where μ_m, K_S, Y, K_d, and V_d are

TABLE I
Estimated Parameters

present work	previous work		
μ_γ=0.34 μ_ξ=0.02	μ_m=0.2945	K_S=0.001	V_d=0.04
β =2.55 K_γ=0.11	Y=7.2	b=0.001	c_1=0.1353
K_ξ=0.001 A_k=0.115	K_R=0.25	K_d=1.15	d=0.385

taken as the same values as determined in a previous report [4], and the evaporation rate A_k is estimated by another experiment. In Table I, parameters in eqs. (1)–(3) are also listed. One example of the comparison between experimental data and the simulation using the estimated parameter is shown in Figures 3 and 4. From parameter estimation, the term $\xi/(\gamma + \nu)$ becomes constant and the following relationship is derived by comparing eqs. (1) and (12):

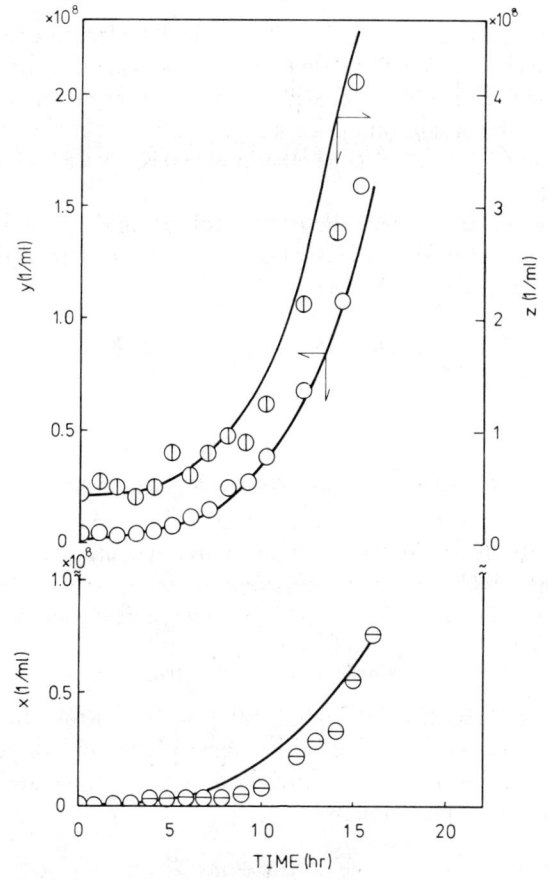

Fig. 4. Comparison of experimental data of x, y, z and calculated data based on population balance. Run 8: (⊖) x; (○) y; (Φ) z.

$$a = \beta(\bar{a} - D) \tag{19}$$

Equation (19) is confirmed by plotting the experimental data. It also means that the state variable $\bar{a}$ is linearly related to the RNA content. This is confirmed by experimental data of the RNA content. Equation (19) is substituted into eq. (14), then

$$\dot{a} = \underbrace{-\mu_M a^2 + \frac{\mu_\gamma S}{K_\gamma + S} a}_{\text{I}} \underbrace{- 2\mu_M D\beta a - \mu_M \beta^2 D^2 + \frac{\mu_\gamma \beta DS}{K_\gamma + S}}_{\text{II}} \tag{20}$$

is deduced where $\mu_M \triangleq \mu/a = \mu_m S/(K_S + S)$. From simulation of the right-hand side of eq. (20) by using various values of S and a, it is shown that the first term of eqn. (3) corresponds to the sum of first and second terms of eq. (20), (I) and that second term of eq. (3) corresponds to the residual terms of eq. (20), (II). Equation (3) can be explained completely by the consideration of the cell population balance.

SIMULATED STUDY OF CONTROL SYSTEM DESIGN

One of the purposes of building a mathematical model representing dynamic behavior is to use the model for design of a control system. As the first step, the following control problems are considered here. This first situation keeps the single-state variable X constant in a continuous stirred-tank reactor by manipulating the input flow rate, while the other problem is to keep both state variables X and S constant.

Considering the case where the output (observable) variable is dissolved oxygen (DO) concentration, the relationship of DO concentration (c) to the specific growth rate can be written

$$\frac{dc}{dt} \triangleq \mu_c = K_{La}(c_* - c) - \gamma_1 X \tag{21}$$

where

$$\gamma_1 = \alpha_1 \mu + \beta_1 \tag{22}$$

Equation (22) is confirmed by experiments as shown in Figure 5. The other cases in which the state variables, e.g., X and S can be observed are also studied to compare the results. Simulated results by a digital computer are based on the assumption that the basic rate process represented by eqs. (1)–(3) is correct. In other words, the real system is assumed to be represented by eqs. (1)–(3).

Single-Variable Control

The following are basic assumptions: i) the aim of control is to keep the state variable X constant in a continuous stirred-tank reactor, and ii) the manipulated variable is the feed flow rate which corresponds to the dilution rate u. The system equations can be written as follows:

$$\frac{dx}{dt} = \mu X - uX \tag{23}$$

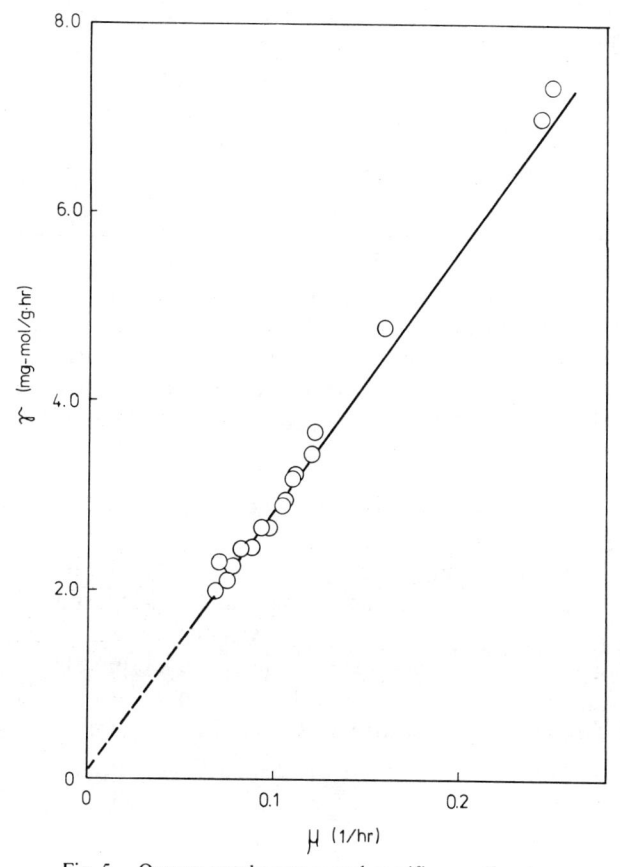

Fig. 5. Oxygen uptake rate γ and specific growth rate μ.

$$\frac{ds}{dt} = \mu_S + u(S_0 - S) \qquad (24)$$

$$\frac{da}{dt} = \mu_a \qquad (25)$$

$$\frac{dc}{dt} = \mu_c \qquad (26)$$

where μ, μ_S, μ_a, and μ_c are given by eqs. (1)–(3) and eq. (21). S_0 and u are input concentration of substrate and dilution rate, respectively.

Case A: State Variable X Is Observable

The performance of control for the following three models is considered by simulation.

i) IA model: This model is represented by eqs. (23)–(26) and is assumed to be a real system.

ii) **Monod model:** The values of parameters of this model can be determined from the dynamics only in the region of a steady state. The rate equation becomes

$$\mu = \mu_{mM}S/(k_m + S), \qquad \mu_S = -(1/Y_m)\mu X \qquad (27)$$

and the system is represented by eqs. (23), (24), and (26) where μ and μ_S are subjected to eq. (27). Parameter values of eq. (27) are given in Table II.

iii) **TF model:** This model is based on the classical transfer function method. The transfer function can be obtained from the response curve of X to an impulse change in the dilution rate u. The results are given in Table II.

The control performance for two kinds of disturbances, i.e., initial disturbance of x, which means that the initial state $[X(0)]$ deviates from the steady state $(\overline{X})$ and the step change of the substrate feed concentration are considered by digital simulation. There are two parameters K_p and T_I in a classical PI control system design as shown in the following equation:

$$\Delta u = K_p \left(\Delta X + \frac{1}{T_I} \int_0^t \Delta X\, d\tau \right) \qquad (28)$$

The values of these two parameters are obtained by computer simulation for each model so as to minimize the time increment associated with ΔX, where ΔX is defined as $\Delta X = X - \overline{X}$ and $\overline{X}$ is the steady-state microogranism concentration which is kept constant. A steady state as a numerical example is shown in Table II. Parameters K_p and T_I of each model are shown in Table III. Based on this TF model, parameters of the control system are used only for initial disturbance of X.

TABLE II
Parameters of Used Models and Given Steady State

Monod model	μ_{mM}=0.15 K_m=0.0476 Y_m=5.4163
TF model	$G(s) = x(s)/u(s) = -14.12(1+11.43s)/(1+8.203s+28.16s^2)$
DO rate eq.	K_{La}=83.53 α_1=879.4 β_1=3.3 c_*=7.6
Steady state	$\overline{X}$=4.5463 $\overline{S}$=0.1693 $\overline{a}$=0.4 $\overline{c}$=1.8153
for single-variable control	$\overline{u}$=0.1171 S_0=1.0
for multi-variable control	S_1^0=0.5 S_2^0=60.0 $\overline{u}_1$=0.116126 $\overline{u}_2$=0.000984

TABLE III
Control Parameters for Each Model

Model & disturbance	K_p [ℓ/ghr]	T_I [hr]	$J = \int_0^{24} \Delta x\, dt$
IA model — initial disturbance of x (Δx=0.1537)	-4.0	∞	8.4067×10^{-2}
IA model — step change of S_0 (ΔS_0=0.1)	-4.0	0.05	3.8013×10^{-4}
Monod model — initial disturbance	-4.0	∞	7.4813×10^{-4}
Monod model — step change of S_0	-4.0	0.06	1.4548×10^{-4}
TF model initial disturbance	-3.0	∞	5.6×10^{-2}

The result of control in the case of the initial disturbance of X is shown in Figure 6. There is no difference between using the IA model and the Monod model because the control parameters K_p and T_I have the same values. The integrated area of the time increment of ΔX for 24 hr is, ERROR_{IA} = $\text{ERROR}_{\text{Monod}}$ = 0.084 (g hr/liter) and ERROR_{TF} = 0.089 (g hr/liter). In the case of a step change of input substrate concentration [ΔS_0 = 0.1 (vol %)], ERROR_{IA} = 0.0003, $\text{ERROR}_{\text{Monod}}$ = 0.0004. It is shown that each control system gives good results if X is observable.

Case B: State Variable X Is Not Directly Observable

It is assumed that only the concentration of dissolved oxygen, c, can be used for control system design. Even if the concentration of dissolved oxygen, c, can be kept constant, the concentration of microorganism, X, cannot always be kept constant. The reason for this is clearly dependent on the fact that the solution that gives the steady state of eq. (26) ($dc/dt = 0$) cannot be obtained uniquely. However, observed data of c can be used for the estimation of X.

The theory of a deterministic state observer can be used if the system equation is fixed. The general theory is summarized in the Appendix, and this general case will be used in the later example. The theory is, of course, based on a linearized model around a steady state. From this theory, the estimated value of the deviation of microorganism concentration ($\Delta \hat{X}$) can be easily obtained. The same control parameters as in Table III are used for control simulation. The simulation results for the case of a step change of input substrate concentration [ΔS_0 = 0.02 (vol. %)] are shown in Figure 7. The result when the direct observation of X is assumed to be possible (there is a 1-hr time lag) is also shown as case 4. The control system based on the IA model with state observer (case 1) is shown to give fairly good results. The state observer does not act effectively in the initial time period; then the deviation in ΔX cannot be improved very much compared to the case without control.

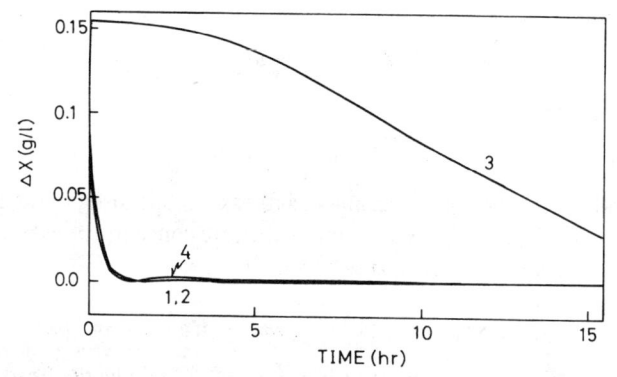

Fig. 6. Result of single-variable (X) control with initial disturbance [$\Delta X(0) = 0.1537$] in case where X is observable. 1, IA model; 2, Monod model; 3, no control; 4, TF model.

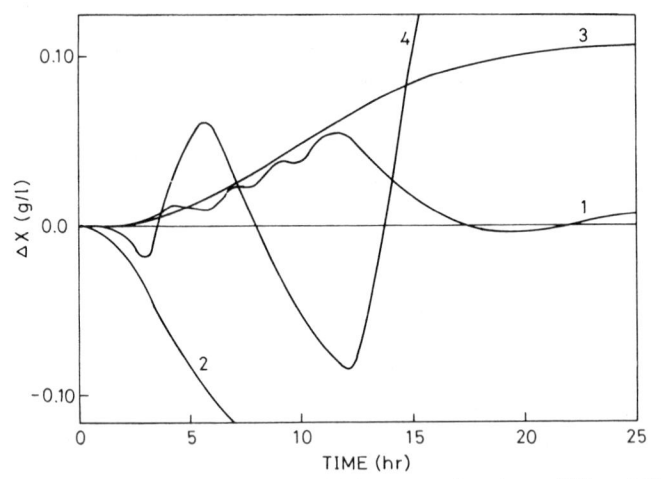

Fig. 7. Result of single-variable (X) control with input step disturbance [$\Delta S_0 = 0.02$ (vol. %)] in case where X is unobservable but the DO concentration is observable. 1, IA model; 2, Monod model; 3, no control; 4, case 4, time lag.

However, the control system based on the Monod model with the state observer is shown to be completely useless for this system. The result for case 4 is also not satisfactory. Therefore, only the control system based on the IA model gives good results when the observed variable is dissolved oxygen.

Multivariable Control

The following are basic assumptions: i) the control objective is to maintain the state variables X and S constant in a continuous stirred-tank reactor, ii) noninteracting control [5] is used, and iii) the manipulated variables are two feed flow rates with different substrate concentrations. From assumption (iii), the basic equation of a continuous stirred-tank system can be easily written by introducing the following bulk flow term to eqs. (23) and (24):
Bulk flow dilution rate

$$\triangleq u = u_1 + u_2 \qquad (29)$$

Input substrate mass per unit time

$$\triangleq u S_0 = u_1 S_1^0 + u_2 S_2^0 \qquad (30)$$

where u_1 and u_2 are the manipulating variables corresponding to the flow rates of different feeds. S_1^0 and S_2^0 are the input substrate concentrations of both feeds and $S_2^0 > S_1^0$. It is also assumed that S_1^0 has disturbance.

Case A: State Variables X and S Are Observable

The theoretical aspect for assumptions (i) and (ii) can be deduced as shown in the Appendix. This theory is also based on a linear model. The control scheme is given in Figure 8. The state estimation mechanism, which is already explained for single-variable control, is also shown in Figure 8. The following three control policies are tested by computer simulation: i) noninteracting control based on a linearized IA model represented by eqs. (23)–(25), (1)–(3), (29), and (30); ii) noninteracting control based on a linearized Monod model represented by eqs. (23), (24), (27), (29), and (30); iii) simple classical control. Input–output relations of a linearized IA model can be written as follows:

$$\begin{pmatrix} \Delta X(s) \\ \Delta S(s) \end{pmatrix} = \begin{pmatrix} G_{11}(s) & G_{12}(s) \\ G_{21}(s) & G_{22}(s) \end{pmatrix} \begin{pmatrix} \Delta u_1(s) \\ \Delta u_2(s) \end{pmatrix} \tag{31}$$

If $G_{11}(s) = G_{22}(s) = 0$ (case I) or $G_{12}(s) = G_{21}(s) = 0$ (case II) is assumed, so-called classical control policy can be applied. Control parameters of PI control of u_1 and u_2 are determined by the Ziegler–Nichols method.

The result of each control policy for a step change in the feed concentration of substrate (ΔS_1^0) is shown in Figure 9. There is no difference between the linearized Monod model and the linearized IA model. Both noninteracting control policies give successful results. However, the control by a linearized Monod model is effective only around a particular steady state, so if the operating steady state is changed to another point, the values for the parameters of the linearized Monod model must be determined for each new steady state. The classical control policy by the simple method employed here does not give good results, but if more suitable control parameters are given, better results will be obtained.

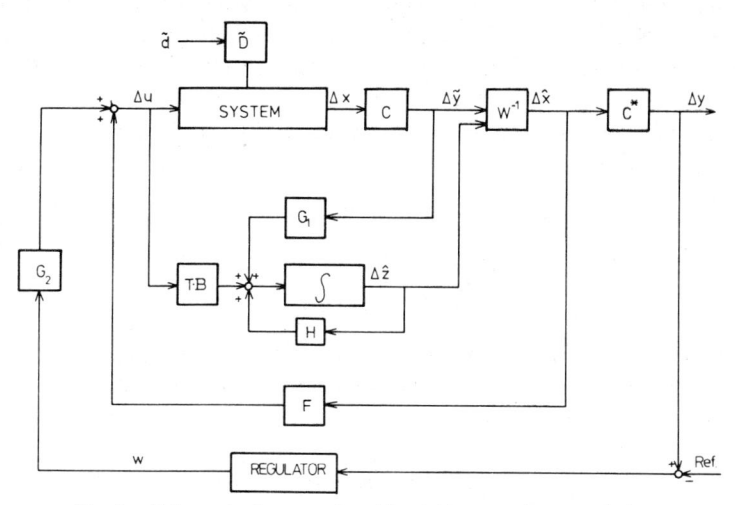

Fig. 8. Schematic diagram of multivariable control system design.

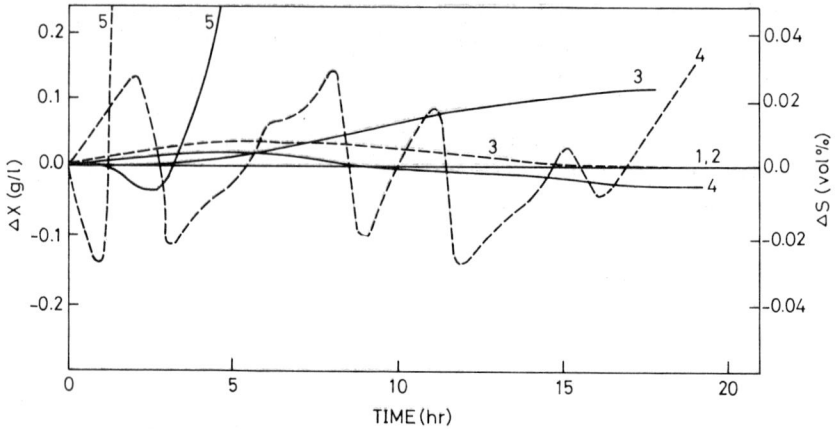

Fig. 9. Result of multivariable control with input step disturbance [$\Delta S_1^0 = 0.0202$ (vol. %)] in case where X and S are observable. (———) X; (- - -) S; 1, IA model; 2, Monod model; 3, no control; 4, case I; 5, case II.

Case B: Only the Dissolved Oxygen Concentration, c, Is Observable

The same observer, as explained in the previous section, is used. One example of observer properties based on a linearized IA model is shown in Figure 10. Noninteracting control policy is considered for a linearized IA model or a Monod model based on the control scheme shown in Figure 8. Some results of each

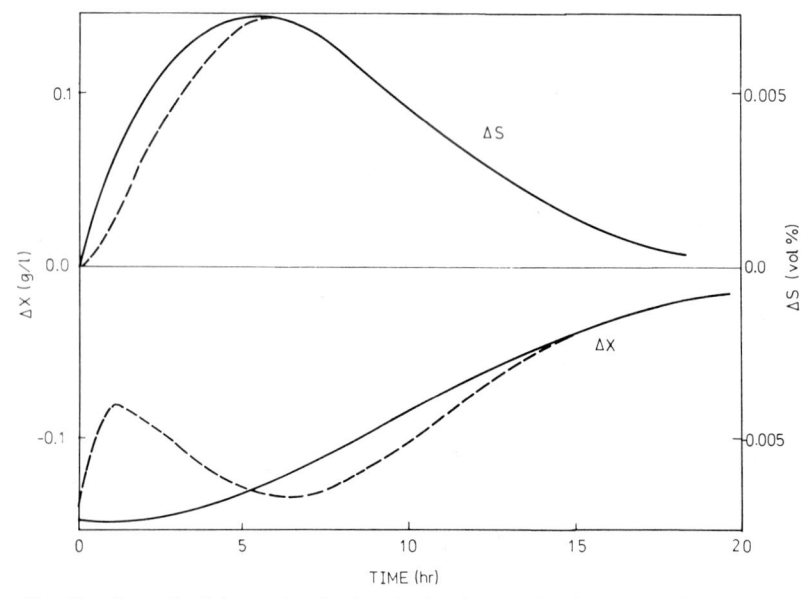

Fig. 10. Example of the results of estimation by observer. (———) Real; (- - -) by observer.

control policy are shown in Figures 11 and 12. The results for the initial disturbance are shown in Figure 11. Figure 12 shows the effect of a step change of input substrate concentration S_1^0. From these figures the IA model, which

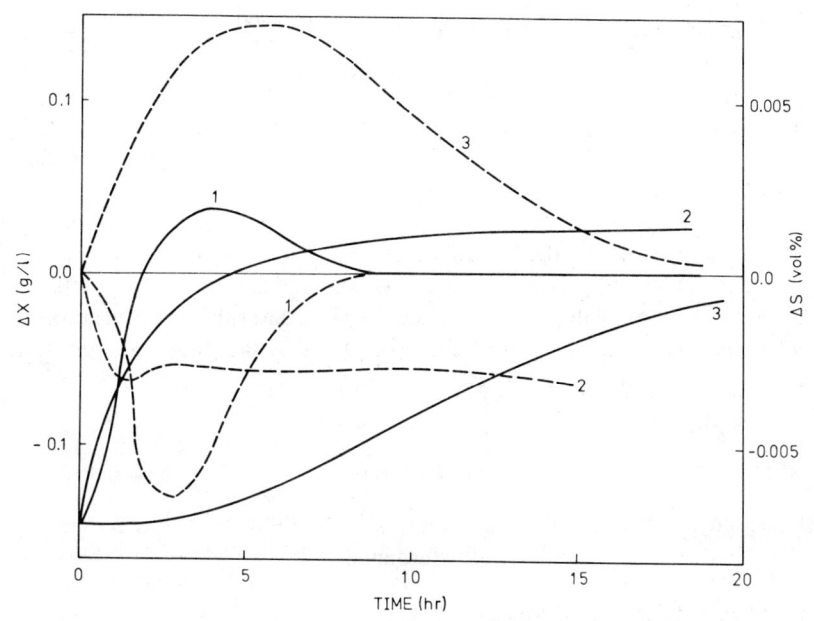

Fig. 11. Result of multi-variable control with initial disturbance [$\Delta X(0) = -0.1463$] by using state observer. (—) ΔX; (- - -) ΔS; 1, IA model; 2, Monod model; 3, no control.

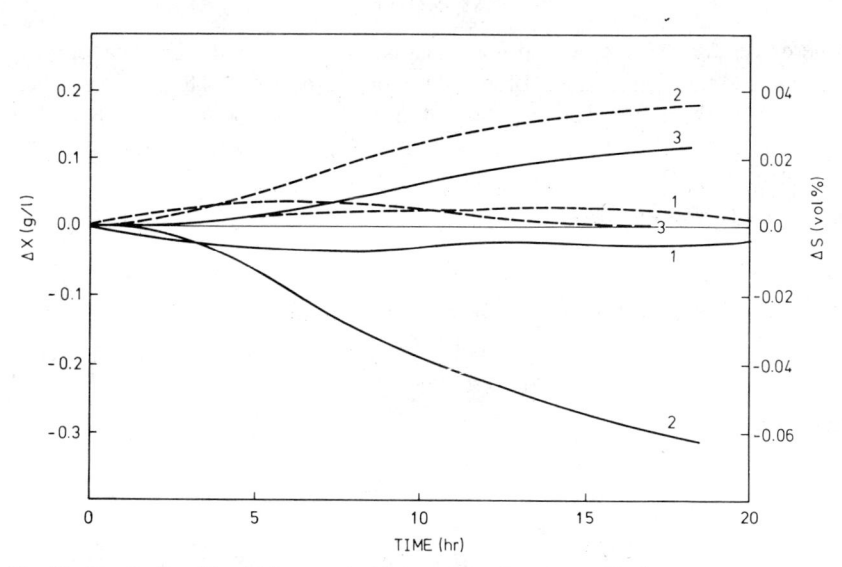

Fig. 12. Result of multivariable control with input step disturbance [$\Delta S_1^0 = 0.0202$ (vol. %)] by using state observer. Symbols are the same as in Figure 11.

represents the dynamic behavior well, is useful for control. The control policy based on a Monod model will lead to another steady state. The reason why the Monod model gives poor control depends on the effectiveness of the state observer. In any event, if we can only observe the DO concentration, the result of control will be good so long as the proper model is used.

DISCUSSION

If the substrate concentration is observable, the performance of control becomes better. The noninteracting control based on a linearized IA model is successful for the case of a setpoint change.

It is shown that a proper mathematical model gives good results for small disturbances even if only the DO concentration is observable. The theory based on a linear model will not give good results for large disturbances. In this case, feed-forward control will be helpful. If the biomass and substrate concentrations can be measured more accurately than the values by the state observer, better control will be obtained. From this viewpoint, indirect estimation methods of microorganism concentration [6] are noticeable.

APPENDIX

Deterministic Observer of Multi-Input, Multi-Output System with Constant Disturbance

The linearized system equations are

$$\Delta \dot{\tilde{x}} = \tilde{A}\Delta \tilde{x} + \tilde{B}\Delta u + \tilde{D}\tilde{d}$$
$$\Delta \tilde{y} = \tilde{C}\Delta \tilde{x} \quad (32)$$

where $\Delta \tilde{x}$, $\Delta \tilde{u}$, $\Delta \tilde{y}$ are the n-dimensional state vectors, r-dimensional control vectors, and the l-dimensional observed vectors, respectively. They are perturbed values from the nominal steady state. $\tilde{A}$, $\tilde{B}$, $\tilde{C}$, $\tilde{D}$ are $n \times n$, $n \times r$, $l \times n$, $n \times m$ constant matrices, respectively. Constant m-dimensional disturbance $\tilde{d}$ exists. The element of $\tilde{D}$ is 1 or 0. All vectors are column vectors. In this case, we treat $\tilde{d}$ as a state variable, then

$$\dot{\tilde{d}} = 0 \quad (33)$$

Then, the improved state equation becomes as follows:

$$\Delta \dot{x} = A\Delta X + B\Delta u$$
$$\Delta y = C\Delta x \quad (32a)$$

where

$$\Delta x^t = [\Delta \tilde{x}^t, \tilde{d}^t]$$

$$A = \begin{pmatrix} \tilde{A}, \tilde{D} \\ 0 \end{pmatrix}, \qquad B = \begin{pmatrix} \tilde{B} \\ 0 \end{pmatrix}, \qquad C = [\tilde{C}, 0]$$

$$(n + m) \times (n + m) \quad (n + m) \times r \quad l \times (n + m)$$

In eq. (32a), a deterministic state observer is constructed as follows. The estimated state is given by

$$\Delta \hat{x} = W^{-1} \begin{pmatrix} \Delta \hat{z} \\ \Delta \tilde{y} \end{pmatrix} \tag{34}$$

where $\Delta \hat{z}$ satisfies the following ordinary differential equation:

$$\Delta \dot{\hat{z}} = H \Delta \hat{z} + G_1 \Delta y + TB \Delta u \tag{35}$$

H is given arbitrarily so as to be negative definite and not to have the same eigenvalues as matrix A. Matrix T is the solution of the following equation:

$$TA - HT = G_1 C \tag{36}$$

where G_1 is given arbitrarily as W being a nonsingular matrix. Matrix W is given as follows:

$$W = \begin{pmatrix} T \\ C \end{pmatrix} \tag{37}$$

If we can find such a T, $\Delta \hat{z}$ approaches $T \Delta x$, and the estimated state $\Delta \hat{x}$ can be given by eq. (34).

Noninteracting Control

According to the above procedure, we can estimate all state variables; then we can rewrite the observation equation as follows:

$$\Delta y = C^* \Delta \tilde{x} \tag{32b}$$

where C^* is an $r \times n$ matrix and Δy is an r-dimensional output column vector to be controlled. The element of C^* is 1 or 0. The necessary and sufficient condition to make noninteracting system control $(\tilde{A}, \tilde{B}, C^*)$ is that B^* be nonsingular, where B^* is defined as

$$B^* = \begin{pmatrix} C_1 \tilde{A}^{e_1} \tilde{B} \\ C_2 \tilde{A}^{e_2} \tilde{B} \\ \vdots \\ C_l \tilde{A}^{e_l} \tilde{B} \end{pmatrix} \tag{38}$$

where C_i is the row vector of C^*; and e_i is given as

$$\begin{aligned} e_i &= \min \{j \mid C_i \tilde{A}^j \tilde{B} \neq 0, j = 0, 1, \cdots, n-1\} \\ &= n-1 \quad \text{if } C_i \tilde{A}^j \tilde{B} = 0 \quad \text{for all} \quad j \end{aligned} \tag{39}$$

In the case when the above condition is satisfied, control Δu can be given by

$$\Delta u = F \Delta \tilde{x} + G_2 \omega \tag{40}$$

$$F = -B^{*-1} A^*, \quad G_2 = B^{*-1} \tag{41}$$

$$A^* = \begin{pmatrix} C_1 \tilde{A}^{e_1+1} \\ \vdots \\ C_l \tilde{A}^{e_l+1} \end{pmatrix} \tag{42}$$

Output Δy_i is noninteracted except for the term of $\tilde{d}$ as follows:

$$\Delta y_i^{(e_i+1)} = \omega_i + (C^* \tilde{A}^{e_i} \tilde{D} \tilde{d})_{i\text{th element}} \tag{43}$$

But in our problems $e_i = 0$, then

$$\Delta \dot{y}_i = \omega_i + d_i \tag{44}$$

where

$$d_i = (C^* \tilde{D} \tilde{d})_{i\text{th element}}$$

The policy of how to operate ω_i is determined as follows as a regulator problem:

$$\Delta \hat{y}_i \triangleq \omega_i + d_i \tag{45}$$

$$\dot{\omega}_i \triangleq \nu$$

then

$$\begin{pmatrix} \Delta \dot{y}_i \\ \Delta \dot{\hat{y}}_i \end{pmatrix} = \begin{pmatrix} 0 & 1 \\ 0 & 0 \end{pmatrix} \cdot \begin{pmatrix} \Delta y_i \\ \Delta \hat{y}_i \end{pmatrix} + \begin{pmatrix} 0 \\ 1 \end{pmatrix} \cdot \nu \tag{46}$$

If the performance index (J) of the regulator is given as

$$J = \int_0^\infty \left\{ [\Delta y_i, \Delta \hat{y}_i] \begin{pmatrix} \rho^2 & 0 \\ 0 & 0 \end{pmatrix} \begin{pmatrix} \Delta y_i \\ \Delta \hat{y}_i \end{pmatrix} + \nu^2 \right\} dt \tag{47}$$

where ρ is a weighting factor, then the problem of finding a ν that minimizes J is solved as follows:

$$\nu = -[0,1] \hat{K} \begin{pmatrix} \Delta y_i \\ \Delta \hat{y}_i \end{pmatrix} \tag{48}$$

where the symmetric positive-definite matrix $\hat{K}$ is the solution of the following Ricatti's equation:

$$\hat{K} \begin{pmatrix} 0 & 0 \\ 0 & 1 \end{pmatrix} \hat{K} - \hat{K} \begin{pmatrix} 0 & 1 \\ 0 & 0 \end{pmatrix} - \begin{pmatrix} 0 & 0 \\ 1 & 0 \end{pmatrix} \hat{K} - \begin{pmatrix} 1 & 0 \\ 0 & 0 \end{pmatrix} = 0 \tag{49}$$

Solving eq. (49) and using eq. (45), we get

$$\omega_i = -(2\rho)^{1/2} \Delta y_i - \rho \int_0^t \Delta y_i \, d\tau \tag{50}$$

Nomenclature

a	activity in eq. (1) (dimensionless)
$\bar{a}$	(average number of cells in a floc)$^{-1}$ [defined by eq. (13)]
A_k	rate constant of ethanol evaporation (hr^{-1})
b	constant in eq. (3) (dimensionless)
c_1	constant in eq. (3) (dimensionless)
c^*	equilibrium DO concentration at atmosphere (ppm)
c	DO concentration (ppm)

d	constant in eq. (3) (dimensionless)
D	constant in eq. (19) $[=\xi/(\gamma+\nu)]$
$f(n,m)$	number of flocs in which total number of bud scars is m and total number of cells is n
$K_i (i = S,d,R,\nu,\gamma,\xi)$	Monod constant in each equation (vol. %)
K_p	parameter in eq. (28)
k_{LA}	overall mass-transfer coefficient (hr^{-1})
s	Laplace operator
S	ethanol concentration (vol. %)
ΔS	deviation of S from steady state (vol. %)
S_0	input substrate concentration (vol %)
S_1^0, S_2^0	input substrate concentration defined in eq. (29) (vol %)
T_I	parameter in eq. (28)
u	dilution rate (hr^{-1})
Δu	deviation of u from steady-state operation (hr^{-1})
u_1, u_2	defined by eq. (29) (hr^{-1})
V_d	maximum endogenous rate in eq. (2) [vol % (hr g/liter)$^{-1}$]
x	number of single cells defined by eq. (6)
ΔX	deviation of X from steady state (g/liter)
X	microorganism concentration (g/liter)
y	number of flocs defined by eq. (6)
Y	yield constant in eq. (2) (g/liter/vol %)
z	number of cells in flocs defined by eq. (6)
α	constant in eq. (11) (g/number)
α_1	constant in eq. (22) [ppm/(g/liter)]
β	constant in eq. (16) (dimensionless)
β_1	constant in eq. (22) [ppm/(g/liter)/hr]
$\mu_i (i = m,\gamma,\xi)$	maximum rate in each equation (hr^{-1})
μ	specific growth rate (hr^{-1})
μ_S	substrate consumption rate defined in eq. (2)
μ_a	changing rate "activity" a defined in eq. (3)
μ_M	defined in eq. (1)
μ_c	defined in eq. (21)
ν_n^m	rate of budding in eq. (4) or eq. (5) (hr^{-1})
γ_n^m	rate of budding in eq. (4) or eq. (5) (hr^{-1})
γ_1	specific oxygen uptake rate defined by eq. (22) (ppm/g/liter/hr)
ξ_n^m	rate of death in eq. (4) or eq. (5) (hr^{-1})

References

[1] P. Shu, *J. Biochem. Microbiol. Technol. Eng.*, 3, 95 (1961).
[2] D. Ramkrishna, A. G. Frederickson, and H. M. Tsuchiya, *J. Gen. Appl. Microbiol.*, 12, 4 (1966).
[3] S. Nagai, Y. Nishizawa, I. Endo, and S. Aiba, *J. Gen. Appl. Microbiol.*, 14, 121 (1968).
[4] T. Takamatsu, S. Shioya, T. Maenaka, and M. Shiota, in *Proceedings of 2nd Pacific Chemical Engineering Conference* (PACHEC, Tokyo, 1977), p. 570.
[5] P. L. Falb and W. A. Wolovich, *Joint Automatic Control Conference*, 791 (1967).
[6] D. W. Zabriskie and A. E. Humphrey, *AIChE J.*, 24, 1 (1978).

Optimization of Erythromycin Biosynthesis by Controlling pH and Temperature: Theoretical Aspects and Practical Application

A. CHERUY

Laboratoire d'Automatique de Grenoble, Institut National Polytechnique de Grenoble, Grenoble, France

A. DURAND

Station de Génie Microbiologique—Institut de Recherche Agronomique, Dijon, France

INTRODUCTION

The progress realized during the past years in the field of sensor as well as in monitoring and automatic analysis make the application of modeling and optimization techniques to fermentation processes possible and valid. Automatic control of the fermentation units with a computer is now possible. Many scientists began their studies in this way and many works in bioengineering have been devoted to this subject in various application fields: yeasts, protein production, amino acid synthesis, etc. [1–18].

In antibiotic production (our field of study), Constantinides et al. [1] showed how to determine a temperature profile in order to maximize the production of penicillin in batch fermentation. In 1976, Giona et al. [14] reported a detailed kinetic analysis on industrial fermentors in order to control the penicillin production in a fed-batch fermentation, but they did not propose a control method.

The purpose of this work is to optimize the erythromycin production in a batch fermentation by controlling the pH and temperature. In this paper, we will present the theoretical study and also the practical applications. First, we will show how to determine the pH and temperature optimal profiles and what are the results expected by simulation, in the theoretical section. The method used here is similar to the one used by Rai and Constantinides [2] for the optimization of the gluconic acid fermentation by the control of both temperature and pH. Then, the experimental part will include the application of the optimal profiles to a fermentation pilot unit and the comparison of the results with those obtained by simulation. The last section will be devoted to the discussion of the results leading to an analysis of the validity of the optimization method and to proposing a practical solution.

MATERIALS AND METHODS

Materials

The antibiotic (erythromycin) is synthesized by *Streptomyces erytheus* grown in a complex industrial medium. The culture is of batch type, and its duration is about 300 hr in standard conditions where the fermentor temperature is kept at 32°C and the pH is adjusted to 7.0 before inoculation and then not controlled during the culture. A calcium carbonate buffer is also used to prevent excessive acidification that is prejudicial to the normal evolution of the process.

All experiments were carried out at the Station de Génie Microbiologique, I.N.R.A. Dijon on "Biolafitte" fermentors of 10-liter capacity. These reactors of a classical design are connected to control racks allowing i) the regulation of the temperature by a PID controller ensuring an accuracy of 0.2°C on the 0–60°C range.

ii) The regulation of the pH by automatic addition of an acidic or basic solution ($2N$ sulfuric acid, $2N$ sodium hydroxide). This PID regulation ensures an accuracy of 0.1 pH unit for the whole range. For the experiments with pH regulation, calcium carbonate was not used.

iii) The automatic monitoring of the dissolved oxygen (DO) concentration in the culture medium. Preliminary experiments showed that this parameter does not influence the process if its value is not lower than 30% (100% corresponds to the saturation of the medium in oxygen). Consequently, the DO concentration was kept above 30% by controlling the agitation speed, the air flow being fixed at 1 v/v/m with a pressure of 0.2 bar. The evolution of the DO concentration was measured with a galvanic probe and continuously recorded.

Each rack is connected to a multichannel recorder allowing an on-line recording of the parameters: temperature, pH, and DO concentration.

Methods

Each series of experiments was carried out in four parallel fermentors. In order to eliminate the possible variations due to the inoculum, each fermentor is inoculated from the same Erlenmeyer flask that is always prepared under the same conditions. For each experiment, biomass and erythromycin concentrations were measured on sterile samples taken every 5 hr.

Biomass Assay

The traditional methods for measuring the biomass [optical density (OD), dry weight (DW), etc.] could not be used in this case for two main reasons: the nature of the microorganism (mycelial pellets of various sizes), and the nature of the culture medium (natural complex medium including nonmycelial solids). To confront these difficulties, an original analysis method was developed which is an indirect evaluation of the biomass based on a differential assay of protein before and after mycelial lysis (unpublished method).

Erythromycin Concentration Measurement

During biosynthesis, several isomeric compounds of erythromycin (A, B, and C erythromycins) and a degradation product are synthesized. Only A erythromycin is active as an antibiotic and is economically interesting. The method used to specifically titrate the A erythromycin is a spectrophotometric one in the uv range. It is based on the formation, in basic medium, of a *"chromophore"* whose absorption at 236 nm is determined in comparison with a blank made under the same conditions but after acidic degradation. This acidic degradation transforms the erythromycin into a hydroerythromycin which does not absorb at the wavelength used. This allows the determination of the parasitic absorption due to compounds of the culture medium. The accuracy of the titration is around 10%.

THEORETICAL STUDY

Determination of a Temperature and pH Optimal Control

In order to determine the optimal control for a process, it is necessary i) to define the objective function to be maximized by this control, ii) to elaborate a mathematical model accounting for the process evolution within experimental range, and iii) to choose and set an optimization method adapted to the problem.

Objective Function

In our case, the definition of the criterion was essentially guided by the process exploitation and the production conditions: the object is to determine a pH and temperature control that maximizes the amount of antibiotic produced at the end of the fermentation. The duration is always 300 hr, the objective function that needs to be maximized is the erythromycin concentration after a fermentation period of 300 hr.

Elaboration of a Mathematical Model

In the literature, there are numerous models, but they are of two main types: semiempirical models (e.g., Monod's model for the bacterial growth), and models that include important biological knowledge (e.g., the model proposed by Peringer [15] for yeast). The elaboration of this second type of model requires a thorough biochemical knowledge of the microorganism metabolism, while the first type consists of fitting the parameters of known laws (e.g., Monod's growth law) in order to account for general measurements such as biomass. In some cases, these parameters can have biological significance.

Taking into account, on the one hand, the lack of knowledge of the erythromycin biosynthesis mechanism, and, on the other hand, the nature and the limited number of our measurements (only biomass and erythromycin concentration are measured), the only possible approach was a semiempirical model.

The object was then to choose the growth and biosynthesis laws best adapted to our problem.

Therefore, to characterize the biomass evolution and the erythromycin production, several growth and biosynthesis laws among those classically used in the literature were considered and one based on the criterion of good fit and complexity was selected. So, for the biomass, the following laws were studied comparatively:

(1) Logarithmic law,

$$\frac{dx_1}{dt} = \mu x_1 \left(1 - \frac{x_1}{x_{1f}}\right) \tag{1}$$

(2) Tessier's law,

$$\frac{dx_1}{dt} = \mu x_1 \{1 - \exp[a_T(x_1 - x_{1f})]\} \tag{2}$$

(3) Monod's law,

$$\frac{dx_1}{dt} = \mu x_1 \left(1 - \frac{1}{1 - a_M(x_1 - x_{1f})}\right) \tag{3}$$

(4) Polynomial law,

$$\frac{dx_1}{dt} = \mu x_1 \left[1 - \left(\frac{x_1}{x_{1f}}\right)^{a_P}\right] \tag{4}$$

where x_1 is the biomass concentration, μ is the growth rate, and x_{1f} is the final biomass concentration; a_T and a_M also have a biochemical meaning,

$$a_T = a_M = S_0/K_S(x_{10} - x_{1f})$$

where S_0 is the initial substrate concentration, x_{10} is the initial biomass concentration, x_{1f} is the final biomass concentration, and K_S is the constant relative to the substrate assimilation. a_P, in the polynomial law, has no biochemical meaning.

In the same way, to characterize the erythromycin biosynthesis, the following laws were compared:

$$\frac{dx_2}{dt} = b_1 x_1 - b_2 x_2 \tag{5}$$

$$\frac{dx_2}{dt} = b_1 x_1 - (b_2 x_2)^{b_3} \tag{6}$$

where x_2 is the antibiotic concentration and b_1, b_2, b_3 are parameters without any biochemical meaning. In these laws (1)–(6), the control variables—temperature and pH—do not appear explicitly, so each parameter depended on the temperature and pH. Indeed, it is known that the variations as a function of temperature or pH of the parameters, such as the growth rate, are Gaussian-like. Therefore they can be approximated by quadratic functions. Consequently, the global model was established in two steps: (1) estimation of the parameters of the growth and biosynthesis laws from experiments carried out with pH and temperature kept constant. (2) Determination of second-order regression functions for each parameter of the laws from the results of the preceding estimation.

In the first estimation, the Marquardt algorithm [21] was used and, in the second step, a classical linear regression method was followed. Statistical tests were done to check the estimation validity. The Student–Fisher test was applied to check whether each regression term was significant, taking into account the dispersion of the measurements. If this test is not satisfied, the concerned term was suppressed and the regression became simplified. If contrary, the term was kept and the accuracy of its estimation and its correlation with the other terms of the regression were calculated.

This identification was carried out from 15 experiments with constant pH and temperature in the ranges 6.4, 7.2, and 28–40°C, respectively. The identification program was written in FORTRAN and run on the 360 IBM computer of the Grenoble University Computing Center [19]. It gave, for each experiment, the estimated value of the parameter laws and their variance analysis (see Table I) and for each parameter the regression analysis as shown in Table II for the b_2 parameter.

For the choice of model, the following criteria were considered:

i) An average error criterion of the approximation by the law, i.e., the average, for all the experiments of the quadratic sum of the errors between measured values and those calculated with the law without considering the parameters as functions of pH and temperature.

ii) A general average error criterion of the approximation by the global model, i.e., the average of the quadratic sum of the errors between measured values and those calculated with the global model including the variations of the law parameters as quadratic functions of pH and temperature.

iii) A model complexity criterion given by the number of the parameters of the global model.

iv) A model validity criterion given by the number of experiments that did not satisfy the F test. The F test allows us to verify whether the error due to the approximation of the measurements by the model is negligible in front of the precision of the measurements.

v) Another model validity criterion given the number of experiments that satisfied the Z test. The Z test indicates whether the errors between the measurements and the simulated values change their sign in a random way.

For the biomass growth laws, the obtained results are shown in Table III.

From these results, the polynomial law appears, in our case, to be the "best" growth model. Thus, it was retained and used to compare the two biosynthesis laws. This comparison was made according to the same criteria as previously. The differences between the two laws was not significant, so, the first one was retained for simplicity.

Therefore, the retained global model was the following:

TABLE I
Identification Results for Three Experiments[a]

Test No.	Temperature (°C)	Acidity (pH)		Criterion	Parameter Values	F-test	Z-test
1	32.0	6.4	Biomass	0.060	$\mu = 0.0707$ $x_f = 15.0558$ $a = 1.4752$	$F_{adm} = 2.240$ $F_{calc} = 1.888$	$Z_{adm} = 1.86$ $Z_{calc} = 2.79$
			Erythromycin	0.075	$b_1 = 0.00195$ $b_2 = 0.09428$	$F_{adm} = 3.122$ $F_{calc} = 7.635$	$Z_{adm} = 1.96$ $Z_{calc} = 2.19$
2	32.0	6.4	Biomass	0.016	$\mu = 0.0760$ $x_f = 14.9152$ $a = 1.3039$	$F_{adm} = 2.240$ $F_{calc} = 1.888$	$Z_{adm} = 1.96$ $Z_{calc} = 0.28$
			Erythromycin	0.133	$b_1 = 0.00198$ $b_2 = 0.08002$	$F_{adm} = 3.122$ $F_{calc} = 7.635$	$Z_{adm} = 1.96$ $Z_{calc} = 3.05$
3	32.0	6.6	Biomass	0.096	$\mu = 0.0744$ $x_f = 15.0201$ $a = 1.0952$	$F_{adm} = 1.982$ $F_{calc} = 10.698$	$Z_{adm} = 1.96$ $Z_{calc} = 3.24$
			Erythromycin	0.137	$b_1 = 0.00227$ $b_2 = 0.08615$	$F_{adm} = 2.139$ $F_{calc} = 1.166$	$Z_{adm} = 1.96$ $Z_{calc} = 2.81$

[a] In this example, for the biomass, the polynomial law is considered.

TABLE II
Regression Analysis for the b_2 Parameter[a] of the Model

Regression coefficients	T-test value
$b_{20} = 0.1356 + 0.0045$	66.43
$b_{21} = 0.2037 + 0.0187$	23.97
$b_{22} = 0.1712 + 0.0179$	21.01
$b_{23} = 0.0250 + 0.0054$	10.20

Correlation Matrix[b]

$$C = \begin{bmatrix} 1.00 & -0.79 & 0.63 & -0.33 \\ -0.79 & 1.00 & -0.94 & -0.09 \\ 0.63 & -0.94 & 1.00 & 0.11 \\ -0.33 & -0.09 & 0.11 & 1.00 \end{bmatrix}$$

Significance test of the correlation for 99% confidence[c]

$$S\text{-test} = \begin{bmatrix} \text{yes} & \text{yes} & - & - \\ \text{yes} & \text{yes} & \text{yes} & - \\ - & \text{yes} & \text{yes} & - \\ - & - & - & \text{yes} \end{bmatrix}$$

[a] $b_2 = b_{20} + b_{21} \cdot T + O \cdot \text{pH} + b_{22} \cdot T^2 + b_{23} \cdot \text{pH}^2 + O \cdot T \cdot \text{pH}$. Degrees of liberty = 11, value of T for the 95% confidence interval = 2.201.

[b] A c_{ij} term is the correlation coefficient between the b_{2i} and b_{2j} parameters.

[c] Yes in the i,j position indicates that the correlation coefficient between the b_{2i} and b_{2j} parameters is significant with a 99% confidence.

TABLE III
Comparison of Several Growth Models

Law	Criterion (law)	Criterion (global model)	No. parameters	No. of not satisfied F test	No. of satisfied Z test
Logarithmic	0.0551	0.832	10	2	8
Teissier	0.0476	4.247	12	1	10
Monod	0.0485	36.142	12	1	8
Polynomial	0.0442	0.816	12	1	8

$$\frac{dx_1}{dt} = \mu(T,\text{pH}) \, x_1 \left[1 - \left(\frac{x_1{}^a}{x_{1f}(T,\text{pH})} \right)^{a(T,\text{pH})} \right]$$

$$\frac{dx_2}{dt} = b_1(T,\text{pH})x_1 - b_2(T,\text{pH})x_2$$

$$\begin{bmatrix}\mu\\x_{1f}\\b_1\\b_2\\a\end{bmatrix} = \begin{bmatrix}0.075 & 0 & -0.065 & 0 & 0.05 & 0.0519\\9.26 & 24.85 & -4.19 & 21.0 & 3.5 & 0\\0.008 & -0.015 & 0 & -0.002 & -0.008 & 0.03\\0.013 & -0.025 & 0 & -0.004 & -0.01 & 0.054\\2.0 & 0 & 0 & 0 & 0.0 & 0\end{bmatrix}\begin{bmatrix}1\\T\\pH\\T^2\\pH^2\end{bmatrix} \quad (7)$$

The numerical values of the model parameters are calculated by using 0–1 range for the 28–40°C temperature ranges and 6.4–7.2 pH ranges.

As an example, Figure 1 shows the experimental points and the simulated curves of biomass and erythromycin concentrations (the dotted lines refer to the curves simulated from the global model, and the solid lines to those simulated from the law alone). Taking into account the dispersion of the measurement points, we can note that the approximation of the growth and biosynthesis curves by the model are satisfied.

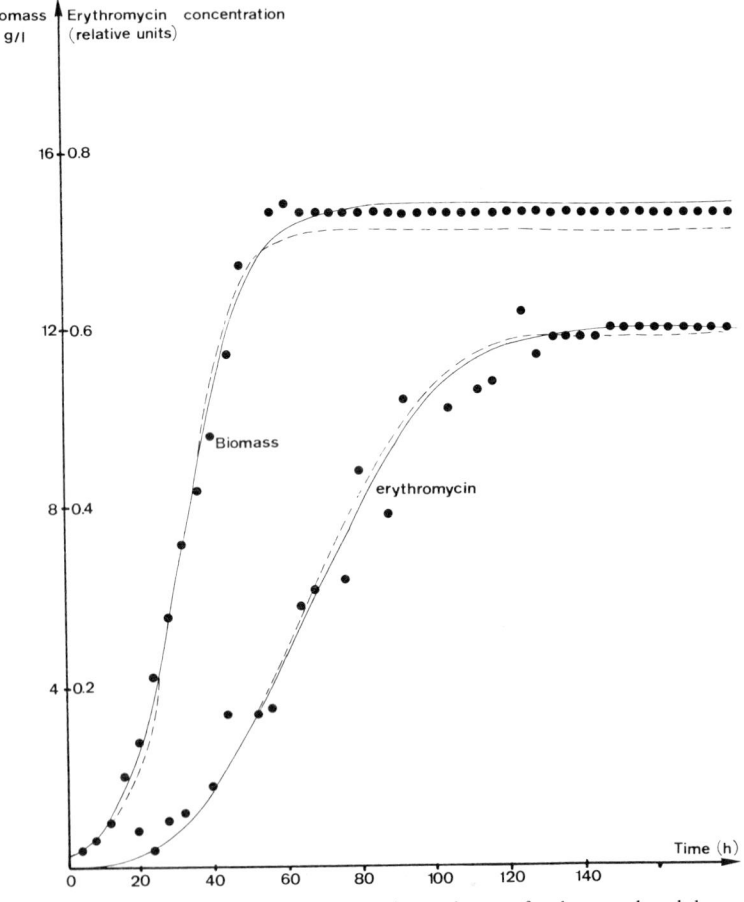

Fig. 1. Comparison between simulated and experimental curves for the growth and the erythromycin biosynthesis.

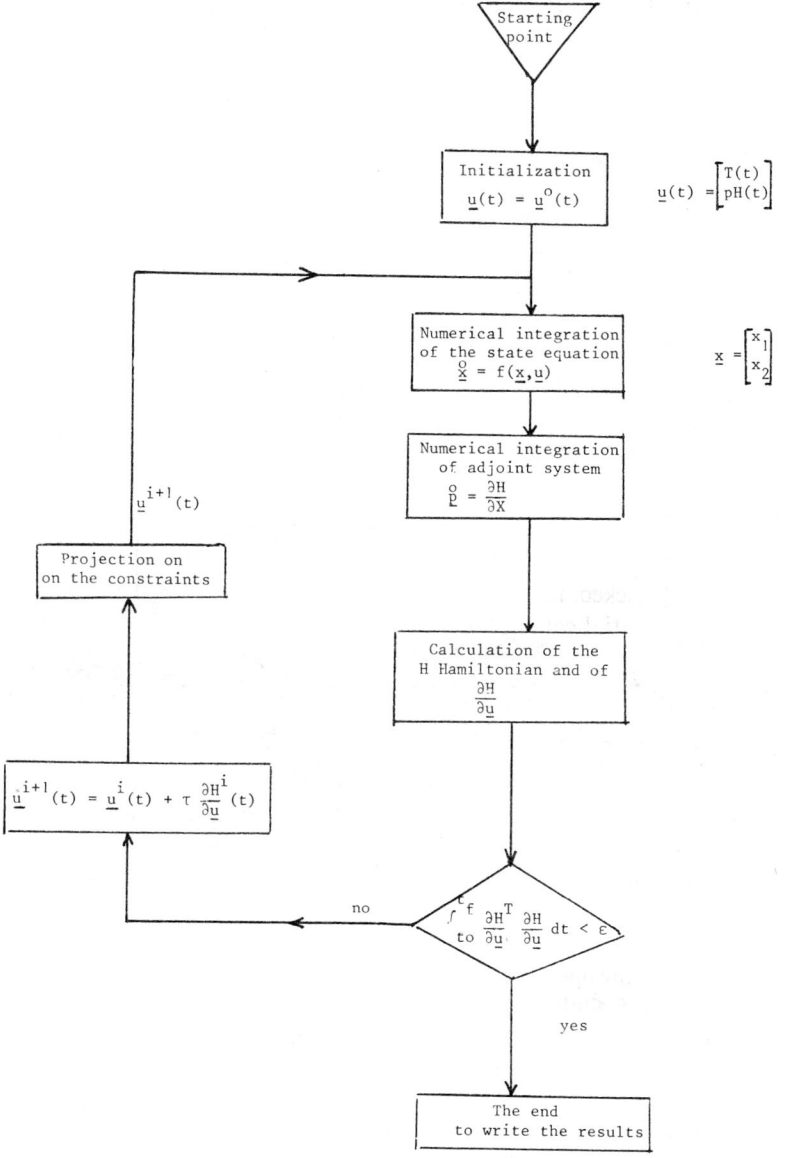

Fig. 2. Flow diagram for the research of the optimal profiles.

Optimization

To determine the profiles of temperature and pH [$T = f(\text{time})$, $\text{pH} = g(\text{time})$] which lead to a maximum erythromycin production after a fermentation period of 300 hr is a problem of dynamic optimization with constraints on the control variables ($28°C \leq T \leq 40°C$, $6.4 \leq \text{pH} \leq 7.2$), the system being defined by the

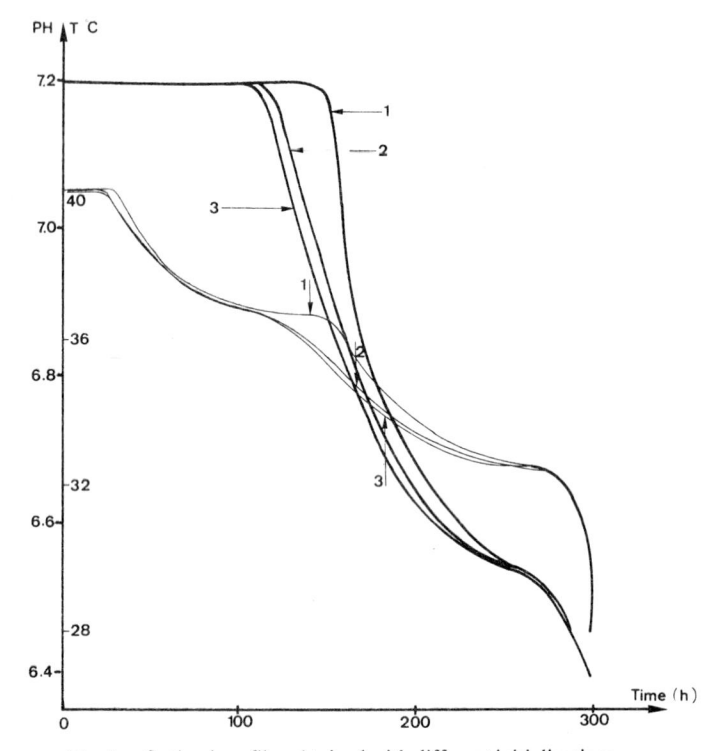

Fig. 3. Optimal profiles obtained with different initializations.

second-order nonlinear differential equations (7) and the criterion to maximize being $J = x_2(300)$. To solve this problem, we used a method based upon the Pontryagin's maximum principle with an optimization algorithm using a gradient method with projection in the range of the admissible control variables [20]. The flow diagram is shown in Figure 2.

As it was not possible to demonstrate the uniqueness of the optimum, the algorithm was checked. It seems that it always converged toward the same solution whatever the initial conditions. During this study, the convergence speed depended on the initial conditions and on the τ gain affecting the gradient in the iterations. Figures 3 and 4 illustrate these convergence problems; in these figures, curve 1 refers to the constant initial profiles $T^0(t) = 32°C$ and $pH^0(t) = 6.8$, and curves 2 and 3 to linear initial profiles with respect to time. Moreover, curve 3 corresponds to another value of the τ gain. The temperature and pH optimal profiles obtained are given in Figure 5.

APPLICATION: EXPERIMENTAL RESULTS

By applying the optimal profiles determined for the admissible range (30°C $\leq T \leq 36°C$, $6.6 \leq pH \leq 7$), the biomass and antibiotic production curves expected by simulation are given in Figure 6 (B_1, A_1 curves). Comparing these

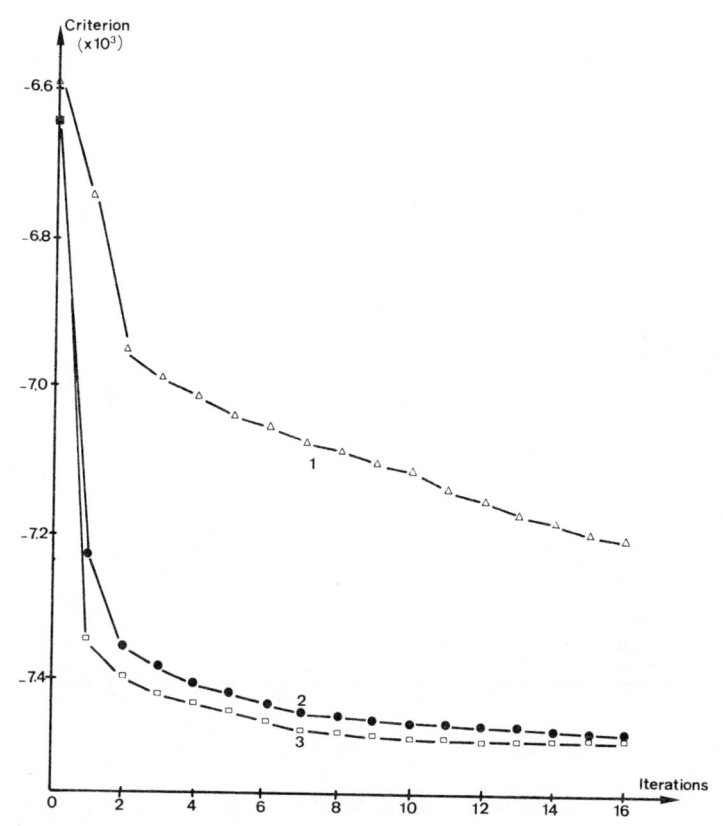

Fig. 4. Convergence of the gradient algorithm for three different initializations.

curves to those also obtained in simulation with the constant temperature and pH of 32°C and 6.8, respectively (the "best" constant conditions according to the experimentor's opinion), it is noticed that an increase of at least 10% in the erythromycin final production is found by using the optimal profiles.

Experiments were performed on the fermentation pilot unit in order to test these optimal profiles. These profiles were realized by piecewise constant signals as shown in Figure 8. An example of the so-obtained experimental results is given on Figure 7, in where you can also find the comparative results of an experiment simultaneously carried out under standard conditions (a constant temperature of 32°C, and free pH). It is noticed that 1) the erythromycin production is nearly the same in both cases (optimal profiles or standard conditions), 2) this production is obtained after a far shorter time with the optimal profiles (200 hr instead of 300 hr), and 3) there are no experimental points in the 200–300-hr range when the optimal profiles are used. In this connection in several experiments it was observed that an erythromycin lysis occurred if the fermentation was allowed to continue (an unexpected phenomenon not taken into account in our model).

Thus, experimentaly, the pH and temperature profiles did not lead to the result

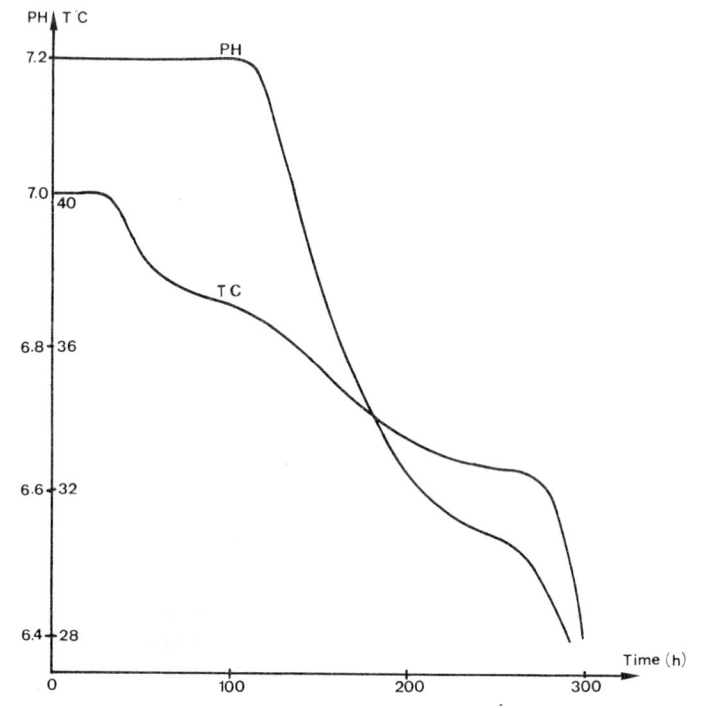

Fig. 5. pH and temperature optimal profiles for 40°C < T < 28°C, 6.4 < pH < 7.2.

expected by simulation (an 10% increase of the antibiotic production), yet they gave an appreciable result: a decrease of about 30% of the fermentation duration, and accordingly about a 30% gain in productivity.

DISCUSSION: A PRACTICAL SOLUTION

When one observes the pH and temperature profiles and the corresponding biomass and erythromycin evolutions (Figs. 7 and 8), it can be noted that pH and temperature profiles lose their constraints only a few hours before the end of the fermentation. So, it was interesting to compare such an experiment with the one carried out with pH and temperature kept constant on their constraint. Taking into account the experimental errors approximately the same results were obtained in both cases. Therefore, in practice, constant values of the control variables will be preferred to optimal profiles which are more difficult to perform.

Moreover, it was interesting to study if it was possible, by using constant control variables, to decrease the fermentation duration further without appreciably modifying the antibiotic production.

This study was theoretically carried out by using the previously established model. For each pH and temperature value, the antibiotic production and the time necessary to reach this production were determined. So, *"isoproduction"*

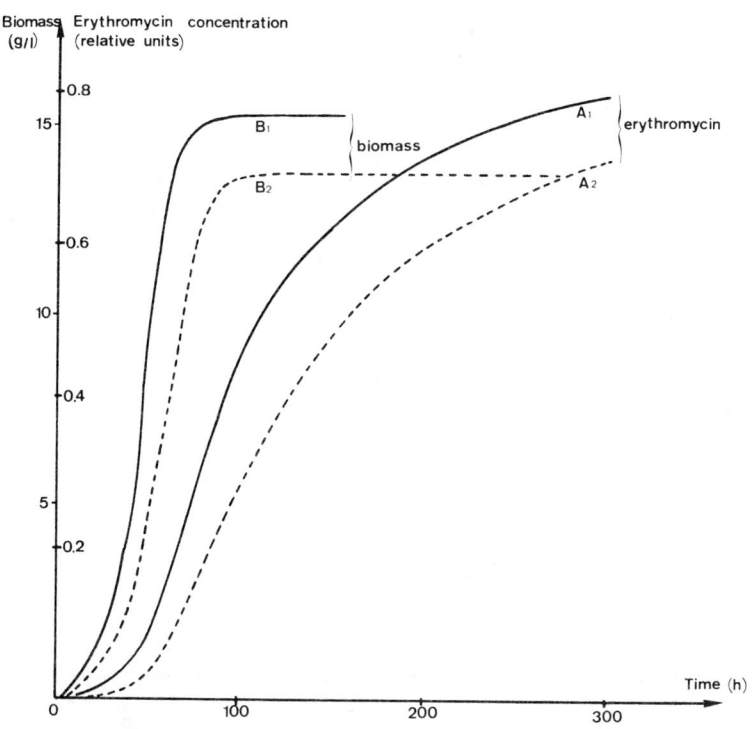

Fig. 6. Simulated results by applying optimal profiles as compared with those obtained when T and pH are kept constant.

and "*isoduration*" curves can be plotted on the (T,pH) plane (see Fig. 9). The "*isoproduction*" curves are ellipsoidal-like; the maximal production is obtained for a pH around 6.85 and a temperature near 35°C, but this maximum is very flat: a production 5% lower than the maximal one can be obtained with a greater range of pH and temperature values (6.65 < pH < 7.15, 32.5°C < T < 37°C). The "*isoduration*" curves are hyperbolic-like and the lowest durations are near the highest temperatures and pH's. So, by using Figure 9, pH and temperature values can be chosen, saving important time in the fermentation duration without appreciably decreasing the antibiotic production. For example, if an 8% decrease of the production in regards to the maximal one is acceptable, the fermentation duration can be reduced to 200 hr by choosing T = 37.5°C and pH = 7.2. An 8% variation of the production was within the range of the experimental error in the operating conditions. A compromise between the production and duration decreases depends on economic considerations, on the equipment available, and also on the measurement quality and experimental reproducibility.

Experiments were carried out to check these theoretical results given by the model. Figure 10 shows an example of the experimental results obtained; the erythromycin production given by a fermentation realized with regulated pH and temperature of 7.2 and 37°C, respectively, is approximately the same as in a fermentation carried out in parallel under classic standard conditions.

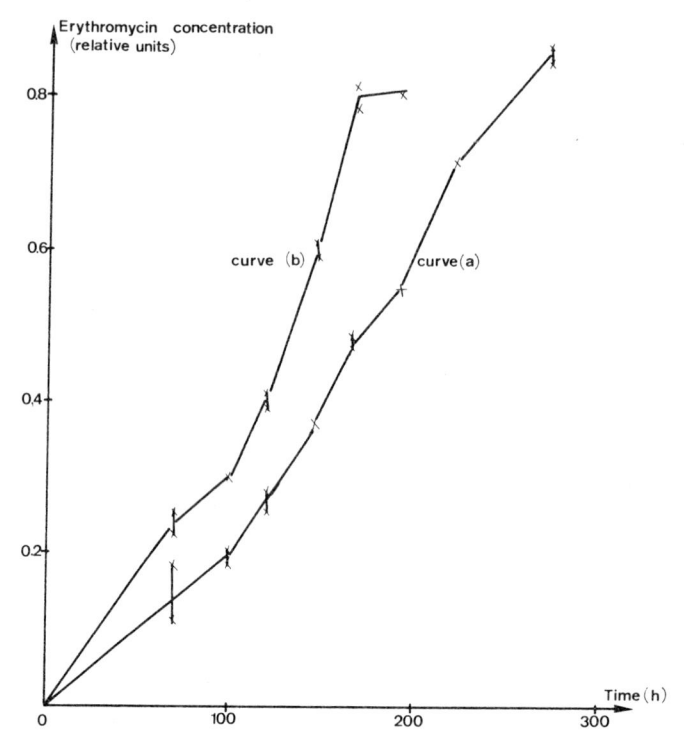

Fig. 7. Experimental results with T and pH optimal profiles [curves (b)] and under standard conditions [curve (a)].

However, to get this production requires 180 hr in one case and 300 hr in the other. This result is more optimistic than those expected by the theory: according to Figure 9, for $T = 37°C$ and pH = 7.2, we were expecting 210 hr for an antibiotic production 7% lower than the maximal one.

Therefore, Figure 9 appears as an interesting tool to choose the pH and temperature values to use according to the possible compromise between the production and duration decreases. The *"isoproductivity"* curves can be plotted on the T,pH plane (production/fermentation duration) is a criterion to consider.

Taking into account the experimental errors, the conditions retained as the best ones were pH = 7.2 and $T = 37°C$. Before adopting these conditions as the operating ones, a practical problem must be solved: how to prevent the lysis that can occur at the end of the fermentation, or, how to detect the end of the fermentation before the lysis occurs. That can easily be done by using the on-line analysis of the fermentor outlet gas.

Therefore, for the erythromycin biosynthesis by controlling pH and temperature, it is not possible to increase the antibiotic production, but, by regulating them at judiciously chosen values, the fermentation duration can be decreased without affecting the production. So, the result is about a 30% gain in productivity.

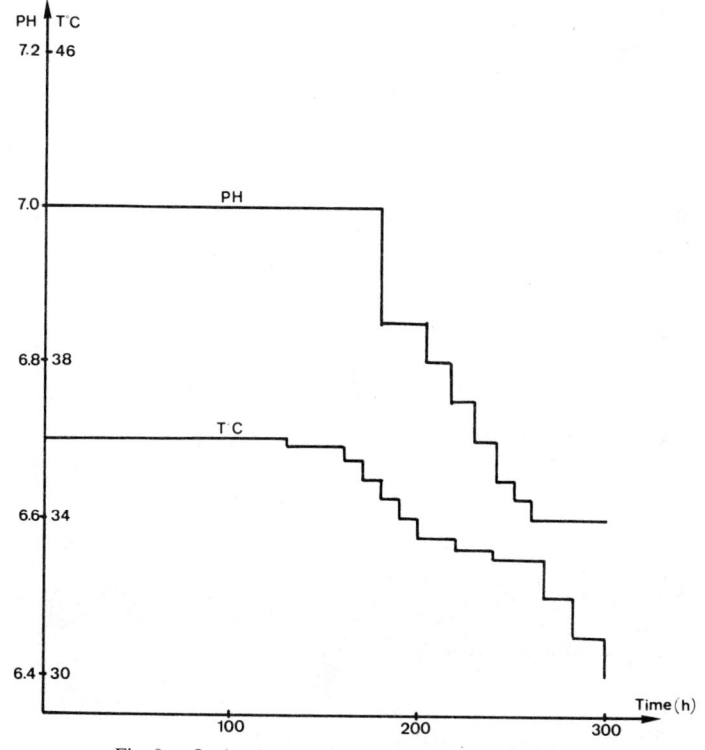

Fig. 8. Optimal profiles used for the curve (b) (Fig. 7).

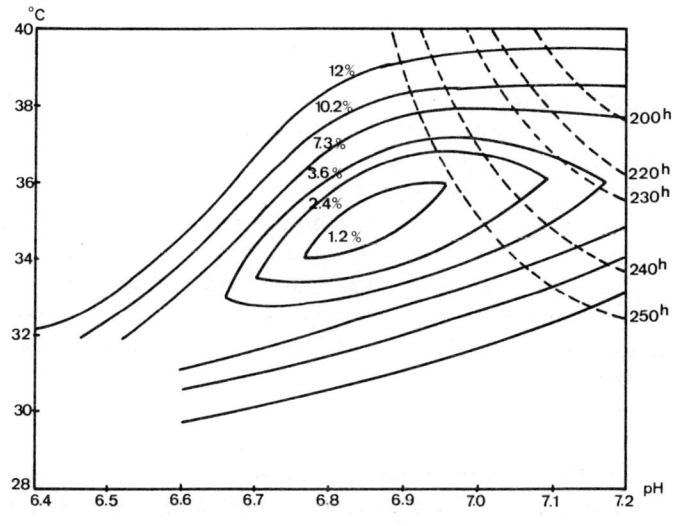

Fig. 9. "Isoproduction" and "isoduration" curves in the (T, pH) plane.

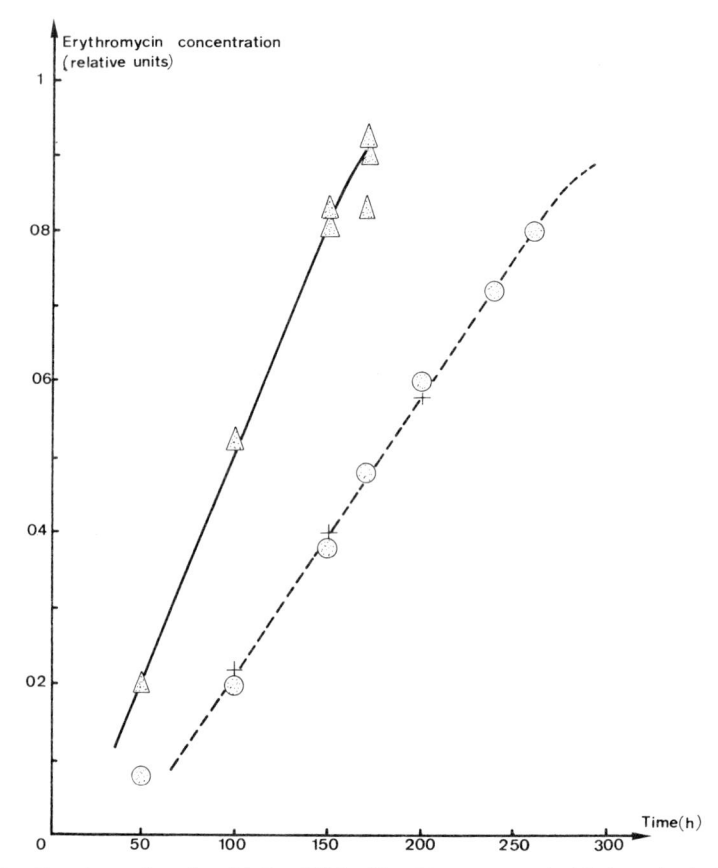

Fig. 10. Experimental results with $T = 37°C$, pH = 7.2 as compared with those obtained under standard conditions.

CONCLUSION

This paper solves, using a concrete example, all the problems connected with the optimal control of a batch antibiotic fermentation by acting on pH and temperature.

The theoretical study shows how to establish a mathematical model of the process, and to calculate the optimal control profiles. The identification and optimization algorithms used are classical in automatic control. They are complex and involve time-consuming computing programs, but a simplified version could certainly be found for implementation on a minicomputer. In the theoretical part, the main difficulty is the mathematical modeling which is the basic starting point.

From a practical viewpoint, the *optimal* control retained is very interesting:

1) It allows about 30% reduction of the fermentation duration without significantly affecting the antibiotic production, which means a 30% improvement of the productivity.

2) It is very simple to implement: pH and temperature must be regulated at constant values.

However, this control does not exactly answer the problem discussed at the beginning of the paper, i.e., the maximization of the erythromycin production. The theoretical study and the simulations forecast an improvement of the production, but the optimal profiles were determined from a *"rough"* model that did not take into account, among others, the substrate consumption, and consequently was unable to predict, for example, the lysis observed at the end of the fermentation, which, in turn, prevented any erythromycin production improvement. So, one could conclude that our model needs to be improved.

Moreover, it was not surprising that, by controlling physiochemical variables such as pH and temperature, we improved the productivity and not the production, which is closely dependent on the amount of substrate supplied by the culture medium at the beginning. The criterion to be optimized (erythromycin production) was not very sensitive to the control variables' variations. In optimal control, the choice of the control variables must be connected with the choice of the optimization criterion.

All these considerations must be taken into account in each optimal control application in batch fermentation.

This study was supported by a grant from the Délégation Générale à la Recherche Scientifique et Technique, with the participation of the Automatic Control Laboratory of the National Polytechnic Institute in Grenoble, the "Station" of microbiological engineering of the I.N.R.A. in Dijon and the Roussel Uclaf Company.

References

[1] A. Constantinides, J. L. Spencer, and E. L. Gaden, *Biotechnol. Bioeng., 12,* 1081 (1970).
[2] V. R. Rai and A. Constantinides, *AICHE Symp. Ser., 132*(69), 114 (1973).
[3] A. Frederickson, R. Megee, and H. Tsuchya, *Advance in Applied Microbiology* (Academic, New York, 1970), Vol. 19.
[4] K. Nyiri, *Advances in Biochemical Engineering* (Springer-Verlag, Berlin, 1972), Vol. 2.
[5] L. Y. Ao, and A. E. Humphrey, *Biotechnol. Bioeng., 12,* 291 (1970).
[6] A. E. Humphrey, "Rational for and principles of computer coupled fermentation," 1st European Conference on Process Control in Fermentation, Dijon, France, September 1973.
[7] R. P. Jefferis, "Control structures for computer coupled fermentation," 1st European Conference on Process Control in Fermentation, Dijon, France, September 1973.
[8] R. P. Jefferis, M. Rems, and F. Wagner, "Applications of computers in process research for SCP," A.F.C.E.T. Symposium on Automatic Control in Fermentation, Paris, 1975.
[9] Y. Nishizawa, S. H. Nagai, and S. Aiba, *J. Gen. Appl. Microbiol., 17,* 131 (1971).
[10] S. Aiba, S. Nagai, and Y. Nishizawa, *Biotechnol. Bioeng., 18,* 1001 (1976).
[11] C. L. Cooney, H. Y. Wang, and D. I. C. Wang, *Biotechnol. Bioeng., 19,* 55 and 69 (1977).
[12] H. Ohno, E. Nakanishi, and T. Takamatsu, *Biotechnol. Bioeng., 18,* 847 (1976).
[13] H. Ohno, E. Nakanishi, and T. Takamatsu, *Biotechnol. Bioeng., 20,* 625 (1978).
[14] A. R. Giona, L. Marrelli, and T. Tora, *Biotechnol. Bioeng., 18,* 473 (1976).
[15] P. Peringer, H. Blachere, G. Corrien, and A. G. Lane, *Biotechnol. Bioeng., 16,* (1974).
[16] H. Blachere, D. Bourdaud, C. Foulard, P. Peringer, A. G. Lane, "Computer control and optimisation in the cultivation of yeast," 4th GIAM Meeting, Sao Paulo, July, 1973.
[17] L. Lukasik, C. Foulard, and A. Cheruy, "Identification des procédés de fermentation discontinus: application au procédé de fabrication de l'erythromycine," A.F.C.E.T. Symposium on Automatic Control in Fermentation, Paris, May, 1975.

[18] D. Bourdaud and C. Foulard, "Optimisation de procédés de fermentation discontinus," A.F.C.E.T. Symposium on Automatic Control in Fermentation, Paris, May, 1975.
[19] A. Lukasik, "Sur l'identification de procédés de fermentation discontinus," Thèse de doctorat 3° Cycle, Université de Grenoble, 1974.
[20] D. Bourdaud, "Contribution à l'identification et à l'optimisation de procédés de fermentation discontinus," Thèse de doctorat 3° Cycle, Université de Grenoble, 1974.
[21] D. Marquardt, *J. Soc. Ind. Appl. Math., 11,* 431 (1963).

Optimal Strategy for Batch Alcoholic Fermentation

ISAO ENDO and TERUYUKI NAGAMUNE
The Institute of Physical and Chemical Research, Wako-shi, Saitama Japan

ICHIRO INOUE
Tokyo Institute of Technology, Ookayama, Tokyo, Japan

INTRODUCTION

It is quite common that many researchers are interested in the cultivation process of a microorganism in their studies on optimization problems in the fermentation processes. However, they should consider how microbial growth reactions differ from ordinary chemical reactions before they deal with the mathematics. How much is known about reaction properties of a microorganism or, in other words, physiological properties of the microorganism with respect to operating conditions? The central dogma of the optimization problems in fermentation engineering may lie in the most effective use of the physiological properties of a microorganism.

The purpose of this presentation is to elucidate the following problems: 1) what types of data about the physiological properties of a microorganism with respect to operating variables are required so as to discuss the optimal strategy for an aerobic batch alcoholic fermentation? 2) How do we determine the initial conditions of a batch culture in order to maximize the objective functions concerning the amount of alcohol and its yield for the substrate and to minimize the fermentation time?

THEORETICAL CONSIDERATIONS

Block Diagram of an Aerobic Batch Alcoholic Fermentation System

The state equations of an aerobic batch alcoholic fermentation systems by brewers' yeast are described in the following set of mass balance equations: glucose:

$$\frac{dG}{dt} = -\nu X$$

oxygen:

$$\frac{dO_2}{dt} = K_L a(O_{2,c} - O_2) - Q_{O_2} X \qquad (1)$$

cell mass:

alcohol:

$$\frac{dX}{dt} = \mu X$$

$$\frac{dA}{dt} = \pi_A X$$

where G, O_2, X, and A are the concentrations of glucose, oxygen, cell mass, and alcohol, respectively, and $O_{2,c}$ is the critical concentration of oxygen. They are the state variables of the systems. ν, Q_{O_2}, μ, and π_A in eq. (1) are the specific rates of glucose consumption, oxygen uptake, cellular growth, and alcoholic production, respectively.

In previous papers [1,2] the authors proposed a transport process for yeast cells in order to relate ν and Q_{O_2} to the glucose concentration in the culture medium, and expressed the functions of the process as follows:

$$\begin{aligned}\nu &= \Phi_G(G, G_0, X_0) \\ Q_{O_2} &= \Phi_{O_2}(G, G_0, X_0)\end{aligned} \quad (2)$$

The metabolic functions of the organism are expressed by a four-terminal network where ν and Q_{O_2} are input terminals and μ and π_A are output terminals [1–3]:

$$\begin{pmatrix} \mu \\ \pi_A \end{pmatrix} = \begin{pmatrix} \alpha & \beta \\ \gamma & \delta \end{pmatrix} \begin{pmatrix} \nu \\ Q_{O_2} \end{pmatrix} \quad (3)$$

μ and π_A, and the metabolic coefficients α, β, γ, and δ are also multivariable functions of the glucose concentration G, the initial glucose concentration, G_0, and the cell mass, X_0.

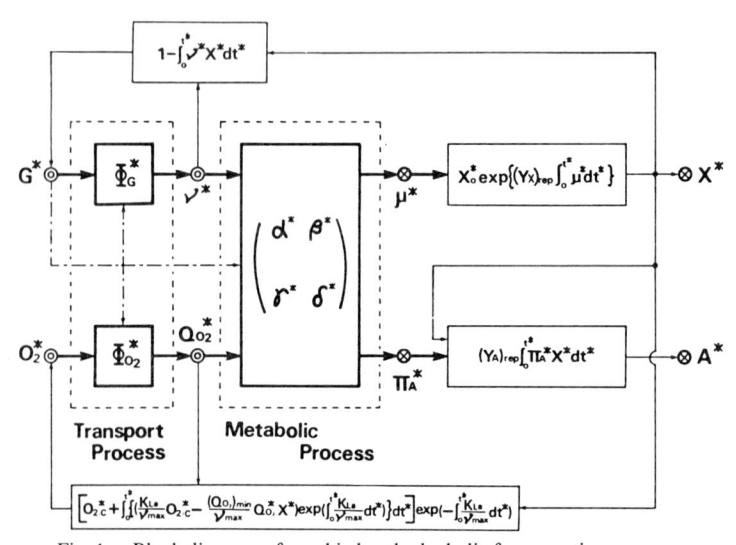

Fig. 1. Block diagram of aerobic batch alcoholic fermentation systems.

OPTIMAL STRATEGY FOR BATCH ALCOHOLIC FERMENTATION   323

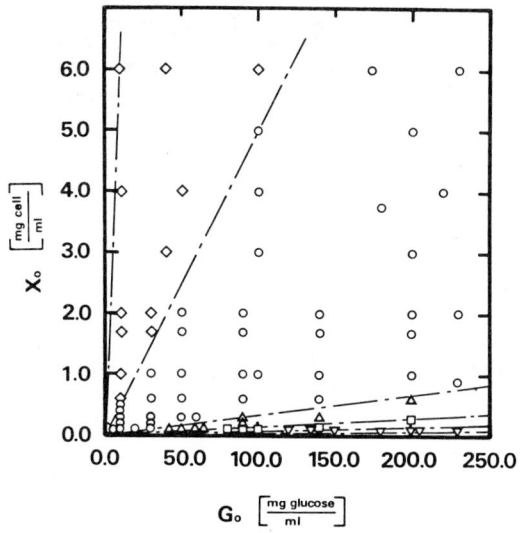

Fig. 2. Experimental regions of G_0 and X_0. ($\diamond,\circ,\triangle,\square,\triangledown$) Values of operating variables tested in this experiment.

In order to obtain both a functional scheme of the batch alcoholic fermentation system and physiological properties of the yeast organism that are independent of the operating variables, we have expressed eqs. (1)–(3) in dimensionless terms [1,2]. These equations are given below:

$$\frac{dG^*}{dt^*} = -\nu^* X^*$$

$$\frac{dO_2^*}{dt^*} = \frac{K_L a}{\nu_{max}}(O_{2,c}^* - O_2^*) - (Y_{O_2})_{rep} Q_{O_2}^* X^*$$

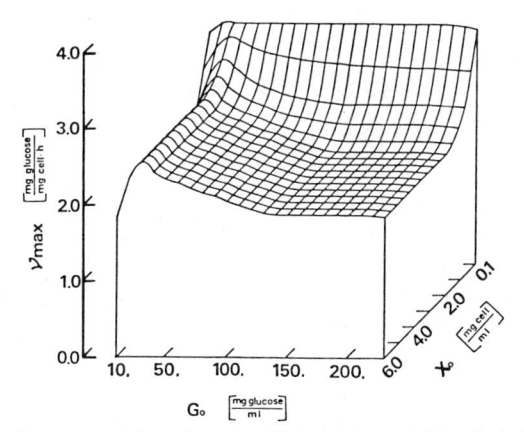

Fig. 3. Stereographic representation of ν_{max} vs. the operating variables G_0 and X_0. Experimental results obtained were omitted for simplicity here and in Figures 4 and 5.

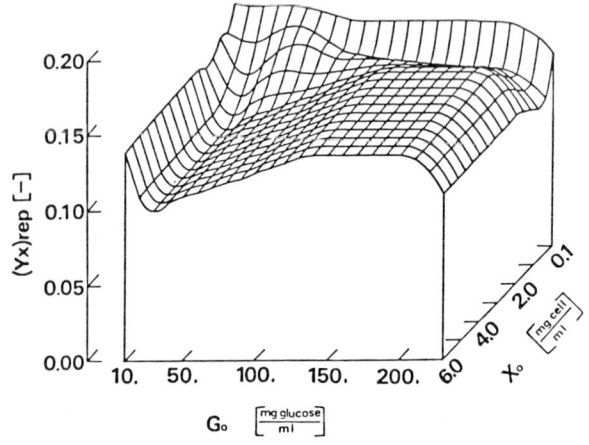

Fig. 4. Stereographic representation of $(Y_X)_{rep}$ vs. G_0 and X_0.

$$\frac{dX^*}{dt^*} = (Y_X)_{rep}\mu^* X^*$$

$$\frac{dA^*}{dt^*} = (Y_A)_{rep}\pi_A^* X^* \qquad (4)$$

where

$G^* = G/G_0, \qquad O_2^* = O_2/G_0, \qquad X^* = X/G_0, \qquad A^* = A/G_0,$
$\nu^* = \nu/\nu_{max}, \qquad Q_{O_2}^* = Q_{O_2}/(Q_{O_2})_{min}, \qquad \mu^* = \mu/\mu_{max},$
$\pi_A^* = \pi_A/(\pi_A)_{max}, \qquad t^* = \nu_{max}t, \qquad (Y_X)_{rep} = \mu_{max}/\nu_{max},$
$(Y_{O_2})_{rep} = (Q_{O_2})_{min}/\nu_{max}, \qquad (Y_A)_{rep} = (\pi_A)_{max}/\nu_{max}$

ν_{max}, μ_{max}, and $(\pi_A)_{max}$ are the maximum specific rates of glucose consumption,

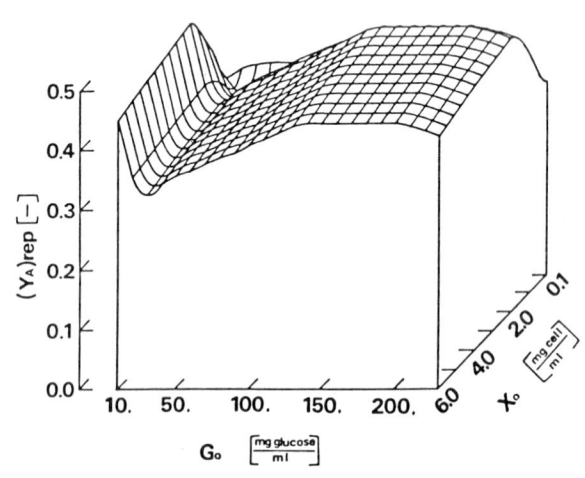

Fig. 5. Stereographic representation of $(Y_A)_{rep}$ vs. G_0 and X_0.

cellular growth, and alcoholic production, respectively. $(Q_{O_2})_{min}$ is the minimum specific rate of oxygen respiration. They are observable in the middle of a batch fermentation and are functions of operating variables G_0 and X_0.

The operation-independent transport and metabolic functions of the yeast cells are restated in the following equations:

$$\nu^* = \Phi_G^*(G^*, X_0^*)$$

$$Q_{O_2}^* = \Phi_{O_2}^*(G^*) \qquad (5)$$

$$\begin{pmatrix} \mu^* \\ \pi_A^* \end{pmatrix} = \begin{pmatrix} \alpha^* & \beta^* \\ \gamma^* & \delta^* \end{pmatrix} \begin{pmatrix} \nu^* \\ Q_{O_2}^* \end{pmatrix} \qquad (6)$$

where,

$X_0^* = X_0/G_0, \qquad \Phi_G^* = \Phi_G/\nu_{max}, \qquad \Phi_{O_2}^* = \Phi_{O_2}/(Q_{O_2})_{min},$

$\alpha^* = (Y_X)_{rep}\alpha \qquad \beta^* = (Y_{O_2})_{rep}\beta, \qquad \gamma^* = (Y_A)_{rep}\gamma, \qquad \delta^* = (Y_{O_2})_{rep}\delta$

The relationship between the state variables and the functions of both transport and metabolic processes of the yeast cells in an aerobic batch alcoholic fermentation system is shown in Figure 1. In Figure 1, the solid arrows indicate the flow of material and the dotted lines indicate the effect of glucose concentration on both the transport and metabolic processes.

Fermentation Time, Alcohol Yield for the Initial Glucose Concentration, and Amount of Alcohol

The time required for one batch fermentation to be complete is expressed

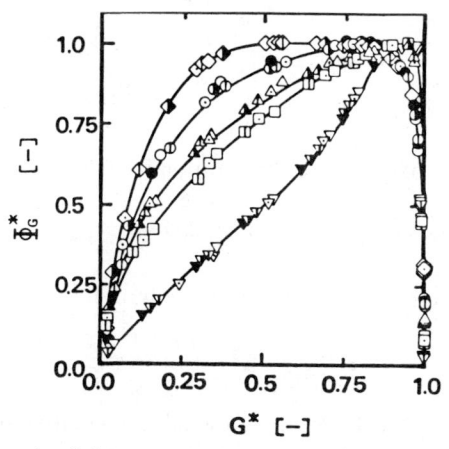

Fig. 6. Functional properties of glucose consumption Φ_G^* vs. G^*. G^* means dimensionless glucose concentration and X_0^* means dimensionless operating variable. X_0^*: (◇) 0.05–0.6; (◊) 0.06; (♦) 0.2; (♦) 0.4; (♦) 0.6. (○) 0.0033–0.05: (○) 0.05; (◎) 0.02; (◐) 0.001; (◑) 0.005; (●) 0.0033. (△) 0.0014–0.0033: (△) 0.0024; (▲) 0.0020; (▲) 0.0017; (▲) 0.0016. (□) 0.00077–0.0014: (□) 0.00125; (▣) 0.0010; (■) 0.00063. (▽) 0.00043–0.00077: (▽) 0.00074; (▼) 0.00067; (▼) 0.00056; (▼) 0.00049; (▼) 0.00043.

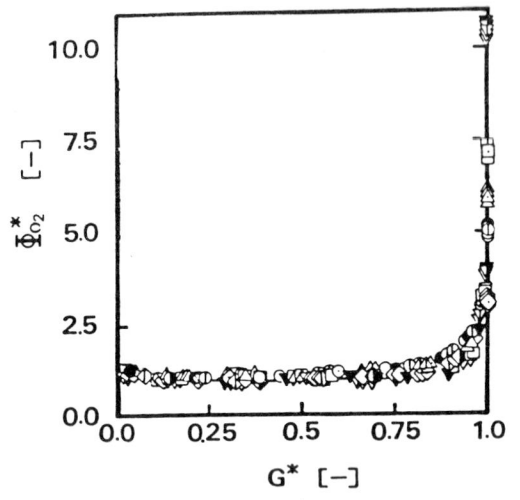

Fig. 7. Functional properties of the transport process for oxygen. Same symbols as in Figure 6.

clearly as a function of initial values of G_0 and X_0. So that, from eqs. (4)–(6), we obtain the following equation:

$$\frac{dX^*}{dG^*} = -(Y_X)_{\text{rep}}\left(\alpha^* + \beta^* \frac{\Phi^*_{O_2}}{\Phi^*_G}\right)$$

Integration of the above equation gives the relationship between the cell mass concentration and glucose concentration:

$$X^* = X_0^* + (Y_X)_{\text{rep}} K_X^*(G^*)$$

where

$$K_X^*(G) = \int_{G^*}^{1}\left(\alpha^* + \beta^* \frac{\Phi^*_{O_2}}{\Phi^*_G}\right) dG^*$$

The time derivative of glucose concentration is rewritten as follows:

$$\frac{dG^*}{dt^*} = -\nu^* X^* = -\Phi_G^*[X_0^* + (Y_X)_{\text{rep}} K_X^*(G^*)]$$

Then, the time required for one batch fermentation t_f^* is given by

$$t_f^* = \int_0^1 \frac{dG^*}{\Phi_G^*[X_0^* + (Y_X)_{\text{rep}} K_X^*(G^*)]} \tag{7}$$

Equation (7) means that the nondimensional fermentation time is constant but a function of the initial conditions of G_0 and X_0. If we express the fermentation time in dimensional terms,

$$t_f \stackrel{\text{def}}{=} \frac{t^*}{\nu_{\max}} = \frac{1}{\nu_{\max}} \int_0^1 \frac{dG^*}{\Phi_G^*[X_0^*(Y_X)_{\text{rep}} K_X^*(G^*)]} \tag{8}$$

TABLE I

Characteristic Values of the Metabolic Coefficients α^*, β^*, γ^*, and δ^* at Three Stages of a Whole Batch Cultivation Process in Five Ranges of Operating Variable X_0^*

Metabolic coefficient Phase X_0^*	α^* lag phase	α^* log phase	α^* stationary phase	β^* lag phase	β^* log phase	β^* stationary phase	γ^* lag phase	γ^* log phase	γ^* stationary phase	δ^* lag phase	δ^* log phase	δ^* stationary phase
$0.05 < X_0^* \leq 0.6$	$1.0 \geq G^* > 0.82$ +0.740	$0.82 \geq G^* > 0.30$ +1.00	$0.30 \geq G^* \geq 0.0$ +0.699	$1.0 \geq G^* > 0.82$ +0.150	$0.82 \geq G^* > 0.30$ 0.0	$0.30 \geq G^* \geq 0.0$ +0.327	$1.0 \geq G^* > 0.82$ +0.958	$0.82 \geq G^* > 0.30$ +1.0	$0.30 \geq G^* \geq 0.0$ +0.893	$1.0 \geq G^* > 0.82$ +0.029	$0.82 \geq G^* > 0.30$ 0.0	$0.30 \geq G^* \geq 0.0$ +0.066
$0.0033 < X_0^* \leq 0.05$	$1.0 \geq G^* > 0.85$ +0.922	$0.85 \geq G^* > 0.45$ +1.33	$0.45 \geq G^* \geq 0.0$ +0.565	$1.0 \geq G^* > 0.85$ +0.055	$0.85 \geq G^* > 0.45$ −0.213	$0.45 \geq G^* \geq 0.0$ +0.493	$1.0 \geq G^* > 0.85$ +1.05	$0.85 \geq G^* > 0.45$ +1.37	$0.45 \geq G^* \geq 0.0$ +1.13	$1.0 \geq G^* > 0.85$ −0.04	$0.85 \geq G^* > 0.45$ −0.244	$0.45 \geq G^* \geq 0.0$ −0.056
$0.0014 < X_0^* \leq 0.0033$	$1.0 \geq G^* > 0.85$ +0.995	$0.85 \geq G^* > 0.45$ +1.80	$0.45 \geq G^* \geq 0.0$ +0.909	$1.0 \geq G^* > 0.85$ +0.003	$0.85 \geq G^* > 0.45$ −0.483	$0.45 \geq G^* \geq 0.0$ +0.196	$1.0 \geq G^* > 0.85$ +1.29	$0.85 \geq G^* > 0.45$ +1.18	$0.45 \geq G^* \geq 0.0$ +1.12	$1.0 \geq G^* > 0.85$ −0.163	$0.85 \geq G^* > 0.45$ −0.094	$0.45 \geq G^* \geq 0.0$ −0.067
$0.00077 < X_0^* \leq 0.0014$	$1.0 \geq G^* > 0.85$ +0.94	$0.85 \geq G^* > 0.50$ +1.01	$0.50 \geq G^* \geq 0.0$ +0.87	$1.0 \geq G^* > 0.85$ +0.031	$0.85 \geq G^* > 0.50$ −0.006	$0.50 \geq G^* \geq 0.0$ +0.106	$1.0 \geq G^* > 0.85$ +1.08	$0.85 \geq G^* > 0.50$ +1.08	$0.50 \geq G^* \geq 0.0$ +1.08	$1.0 \geq G^* > 0.85$ −0.048	$0.85 \geq G^* > 0.50$ −0.058	$0.50 \geq G^* \geq 0.0$ −0.058
$0.00043 < X_0^* \leq 0.00077$	$1.0 \geq G^* > 0.85$ +0.98	$0.85 \geq G^* > 0.50$ +2.22	$0.50 \geq G^* \geq 0.0$ +0.58	$1.0 \geq G^* > 0.85$ +0.017	$0.85 \geq G^* > 0.50$ −0.69	$0.50 \geq G^* \geq 0.0$ +0.04	$1.0 \geq G^* > 0.85$ +1.03	$0.85 \geq G^* > 0.50$ +1.0	$0.50 \geq G^* \geq 0.0$ +1.01	$1.0 \geq G^* > 0.85$ −0.021	$0.85 \geq G^* > 0.50$ 0.0	$0.50 \geq G^* \geq 0.0$ −0.013

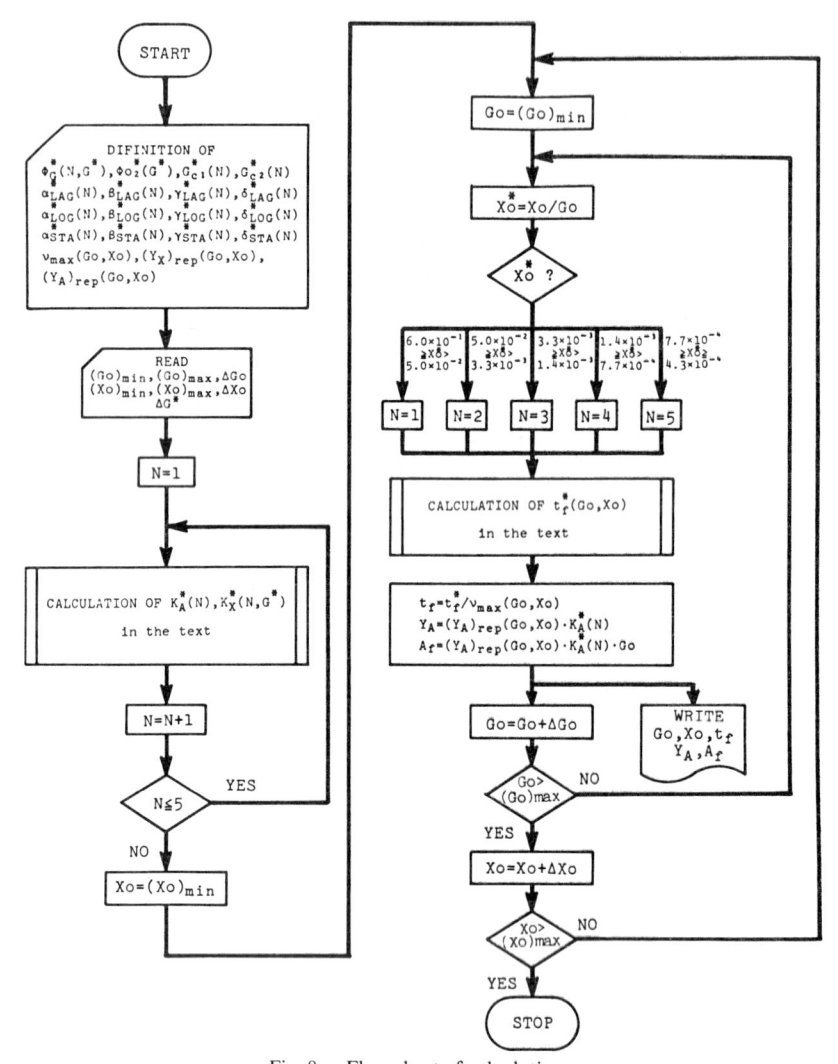

Fig. 8. Flow chart of calculations.

is obtainable. From eq. (8), the fermentation time t_f is a function of the maximum specific rate of glucose consumption (ν_{max}) and the initial conditions G_0 and X_0.

The alcohol yield for the initial glucose concentration is given in a similar way as in eqs. (7) and (8). Namely, from eqs. (4)–(6), we obtain

$$\frac{dA^*}{dG^*} = (Y_A)_{\text{rep}} \left(\gamma^* + \delta^* \frac{\Phi_{O_2}^*}{\Phi_G^*} \right)$$

$$A_f^* = (Y_A)_{\text{rep}} \int_0^1 \left(\gamma^* + \delta^* \frac{\Phi_{O_2}^*}{\Phi_G^*} \right) dG^* \qquad (9)$$

OPTIMAL STRATEGY FOR BATCH ALCOHOLIC FERMENTATION 329

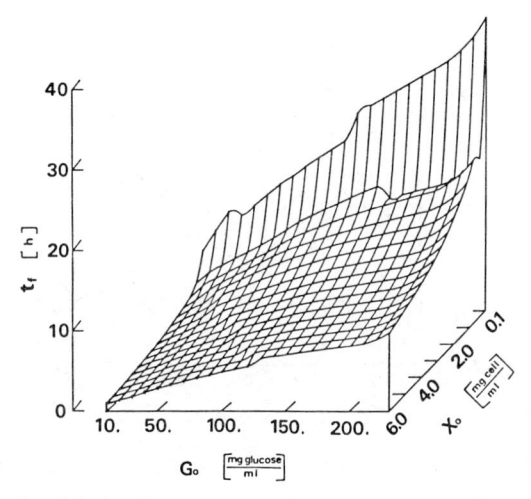

Fig. 9. Calculated results for fermentation time t_f vs. G_0 and X_0.

here, we have assumed that $A_0^* = 0$ at $t^* = 0$.

Then, the alcohol yield is given as follows

$$Y_A \stackrel{\text{def}}{=} \frac{A_f}{G_0} \stackrel{\text{def}}{=} A_f^* = (Y_A)_{\text{rep}} K_A^* \qquad (10)$$

where

$$K_A^* = \int_0^1 \left(\gamma^* + \delta^* \frac{\Phi_{O_2}^*}{\Phi_G^*} \right) dG^*$$

The amount of alcohol A_f that is produced during a batch fermentation is obtained from eq. (10):

$$A_f = (Y_A)_{\text{rep}} K_A^* G_0 \qquad (11)$$

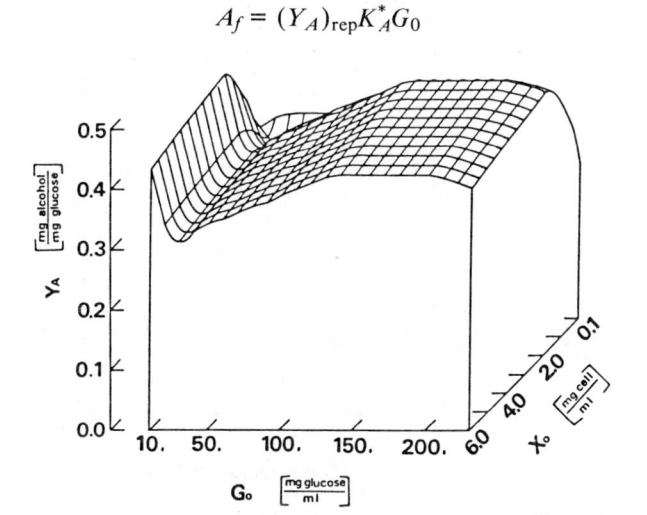

Fig. 10. Calculated results for yield Y_A vs. G_0 and X_0.

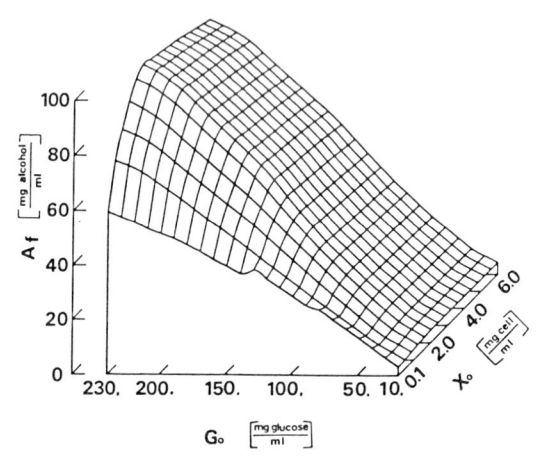

Fig. 11. Calculated results for the amount of alcohol production A_f vs. G_0 and X_0.

From the previous discussion we can conclude that the following nine physiological properties of the microorganism are the most important to discuss an optimal strategy for an aerobic batch alcoholic fermentation: $\nu_{\max}(G_0, X_0)$, $(Y_X)_{\rm rep}(G_0, X_0)$, $(Y_A)_{\rm rep}(G_0, X_0)$, $\Phi_G^*(G^*, X_0^*)$, $\Phi_{O_2}^*(G^*)$, $\alpha^*(G^*, X_0^*)$, $\beta^*(G^*, X_0^*)$, $\gamma^*(G^*, X_0^*)$, and $\delta^*(G^*, X_0^*)$.

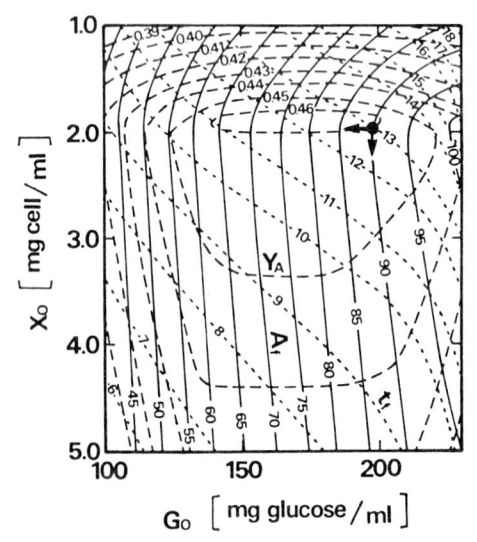

Fig. 12. Contour map of t_f, Y_a, and A_f vs. the operating variables G_0 and X_0. Optimal operational conditions can be determined from this figure.

EXPERIMENTAL

Cultivation Method

Brewers' yeast (*S. cerevisiae*) was cultivated batchwise and aerobically using Hayduck's synthetic medium where glucose was the sole carbon source and the growth-limiting factor. The initial glucose concentration, G_0, and that of cell mass, X_0, were selected as operating variables and were allowed to fluctuate as shown in Figure 2. All the experiments were carried out mainly with a 10 liter jar fermentor. The working volume of the fermentor and agitation speed were set at 5 liter and 450 rpm, respectively. The air flow rate was controlled at 5 liter/min. The cultivation temperature and pH were adjusted at 30°C and 4.5, respectively. Prior to a jar cultivation, the yeast organisms were grown overnight at 30°C in 300 ml shake flasks and were transferred to the fermentor as a seed culture.

Analytical Method

The determination of dry cell mass X, glucose concentration, G, and ethyl alcohol concentration, A, was described in the previous paper [3]. Dissolved oxygen concentration in the culture medium was measured with a NBS oxygen analyzer, and the specific oxygen uptake rate Q_{O_2} was calculated by the following equation and is expressed in mg oxygen uptake/mg cells [dry weight (DW)]/hr:

$$Q_{O_2} = [K_L a(O_{2,c} - O_2) - \dot{O}_2]/X$$

where $K_L a$ is the volumetric liquid-phase oxygen mass transfer coefficient and $O_{2,c}$ is the critical concentration of dissolved oxygen at 30°C.

Experimental Results

Almost all the experimental results that are explained hereafter were checked by using various scales of jar fermentors (2, 30, 200, and 600 liter) and found that they were not affected by the fermentor volume.

Experimental results concerning the maximum specific rate of glucose consumption [$\nu_{max}(G_0, X_0)$], the yield factor of cell mass for glucose [$(Y_X)_{rep}(G_0, X_0)$], and yield factor of alcohol for glucose [$(Y_A)_{rep}(G_0, X_0)$] are shown stereographically in Figures 3–5, respectively. The experimental results obtained were omitted for simplicity.

From Figure 3 we can see that ν_{max} has its peak along $X_0 = 0.1$ mg/ml and $G_0 = 30$ mg/ml, but the value decreases sharply as X_0 and G_0 increase. ν_{max} remains almost constant in the region where $2.0 \leq X_0 \leq 6.0$, $130.0 \leq G_0 \leq 230.0$. Although the highest values of $(Y_X)_{rep}$ lie along $X_0 = 0.1$ mg/ml or $G_0 = 10.0$ mg/ml, they fluctuate easily with a slight change of the operating variables (X_0 and G_0) as shown in Figure 4. Not only Figure 4 but also Figure 5 show that the higher but metastable values of $(Y_X)_{rep}$ and $(Y_A)_{rep}$ lie in the region of $X_0 = 2.0$ and $130.0 \leq G_0 \leq 200.0$.

The functional properties of the transport process of the yeast cells, namely, the dimensionless specific consumption rate of glucose $\Phi_G^*(G^*, X_0^*)$ and the oxygen uptake rate $\Phi_{O_2}^*(G^*)$ are shown in Figures 6 and 7, respectively. It is clear from Figure 6 that Φ_G^* showed five kinds of patterns against G^* corresponding to five regions of operating variables X_0^*. The other physiological properties of the yeast organisms, namely, μ^* and π_A^*, although they are not shown in the figures, showed the same variation according to five regions of the operating variables as are seen in Figure 6. The dotted lines in Figure 2 are nothing more than such borders of the operating variables.

Meanwhile, it is evident from Figure 7 that $\Phi_{O_2}^*$ was not influenced by the operating variables and showed a uniform pattern against G^*. That is, $\Phi_{O_2}^*$ reached its maximum just after the batch operation began, and decreased sharply until it reached its minimum in the initial stage of batch cultivation. At the end of cultivation, the respiratory activity recovered slightly because glucose was consumed.

The functional properties of the metabolic processes of the yeast organism that were represented by the metabolic coefficients α^*, β^*, γ^*, and δ^* in eq. (6) are shown in Table I. Their constant values changed during the three stages of batch cultivation. Those stages approximately corresponded to growth phases of the cells, namely, the lag, logarithmic growth, and stationary phases. It is also evident from the relationship expressed by eq. (6) that a combination of the metabolic coefficients differ significantly in the five regions of the operating variables X_0^*.

RESULTS AND DISCUSSIONS

We will discuss here the optimization strategy for an aerobic batch alcoholic fermentation on the basis of the eqs. (8), (10), and (11), and the experimental results shown in Figures 3–7. We have shown the flow chart of these calculations for Figure 8 and calculated results for Figures 9–11.

It is evident from Figure 9 that fermentation takes much time in accordance with a small value of X_0 and a large value of G_0. Especially, Figure 9 leads one to the conclusion that the influence of the initial concentration of cell mass upon the time required for one batch operation is great.

As may be seen from Figure 10, the alcohol yield for the initial glucose concentration has a peak in a region where $X_0 = 2.0$ mg/ml and G_0 is between 130.0 and 200.0 mg/ml except for the unstable operating region where $G_0 = 10.0$ mg/ml. In Figure 11, we have drawn the direction of G_0 in an opposite way to easily represent the changing mode of alcoholic production A_f against the operating variables G_0 and X_0. From Figure 11, we can see that the higher the glucose concentration G_0 becomes, the greater amount of alcohol is produced. This result coincides with the usual techniques in Sake brewing where the yeast cells are harvested under thick mash. Moreover, Figure 11 shows that the influence of a smaller value of X_0 upon A_f was found to be considerable in the region where $G_0 > 150$ mg/ml.

If we draw the values of t_f, Y_A, and A_f against G_0 and X_0 in a contour map, we obtain Figure 12. This figure is useful as a chart for deciding the optimal strategy for a batch alcoholic fermentation. Namely, the following conclusion can be reached from this chart: To make a decision for the highest yield of alcohol for the mash concentration of glucose and the highest concentration of alcohol with minimum fermentation time, select the initial glucose concentration (G_0) to be 200 mg/ml and that of cell mass (X_0) to be 2.0 mg/ml. Then it is possible to obtain about 90 mg/ml alcohol after 13 hr of fermentation. Experimental values agreed quite well with the calculated results illustrated by this chart.

CONCLUSION

To summarize, the interpretation of our results is as follows: 1) The functional scheme of an aerobic batch fermentation systems by the brewers' yeast has been elucidated in a block diagram. 2) The physiological properties of the yeast cells have been identified experimentally. 3) The optimal strategy for batch alcoholic fermentation has been obtained, under which we obtain about 90 mg/ml alcohol, and its yield for G_0 equals 46% after 13 hr of fermentation if $G_0 = 200.0$ mg/ml and $X_0 = 2.0$ mg/ml.

Nomenclature

A	alcohol concentration in the medium (mg/ml)
A_f	amount of alcohol produced in one batch of fermentation (mg/ml)
G	glucose concentration in the medium (mg/ml)
$K_L a$	volumetric liquid-phase oxygen mass transfer coefficient (liter/hr)
O_2	dissolved oxygen concentration (mg O_2/ml)
Q_{O_2}	specific consumption rate of oxygen (mg O_2/mg cell hr)
t	culture time (hr)
t_f	time required for one batch to complete fermentation (hr)
X	cell mass concentration in the medium (mg/ml)
Y_A	production yield of alcohol for glucose π_A/ν (mg alcohol/mg glucose)
Y_{O_2}	growth yield for oxygen μ/Q_{O_2} (mg cell/mg O_2)
Y_X	growth yield for glucose μ/ν (mg cell/mg glucose)
$\alpha, \beta, \gamma, \delta$	metabolic coefficients
μ	specific growth rate of cells (mg cell/mg cell hr)
ν	specific consumption rate of glucose (mg glucose/mg cell hr)
π_A	specific production rate of alcohol (mg alcohol/mg cell hr)
Φ_G	function of transport process for glucose (liters/hr)
Φ_{O_2}	function of transport process for oxygen (liters/hr)

Subscripts

c	critical concentration
max	maximum value
min	minimum value
rep	representative value
0	initial value
*	dimensionless value

Optimization of a Repeated Fed-Batch Reactor for Maximum Cell Productivity

W. A. WEIGAND, H. C. LIM, C. C. CREAGAN, and R. D. MOHLER

School of Chemical Engineering, Purdue University, West Lafayette, Indiana 47907

INTRODUCTION

The objective of this study was to develop an optimal feeding policy for a fed-batch fermentation where the product is cell mass rather than a metabolite. Examples of this type of fermentation are the production of single-cell protein (SCP) by the growth of methonal utilizers [1], and the growth of bakers' yeast on glucose [2].

The optimization of fed-batch fermentations producing a nonbiomass product has been studied by the methods of path optimization. The maximum principle was utilized by Andreyeva and Biryukov [3] and Fishman and Biryukov [4] to find the optimal rate of substrate addition and optimal pH profile that maximize the penicillin concentration for a fixed processing time for a fermentation of *Penicillium chrysogenum* growing on glucose. Another case of fed-batch optimization is given by Ohno et al. [5]. Their problem was formulated to maximize the rate of production of a metabolite by manipulating the volumetric flow rate of substrate into the fermentor. The system equations were reduced to two state variables, and (after relating their performance index to a line integral) a method developed by Miele [6] based on the use of Green's theorem was used to determine the optimal functions. In a more recent paper, Ohno et al. [7] generalized the volume balance equation by considering the outflow from the fermentor, in addition to the inflow, as a manipulated variable for optimization. The maximum principle was utilized after first using a transformation developed by Kelley [8] that reduces the dimensionality of the state variables to two and also eliminates the singular problem which is common for fermentor path optimization. The results were illustrated by the example of a lysine fermentation. A simplified procedure for optimization of a fed-batch fermentation was given by Yamané et al. [9]. The approach was to manipulate the system equations into a form where control of the system occurred by optimizing the trajectory of the specific growth rate. The inlet volumetric flow policy was then related to the specific growth rate trajectory and optimal and suboptimal flow trajectories were developed and illustrated by an example of metabolite production.

In this work the operation of a repeated fed-batch bioreactor that produces cell mass as product was examined by the maximum principle with initial reactor volume as a parameter. The final cell concentration and feed substrate concentration were prescribed and the temperature, pH, and dissolved oxygen (DO) were considered constant during the course of the fed-batch cycle. The optimal cell productivity was formulated as a minimum time problem and the substrate feed policy was optimized subject to constraints on the inlet substrate flow rate and the maximum fermentor volume. The productivity is also dependent on initial volume and its optimal value was obtained by numerical search.

FORMULATION

The fed-batch production of cell mass can be represented by balance equations for cell, substrate, and reactor volume as

$$\frac{d(VX)}{dt} = Vr_x, \qquad X(0) = X_0 \tag{1}$$

$$\frac{d(VS)}{dt} = FS_F + Vr_s, \qquad S(0) = S_0 \tag{2}$$

$$\frac{dV}{dt} = F, \qquad V(0) = V_0 \tag{3}$$

where V, X, and S are the reactor volume, cell concentration, and substrate concentration, respectively, F is the volumetric feed rate, S_F is the feed substrate concentration, and r_x and r_s are the rate of generation of cells and substrate, respectively.

The rates of cell growth and substrate utilization are considered to have the form

$$r_x = \mu(S)X \tag{4}$$

and

$$r_s = -r_x/Y(S) = -\mu(S)X/Y(S) \tag{5}$$

Note that in eq. (5) the biomass yield is considered to be a function of the substrate concentration.

Combining eqs. (1)–(5), the system can be represented as

$$\frac{dX}{dt} = \mu(S)X - \frac{FX}{V}, \qquad X(0) = X_0 \tag{6}$$

$$\frac{dS}{dt} = \frac{F}{V}(S_F - S) - \frac{\mu(S)X}{Y(S)}, \qquad S(0) = S_0 \tag{7}$$

$$\frac{dV}{dt} = F, \qquad V(0) = V_0 \tag{8}$$

The process of repeated fed-batch operation consists of three stages. The first is the filling stage where $F(t) > 0$ for $0 < t < t_f$ so that the volume increases from V_0 to V_{max} and the cell and substrate concentrations change from X_0, S_0, to X_{fill} and S_{fill}, respectively. The second stage is the batch portion where $F(t) = 0$ for $t_f < t < T_f$ and growth continues so that X increases to X_f and S decreases to S_f. The third stage is rapid draw off where V is quickly reduced from V_{max} to V_0 so that X and S remain practically unchanged at X_f and S_f. The process composed of these three stages is then repeated, and, after a steady-state cyclic operation has been achieved, the initial conditions of cell and substrate concentrations at the beginning of the fed-batch cycle will be equal to those at the end of the previous cycle, i.e., $X_0 = X_f$ and $S_0 = S_f$.

For this type of process the production rate of cell mass can be represented by

$$P = V_f(V_{max} - V_0)/T_f \tag{9}$$

where T_f is the time required to perform one cycle. If we further consider X_f, V_{max}, and V_0 to be prescribed, then the maximum value of P is obtained when T_f is minimized. Therefore, the problem is to determine the optimum feed flow profile, $F(t)$, that transforms the system given by eqs. (6)–(8) from V_0, X_f, S_f to V_{max}, X_f, S_f in the minimum time subject to the flow rate constraint

$$0 \leq F(t) \leq F_{max} \tag{10}$$

and also a fermentor volume constraint

$$0 \leq V_0 \leq V(t) \leq V_{max} \tag{11}$$

In this formulation, V_0 is a parameter that significantly affects productivity. Therefore maximum productivity is determined by a time optimal solution with the embedded parameter optimization on V_0.

The objective function to minimize is then

$$\min_{F(t)} \left(T_f = \int_0^{T_f} dt \right) \tag{12}$$

SOLUTION

In order to use the maximum principle, eq. (12) is written

$$\min T_f = \max \int_0^{T_f} (-1) \, dt \tag{13}$$

With the system given by eqs. (6)–(8) and the objective function given by eq. (13), the Hamiltonian can be written

$$H = \lambda_x \mu(S) X - \frac{\lambda_s \mu(S) X}{Y(S)} + \left(\lambda_v + \lambda_s \frac{S_F - S}{V} - \frac{X}{V} \lambda_x \right) F - 1 \tag{14}$$

where the adjoint variables must satisfy the equations given by

$$\frac{\partial H}{\partial X} = -\dot{\lambda}_x = \lambda_x \left(\mu - \frac{F}{V} \right) - \frac{\lambda_s \mu}{Y} \tag{15}$$

$$\frac{\partial H}{\partial S} = -\dot{\lambda}_s = \lambda_x X \mu' - \lambda_s \left[X \left(\frac{\mu}{Y} \right)' + \frac{F}{V} \right] \quad (16)$$

$$\frac{\partial H}{\partial V} = -\dot{\lambda}_v = -\frac{\lambda_s (S_F - S) F}{V^2} + \frac{\lambda_x X F}{V^2} \quad (17)$$

where the primes denote differentiation with respect to S. Also by transversality the Hamiltonian on the optimal path vanishes identically to zero,

$$H^*(X^*, S^*, V^*, \lambda_x^*, \lambda_s^*, \lambda_v^*, F^*) = 0, \quad 0 \leq t \leq T_f \quad (18)$$

Maximization of H by $F(t)$ yields an optimal input function of the type

$$F(t) = \begin{cases} F_{\max}, & \phi_s(t) > 0 \\ F_S, & \phi_s(t) = 0, \quad t_1 < t < t_2 \\ 0, & \phi_s(t) < 0 \end{cases} \quad (19)$$

where $\phi_s(t)$, the switching function is

$$\phi_s(t) = \frac{\partial H}{\partial F} = \lambda_v + \frac{\lambda_s (S_F - S)}{V} - \frac{\lambda_x X}{V} \quad (20)$$

and F_S denotes a singular control policy yet to be determined for the interval (t_1, t_2) in which the switching function is identically zero,

$$\phi_s(t) = 0 = \lambda_v + \frac{\lambda_s (S_F - S)}{V} - \frac{\lambda_x X}{V}, \quad t_1 < t < t_2 \quad (21)$$

The fact that $\phi_s(t)$ is identically zero over the entire interval implies also that all of its time derivatives of any order must also vanish over the same interval. Thus differentiation of eq. (21) and substitution of eqs. (6)–(8), (15)–(17) into it yield

$$\frac{d\phi_s}{dt} = \left[\frac{X(S_F - S)}{V} \right] \left[\lambda_s \left(\frac{\mu}{Y} \right)' - \lambda_x \mu' \right] = 0 \quad (22)$$

Equation (22) implies

$$\lambda_s (\mu/Y)' - \lambda_x \mu' = 0 \quad (23)$$

Once more $d\phi/dt$ is differentiated, set to zero, and rearranged using eqs. (6)–(8), (15)–(17), and (23) to obtain

$$\mu \left[\left(\frac{\mu}{Y} \right)' - \frac{\mu'}{Y} \right] + \frac{d}{dt} \left[\left(\frac{\mu}{Y} \right)' \right] - \frac{(\mu/Y)'}{\mu'} \frac{d}{dt} (\mu') = 0 \quad (24)$$

Carrying out the indicated time differentiation $[(d/dt) g(S) = (dg/dS)(dS/dt)]$ and substituting eq. (7) into the result yield

$$F_S = \frac{V}{S_F - S} \left(\frac{\mu X}{Y} - \frac{\mu \mu' [(\mu/Y)' - \mu'/Y]}{\mu'(\mu/Y)'' - (\mu/Y)' \mu''} \right) \quad (25)$$

where the primes denote differentiation with respect to S. Therefore, the optimal feed policy is

$$F(t) = \begin{cases} F_{max} & \text{if } \phi_s(t) > 0, \quad 0 < t < t_1 \\ F_{si} = \begin{cases} F_{max} & \text{if } F_S > F_{max} \\ F_S & \text{if } F_S < F_{max} \\ 0 & \text{if } F_S < 0 \end{cases} & \text{if } \phi_s(t) = 0, \quad t_1 < t < t_2 \\ 0 & \text{if } \phi_s(t) < 0, \quad t_f < t < T_F \end{cases} \quad (26)$$

According to eq. (26) the general optimal feed rate profile consists of three stages; the initial filling stage, $0 < t < t_1$, in which the feed rate is held at the maximum value allowed, F_{max}, followed by the second stage, $t_1 < t < t_2$, in which the feed policy is singular, F_S, and finally followed by the third stage, a batch operation where the cells are allowed to grow to final desired cell concentration. The cells are rapidly pumped out of the reactor to an optimum volume and the whole process is repeated. Hence the problem is reduced to finding the switching times t_1 and t_2 and the batch times t_f and T_f. At this point, it is convenient to treat the simple case of constant yield first.

Constant Yield

In this case eq. (23) reduces to

$$(\lambda_s/Y - \lambda_x)\mu' = 0 \quad (27)$$

or

$$\frac{d\mu}{dS} = 0 \quad (28)$$

and eq. (25) reduces to

$$F_S = V\mu X/Y(S_F - S) \quad (29)$$

Equation (28) implies that during the singular control the substrate concentration should be held constant at $S = S_c$, the value at which the specific growth rate is at its maximum. A constant substrate concentration implies a corresponding constant value of $\mu = \mu(S_c)$ so that eq. (1) can be integrated to yield

$$XV = X(t_1) V(t_1) \exp[\mu(S_c)(t - t_1)] \quad (30)$$

where t_1 denotes the time at which $S(t_1) = S_c$. Equation (30) is now substituted into eq. (29) to obtain

$$F_S = V(t_1) \mu(S_c)[X(t_1)/Y(S_F - S_c)] \exp[\mu(S_c)(t - t_1)] \quad (31)$$

according to which the singular control is an exponential feed rate. Note that for the constant yield case

$$X(t_1) = Y(S_F - S_c) = Y[S_F - S(t_1)] \quad (32)$$

provided that

$$X_0 = Y(S_F - S_0) \tag{33}$$

Therefore, eq. (31) simplifies to

$$F_S = V(t_1)\,\mu(S_c)\,\exp[\mu(S_c)(t - t_1)] \tag{34}$$

Therefore,

$$F_S/V = D = \mu(S_c) \tag{35}$$

According to which the dilution rate is also kept constant. The cell balance equation, eq. (6), reduces to

$$\frac{dX}{dt} = 0 \tag{36}$$

Equation (36) states that the cell concentration is held constant at $X = X_c = Y(S_F - S_c)$.

Thus for the constant yield case the singular solution is an exponential feed rate that forces not only the substrate concentration but also the cell concentration to remain constant at S_c and X_c, respectively, for the duration of time $t_1 < t < t_2$. As shown by Lim et al. [10] this situation described by eqs. (28)–(36) is entirely equivalent to an extended culture in which the substrate and cell concentrations are kept constant.

The general form of optimal policy is then, with $S(t_1) = S_c$,

$$F(t) = \begin{cases} F_{max}, & 0 < t < t_1 \quad (37a) \\ V(t_1)\,\mu(S_c)\exp[\mu(S_c)(t - t_1)] = \mu(S_c)V \le F_{max}, & t_1 \le t \le t_2 \quad (37b) \\ 0, & t_f \le t \le T_f \quad (37c) \end{cases}$$

The reactor should be fed at the maximum allowed flow rate until the substrate concentration reaches the value at which the specific growth rate is at the maximum. This is the switching time, t_1, at which the flow rate is changed to the exponential form given by eq. (37b) and this form is continued until the reactor is full provided that it never exceeds the allowable maximum, F_{max}. If the exponential flow rate reaches the maximum allowable F_{max} during the interval (t_1, t_2), then the flow rate is set at F_{max} and the reactor is fed at this rate until it is full, $t = t_f$. Then, the reactor is operated in batch mode until the specified cell concentration, X_f, is realized ($t = T_f$), at which point the contents are pumped out as rapidly as possible to a specified volume, V_0, for another cycle. This is then the optimal fed-batch operation. This policy is illustrated in Figure 1. Note that all switching times have been determined in terms of measurable

Fig. 1. Optimal feed profiles.

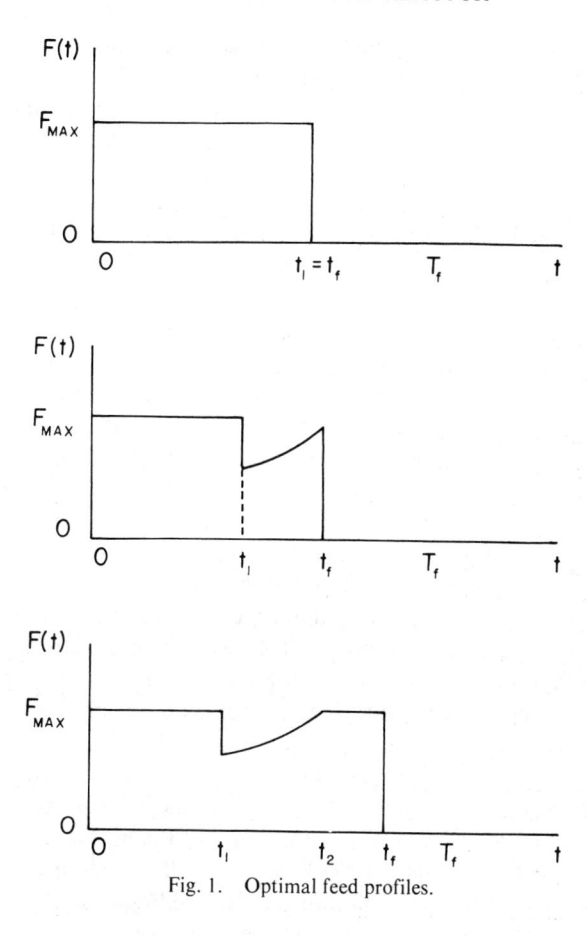

state variables. Since the cell productivity depends on V_0, the value of V_0 that gives the highest cell productivity must be found by numerical search.

From a practical standpoint the policy given by eq. (37) requires a continuous monitoring of S to obtain t_1, of S and V to implement the exponential feed rate, t_2 and t_f, and X to obtain T_f. In other words a complete monitoring of the state variables is needed. In this sense it is a form of feedback policy. However, in practice continuous measurements of all state variables may be difficult, particularly the substrate concentration. Open-loop policy may be used with a computer by having the computer integrate the state equations with only the known initial conditions, X_0, S_0, and V_0 so that it can generate t_1, the exponential feed rate [eq. (37b)], t_2, and T_f. Often it is possible to monitor X (and V). In this situation it may be possible to replace the exponential feed rate of eq. (37b) by a simple feedback of X to maintain the turbidity constant as indicated by eq. (36). This should result in the required exponential feed rate and also maintain the specific growth rate at the highest possible value.

In many cases the maximum rate at which the medium can be pumped into the reactor may be sufficiently high in comparison to the maximum growth rate so that it may be assumed that practically no growth takes place during filling at the maximum feed rate. In this case feed rate approaches the impulse function. The input policy is then an instantaneous fill to bring the substrate concentration to S_c followed by the singular control and batch mode operation. It should be noted that for cultures having monotonic specific growth rates, such as Monod growth, the optimal policy is to fill up the reactor rapidly as possible, run in batch mode, and then to pump out to the optimum volume V_0. Numerical results for the Monod equation with parameter values given in Table I are shown in Figures 2 and 3. The effects of the maximum feed rate on the optimum productivity and the optimum initial volume ratio are shown in Figure 2 for two values of substrate feed concentration. As expected, productivity is enhanced by larger values of feed rate and substrate feed concentration, with the limiting values of feed rate representing the case of instantaneous fill, or impulse input. The effect of maximum feed rate, F_{max}, on the optimum volume ratio can be explained by examining eq. (9). As F_{max} decreases the filling time and the total cycle time increase. This increases the value of the denominator. To compensate for this, V_0 decreases so as to increase the numerator of eq. (9). The effect of the initial volume ratio on productivity is shown in Figure 3 for the case of instantaneous fill. The initial volume has a dramatic effect, with the productivity changing by as much as a factor of 5. Operation with V_0/V_{max} close to zero might be an intuitive mode of operation since it utilizes most of the fermentor volume. However, the cell productivity is only 20% of the optimum operation in which only 24% of the reactor volume is withdrawn ($V_0/V_{max} = 0.76$). Although the reactor volume available is larger with smaller V_0/V_{max}, the cell density is much lower after the rapid fill when V_0/V_{max} is small so that the net result is a lower cell production. With higher values of V_0/V_{max} more cells are retained and therefore the cell concentrations after instantaneous fill are higher so that the cell production rate per unit volume is higher. However, the reactor volume withdrawn is also smaller so that there is an optimum in terms of the total cell production rate.

The results for the case of inhibited growth and instantaneous fill are given in Figures 4 and 5 for Andrews' growth model [11] with parameter values given

TABLE I
Growth Models

1. Monod
$$\mu = \frac{S}{0.1 + S}$$
$Y = 0.473$
$S_F = 0.1$
$V_M = 10$ liters

2. Andrews
$$\mu = \frac{S}{0.03 + S + 0.5S^2}$$
$Y = 0.45$
$S_F = 0.001$ g/liter
$V_M = 10$ liters

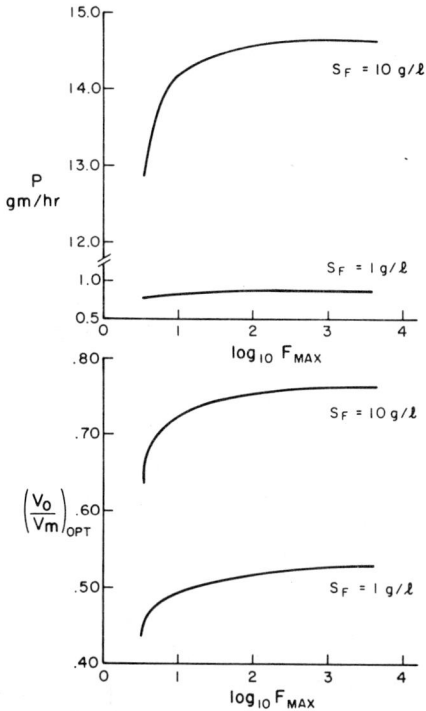

Fig. 2. Effects of F_{max} on performance index and optimal initial volume for Monod growth.

in Table I. In this case the time optimal filling policy involves an instantaneous fill up to the substrate concentration which gives the maximum specific growth rate, followed by an exponential filling policy that maintains μ at its maximum value until the vessel is full. Then the reactor is operated in batch mode until the final desired cell concentration is achieved and then the contents are withdrawn rapidly to an optimum volume. As shown in Figure 4 the optimum volume ratio decreases as the final substrate concentration is reduced. Once again, referring to eq. (9), as the final substrate concentration decreases the batch time increases, causing the total cycle time to increase, and therefore the numerator must increase to improve productivity. Figure 5 shows the strong effect of initial volume on the productivity. It is seen that the productivity is greatly reduced, even with the time optimal input of feed rate, if the initial volume differs appreciably from the optimal ratio of 0.873. Also shown in Figure 5 is the performance of the suboptimal policy in which the reactor is filled instantaneously to its maximum volume, operated in batch mode until the desired cell concentration is achieved, and then pumped out rapidly to a prescribed volume. In all situations the suboptimal policy is considerably inferior to the optimal policy. Note that the optimal initial volume for the suboptimal policy is quite close to that of the optimal policy. The superiority of the optimal feed policy over the suboptimal policy is not as great at the optimal initial volume as at other initial volumes,

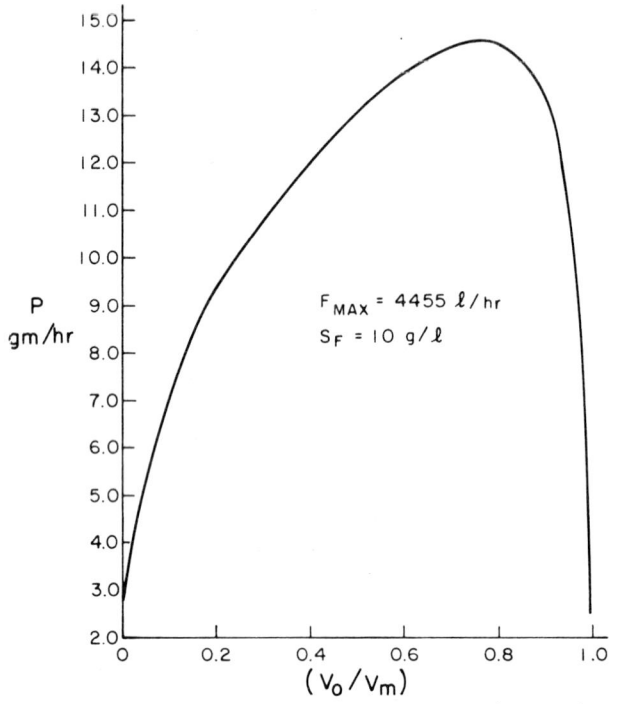

Fig. 3. Dependence of performance index on initial volume for Monod growth.

indicating that for this example the effect of V_0 overshadows the effect of optimal feed policy.

Variable Yield

The general form of the optimal policy was given by eq. (26). The singular control is a complex exponential function which can be obtained from eqs. (25) and (3),

$$F_S(t) = V(t_1) f \exp\left(\int_{t_1}^{t} f \, d\tau\right) \tag{38}$$

where

$$f = \left(\frac{\mu X}{Y} - \frac{\mu\mu'[(\mu/Y)' - \mu'/Y]}{\mu'(\mu/Y)'' - (\mu/Y)'\mu''}\right) \Big/ (S_F - S) \tag{39}$$

According to eq. (38) the preexponential factor as well as the exponent vary with time. It also reveals that if $f < 0$ the singular control cannot be realized so that F_S must be set at zero. Likewise, if $F_S > F_{\max}$, it must be set at $F_{\max}$. Let us see the fate of the substrate and cell concentration during the singular interval. Substitution of eq. (25) into eq. (7) yields

$$\frac{dS}{dt} = \frac{-\mu\mu'[(\mu/Y)' - \mu'/Y]}{\mu'(\mu/Y)'' - (\mu/Y)'\mu''}, \qquad t_1 \le t \le t_2 \tag{40}$$

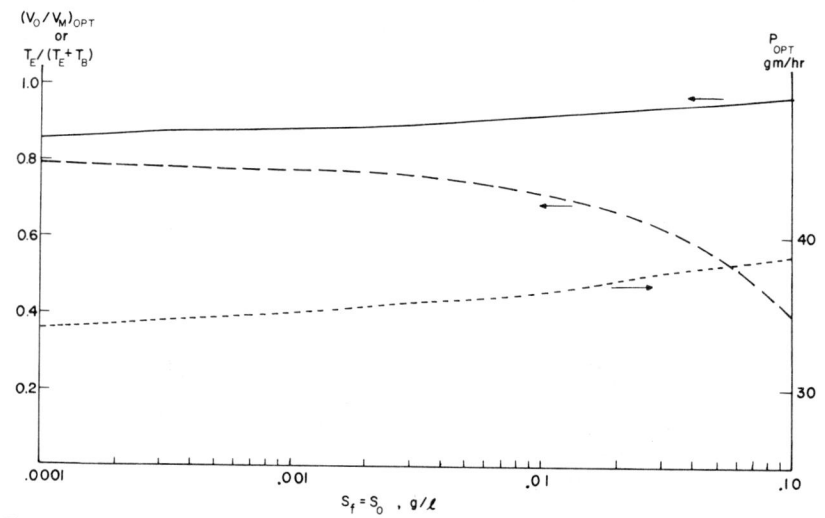

Fig. 4. Effects of final substrate concentration on performance for Andrews' growth model. (—) $(V_0/V_m)_{opt}$; (— —) $T_E/(T_E + T_B)$; (- - -) P_{opt}.

Inspection of eq. (40) reveals that in general the substrate concentration varies, in contrast to the constant yield case in which the substrate concentration is held constant at the value that gives the maximum specific growth rate. Substitution of eq. (25) into eq. (6) and rearrangement using eq. (40) yield the information regarding the fate of the cell concentration,

$$\frac{dX}{dt} = \mu X - \frac{\mu X^2}{Y(S_F - S)} - \frac{X}{S_F - S}\frac{dS}{dt}, \quad t_1 \leq t \leq t_2 \quad (41)$$

that also predicts that the cell concentration varies during the interval again in contrast to the constant yield case. In summary, the optimal policy consists of a rapid fill, followed by a complex exponential feed policy, and a batch mode of operation.

Unlike the case of constant yield in which the switching time, t_1, was predetermined by the fact that at $t = t_1$, $S = S_c$, in this situation no such information is available and hence an iterative technique must be used to solve for the optimum t_1. A procedure would be to integrate eqs. (6)–(8) with $F = F_{max}$ starting from the known $X_0 = X_f$ with an assumed S_0 and V_0 up to switching time, t_{1g}. At this point we implement the singular control F_S given by eq. (25) provided that $0 \leq F_S \leq F_{max}$. If $F_S \geq F_{max}$, we must set $F_S = F_{max}$ and there is no singular control. Let us suppose that $F_S < F_{max}$ at $t = t_{1g}$ so that we implement $F = F_S$ and continue to integrate eqs. (6)–(8) as long as $F_S < F_{max}$ and $V < V_{max}$. When the magnitude of the singular control reaches F_{max} ($F_S \geq F_{max}$), the time should be $t = t_2$, at which point we set $F = F_{max}$ and integrate until $V = V_{max}$. At this point in time, $t = t_f$ and batch operation takes place until the final time $X(T_f) = X_f$. At this point $S(T_f) = S_f$. Because of cyclic operation we must also meet the cyclic condition

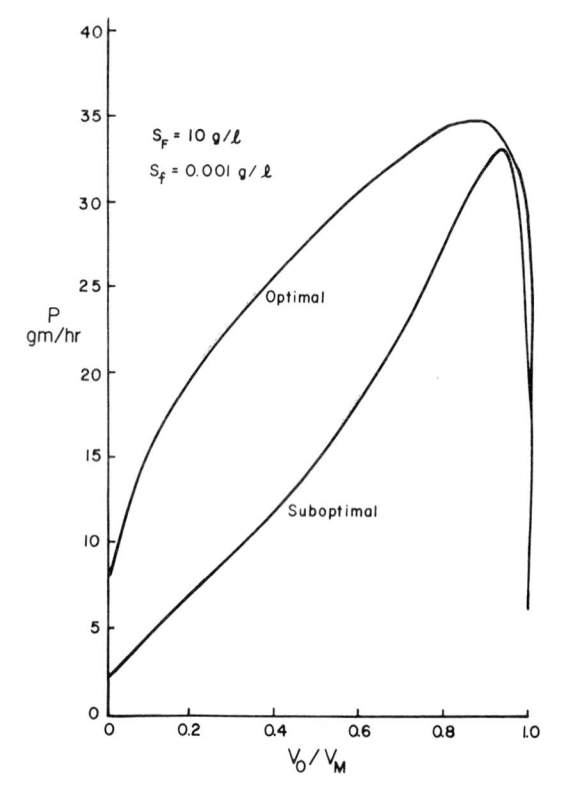

Fig. 5. Comparison of optimal and suboptimal feed policies.

$$X(0) = X_0 = X_f = X(T_f) \tag{42}$$

$$S(0) = S_0 = S_f = S(T_f) \tag{43}$$

In general, the above procedure does not yield the final substrate concentration, S_f, which agrees with the assumed initial value, S_0. Therefore, it is now necessary to iterate on S_0 until it agrees with S_f. Then, we evaluate the performance given by eq. (9). The above procedure must now be repeated for different values of t_1 until there is found the best t_1 that gives the highest productivity defined by eq. (9). Since the optimum t_1 is now found for an arbitrary culture volume, V_0, the entire procedure must be repeated for different values of V_0 until the one that gives the highest cell productivity is found. Thus an extensive numerical computation is required to obtain the optimum starting volume V_0 and the optimum feed policy that gives the highest cell productivity.

SUMMARY AND CONCLUSIONS

The optimal operation of a repeated fed-batch reactor for maximal biomass productivity was determined by using the maximum principle with the initial reactor volume as a parameter. The final cell concentration and feed substrate concentration were prescribed. The optimal cell productivity by manipulation of the substrate feed rate was formulated as a minimum time problem under constrained inlet substrate flow rate and maximum fermentor volume. The productivity is also dependent on initial volume and its optimal value was obtained by numerical search.

Analyses were performed for both the constant and variable yield cases. The general optimal substrate feed policy consists of an initial filling portion where the feed rate is at the maximum value permitted, followed by a singular portion in which the feed rate is a complex exponential function, and finally followed by a batch portion of the cycle to the final desired cell concentration. Finally, the cells in the reactor are rapidly pumped out to a specified optimum volume. Then, the process is repeated. For the constant yield case the singular control keeps the cell and substrate concentrations and dilution rate at constant values, while all these three generally vary during the singular interval for the case of variable yield. For Monod growth it is shown that optimal fed-batch operation consists of a instantaneous fill to the maximum reactor volume followed by batch operation until the final desired cell concentration is achieved. For growth exhibiting a maximum value of specific growth rate, the optimum policy would be an instantaneous fill to the cell and substrate concentration corresponding to this maximum μ, followed by the singular exponential filling function, finally followed by batch operation to final conversion. The case of finite volumetric flow rate has the possibility of three filling policies, depending on the magnitude of the upper bound on the substrate mass flow. Optimal productivity depends on the initial fermentor volume at the beginning of the fed-batch cycle. A simple means to implement the policies for inhibited growth with a constant biomass yield would be a rapid fill to the maximum specific growth rate, followed by the operation in which the reactor turbidity is held constant with the feedback control of the substrate addition during the remaining of the filling portion, followed by the batch portion, and finally by draw-off to the optimum initial volume.

Nomenclature

F	volumetric flow rate (liter/hr)
F_s	volumetric flow rate during the singular control, eq. (25)
f	defined in eq. (39)
H	Hamiltonian
r_s	rate of generation of substrate [g/(liter)(hr)]
r_x	rate of generation of cells [g/(liter)(hr)]
S	substrate concentration (g/liter)
S_c	substrate concentration at t_1 (g/liter)
S_F	feed substrate concentration (g/liter)
t	time (hr)

t_1	time at which $\phi_s = 0$
t_{1g}	estimated t_1
t_2	time at which $\phi_s < 0$
t_f	time at which the reactor is full
T_f	cycle time
V	reactor volume (liter)
X	cell concentration (g/liter)
X_c	cell concentration at t_1
Y	biomass yield

Greek

ϕ_s	switching function
λ_s	adjoint variable associated with S
λ_x	adjoint variable associated with X
λ_V	adjoint variable associated with V
μ	specific growth rate (hr^{-1})

Subscripts

f	final
max	maximum
0	initial

This work was supported in part by National Science Foundation grants No. ENG 75-17796 and No. ENG 76-22309.

References

[1] B. J. Chen, W. Hirt, H. C. Lim, and G. T. Tsao, *Appl. Environ. Microbiol.*, 33, 269 (1977).
[2] H. Y. Wang, C. L. Cooney, and D. I. C. Wang, *Biotechnol. Bioeng.*, 19, 69 (1977).
[3] L. N. Andreyeva and V. V. Biryukov, *Biotechnol. Bioeng. Symp.*, 4, 61 (1973).
[4] V. M. Fishman and V. V. Biryukov, *Biotechnol. Bioeng. Symp.*, 4, 647 (1974).
[5] H. Ohno, E. Nakanishi, and T. Takamatsu, *Biotechnol. Bioeng.* 18, 847 (1976).
[6] A. Miele, *Optimization Techniques*, G. Leitman, Ed. (Academic, New York, 1962).
[7] H. Ohno, E. Nakaniski, and T. Takamatsu, *Biotechnol. Bioeng.*, 20, 625 (1978).
[8] J. H. Kelley, *J. SIAM Control*, 2, 234 (1965).
[9] T. Yamané, T. Kume, E. Sada, and T. Takamatsu, *J. Ferment. Technol.*, 55, 587 (1977).
[10] H. C. Lim, B. J. Chen, and C. C. Creagan, *Biotechnol. Bioeng.*, 19, 425 (1977).
[11] J. F. Andrews, *Biotechnol. Bioeng.*, 10, 707 (1968).

Control of the Quasi-Steady-State in Fed-Batch Fermentation

THOMAS J. BOYLE

Department of Chemical Engineering, McGill University, Montreal, Quebec, Canada H3A 2A7

INTRODUCTION

In fed-batch fermentation the volume of material in the fermentor and the cell mass increase with time until the process is terminated. The fed-batch approach makes a high yield of cells on substrate possible by keeping substrate concentration low at all times. Normally both cell and substrate concentration change with time during the run. Theoretical analyses [1,2] have shown that under certain conditions cell and substrate concentration will be time invariant even though the cell mass and reaction volume are growing. This has been called a quasi-steady-state (QSS). It would certainly be desirable to investigate this phenomenon experimentally. The present work extends the previous efforts by elucidating the question of the rate of approach to QSS and by using the techniques of control systems engineering to propose a linear control law that will make the state accessible to experimenters.

PROCESS MODEL

The process model is based on material balances on limiting substrate and cells and a volume balance on the fermentor as in equations (1)–(3):

$$\frac{ds}{dt} = D(s_1 - s) - \sigma x \tag{1}$$

$$\frac{dx}{dt} = (\mu - D)x \tag{2}$$

$$\frac{dV}{dt} = DV \tag{3}$$

Although there are other possible interpretations, here QSS is taken to be

$$\frac{ds}{dt} = \frac{dx}{dt} = 0 \tag{4}$$

from which the basic conditions for QSS are obtained:

$$D = \sigma x/(s_1 - s) \tag{5}$$

$$\mu = D \tag{6}$$

In principle, if D and s_1 are fixed at suitable values $\bar{D}$ and $\bar{s}_1$, a QSS will be reached in which substrate concentration is defined by

$$\bar{\mu} = \bar{D} \tag{7}$$

and cell concentration by

$$\bar{x} = (\bar{\mu}/\bar{\sigma})(\bar{s}_1 - \bar{s}) \tag{8}$$

Here, $\bar{s}$ and $\bar{x}$ represent the state variables at QSS; it is assumed that μ and σ are functions only of the substrate concentration.

Neither the foregoing nor what follows is dependent on the assumption of Monod kinetics; however, such can be used to provide a convenient numerical example appropriate to bakers' yeast fermentation. Let

$$\mu = 0.3s/(0.1 + s) \tag{9}$$

$$\sigma = \mu/Y \tag{10}$$

$$Y = 0.25, \qquad \bar{D} = 0.1, \qquad \bar{s}_1 = 40.05$$

Then from eqs. (7) and (8),

$$\bar{s} = 0.05, \qquad \bar{x} = 10$$

NEED FOR CONTROL

The feasibility of observing the QSS depends on the rate of approach to equilibrium of the system defined by eqs. (1) and (2). Since these equations are nonlinear, it is appropriate to address the issue of rate of approach to equilibrium by linearizing the equations about the QSS operating point and investigating the transient behavior in this vicinity. For this purpose, define normalized incremental variables as follows:

$$\begin{aligned} s^* &= (s - \bar{s})/\bar{s}, & x^* &= (x - \bar{x})/\bar{x} \\ D^* &= (D - \bar{D})/\bar{D}, & s_1^* &= (s_1 - \bar{s}_1)/\bar{s}_1 \end{aligned} \tag{11}$$

and introduce two parameters

$$\begin{aligned} a &= \left.\frac{\partial \sigma}{\partial s}\right|_{s=\bar{s}} \\ b &= \left.\frac{\partial \mu}{\partial s}\right|_{s=\bar{s}} \end{aligned} \tag{12}$$

Then, the linear incremental model is

$$\frac{ds^*}{dt} = -(\bar{D} + a\bar{x})s^* - \frac{\bar{D}(\bar{s}_1 - \bar{s})}{\bar{s}}x^* + \frac{\bar{D}(\bar{s}_1 - \bar{s})}{\bar{s}}D^* + \frac{\bar{D}\bar{s}_1}{\bar{s}}s_1^* \tag{13}$$

$$\frac{dx^*}{dt} = b\bar{s}s^* - \bar{D}D^* \tag{14}$$

If
$$b/a = \bar{\mu}/\bar{\sigma} \tag{15}$$

i.e., if incremental yield equals total yield these equations can be simplified and put in matrix form as

$$\begin{bmatrix} s^{*\prime} \\ x^{*\prime} \end{bmatrix} = \begin{bmatrix} -(\bar{D} + a\bar{x}) & -\dfrac{\overline{Dx}}{Y\bar{s}} \\ aY\bar{s} & 0 \end{bmatrix} \begin{bmatrix} s^* \\ x^* \end{bmatrix} + \begin{bmatrix} \dfrac{\overline{Dx}}{Y\bar{s}} & \dfrac{\overline{Ds_1}}{\bar{s}} \\ -\bar{D} & 0 \end{bmatrix} \begin{bmatrix} D^* \\ s_1^* \end{bmatrix} \tag{16}$$

which can be stated more compactly as

$$\mathbf{z}' = \mathbf{Az} + \mathbf{Bu} \tag{17}$$

Key information concerning the rate of approach to equilibrium can be found from the eigenvalues of the matrix A

$$\begin{aligned} \lambda_s &= -\bar{D} \\ \lambda_L &= -a\bar{x} \end{aligned} \tag{18}$$

where λ_s is the smaller eigenvalue, as can be seen from our numerical example, where $\lambda_s = -0.1$ and $\lambda_L = -53.33$. The approach to equilibrium is by way of decaying exponentials having time constants given by the negative reciprocals of the eigenvalues

$$\begin{aligned} \tau_L &= -1/\lambda_s = 1/\bar{D} \\ \tau_s &= -1/\lambda_L = 1/a\bar{x} \end{aligned} \tag{19}$$

or for the numerical case $\tau_L = 10$ hr, $\tau_s = 0.01875$ hr. Systems having such widely separated eigenvalues are known as stiff systems and can present special problems in simulation and control.

From the numerical example, it is seen that the rate of approach to equilibrium is quite slow. If we assume that three to five time constants of the response are required, an experiment to achieve QSS would extend to days. Actually, it is virtually infeasible to observe the state without a control system, as can be seen by the following argument. A given apparatus will be characterized by the maximum volume it can handle V_M and by an appropriate initial volume V_0. These values are dictated by the practical aspects of size, mixing, heat transfer, etc. For operation at constant dilution factor, eq. (3) may be integrated to give the maximum time the apparatus can run

$$t_M = (1/\bar{D})\ln(V_M/V_0) \tag{20}$$

The number of time constants of the response which can be observed is given by

$$t_M/\tau_L = \ln(V_M/V_0) \tag{21}$$

Note that this ratio is independent of dilution factor and other system parameters. For example, if a laboratory apparatus has $V_M = 12$, $V_0 = 3$, then only one time constant of the response can be observed. With $V_0 = 3$, the apparatus would have to accommodate 148 liter to observe five time constants. Equation (21) clearly establishes the need for a control system if QSS is to be observed.

ANALYSIS OF THE UNCONTROLLED PROCESS

According to the process model [eqs. (1) and (2)] there are two state variables whose value is to be set, s and x, and two variables that can be manipulated, D and s_1. The mathematical aspect of control systems design is to specify a rule for manipulating D and s_1 to cause s and x to take on desired values s_D and x_D. The simplest design tool available is the influence coefficient matrix relating the sustained change in s and x to applied sustained changes in D and s_1. In terms of eq. (17) this is given as $-\mathbf{A}^{-1}\mathbf{B}$, which yields

$$\begin{bmatrix} s^* \\ x^* \end{bmatrix} = \begin{bmatrix} \dfrac{\bar{D}}{aY\bar{s}} & 0 \\ \dfrac{\bar{D}}{a\bar{x}} & -\dfrac{Y\bar{s}_1}{\bar{x}} \end{bmatrix} \begin{bmatrix} D^* \\ s_1^* \end{bmatrix} \qquad (22)$$

It can be seen that the only nonzero off-diagonal element is quite small. For the numerical case,

$$\begin{bmatrix} s^* \\ x^* \end{bmatrix} = \begin{bmatrix} 1.5 & 0 \\ -0.002 & 1.001 \end{bmatrix} \begin{bmatrix} D^* \\ s_1^* \end{bmatrix} \qquad (23)$$

In terms of sustained changes the system may be said to be decoupled. It appears that the control strategy should use D to control s and s_1 to control x.

Given an arbitrary initial state, $s^*(0)$ and $x^*(0)$, and constant inputs, $D^*(t)$ and $s_1^*(t)$, the transient response of the system described by eq. (17) can be formally written

$$\mathbf{z}(t) = \mathbf{E}(t)\mathbf{z}(0) + \mathbf{F}(t)\bar{u} \qquad (24)$$

which in this case implies

$$\begin{bmatrix} s^*(t) \\ x^*(t) \end{bmatrix} = \frac{1}{1 - \bar{D}/a\bar{x}} \begin{bmatrix} e^{-\bar{a}\bar{x}t} - \dfrac{\bar{D}}{a\bar{x}}e^{-\bar{D}t} & \dfrac{\bar{D}}{aY\bar{s}}(e^{-\bar{a}\bar{x}t} - e^{-\bar{D}t}) \\ \dfrac{Y\bar{s}}{\bar{x}}(e^{-\bar{D}t} - e^{-\bar{a}\bar{x}t}) & e^{-\bar{D}t} - \dfrac{\bar{D}}{a\bar{x}}e^{-\bar{a}\bar{x}t} \end{bmatrix} \begin{bmatrix} s^*(0) \\ x^*(0) \end{bmatrix}$$

$$+ \begin{bmatrix} \dfrac{\bar{D}}{aY\bar{s}}(1 - e^{-\bar{a}\bar{x}t}) & \dfrac{\bar{D}}{a\bar{x}}(e^{-\bar{D}t} - e^{-\bar{a}\bar{x}t}) \\ -\dfrac{\bar{D}}{a\bar{x}}(1 - e^{-\bar{a}\bar{x}t}) & \dfrac{Y\bar{s}_1}{a}\left(1 - \dfrac{1}{1 - \bar{D}/a\bar{x}}e^{-\bar{D}t} + \dfrac{\bar{D}/a\bar{x}}{1 - \bar{D}/a\bar{x}}e^{-\bar{a}\bar{x}t}\right) \end{bmatrix}$$

$$\times \begin{bmatrix} D^* \\ s_1^* \end{bmatrix} \qquad (25)$$

Since the first column of the matrix **F** contains no exponential terms involving the smaller eigenvalue, it is seen that the process will respond rapidly to changes in dilution factor. The influence coefficient matrix showed that changing dilution factor produces large changes in s, but only small changes in x. It is, therefore, anticipated that it will be easy to achieve fast noninteracting control of substrate concentration using dilution factor as the manipulated variable.

Much more can be gleaned from eq. (25) if it is evaluated at the base case,

$$\begin{bmatrix} s^*(t) \\ x^*(t) \end{bmatrix} = 1.002 \begin{bmatrix} e^{-50t} - 0.002e^{-0.1t} & 1.5(e^{-50t} - e^{-0.1t}) \\ 0.00125(e^{-0.1t} - e^{-50t}) & e^{-0.1t} - 0.002e^{-50t} \end{bmatrix}$$

$$\times \begin{bmatrix} s^*(0) \\ x^*(0) \end{bmatrix}$$

$$+ \begin{bmatrix} 1.5(1 - e^{-50t}) & 1.502(e^{-0.1t} - e^{-50t}) \\ -0.002(1 - e^{-50t}) & 1.001(1 - 1.002e^{-0.1t} + 0.002e^{-50t}) \end{bmatrix}$$

$$\times \begin{bmatrix} D^* \\ s_1^* \end{bmatrix} \qquad (26)$$

By considering each of the elements of the matrix **E** in turn the nature of transients in the uncontrolled system emerges. From E_{11} in eq. (26), it is seen that if the substrate concentration is displaced from equilibrium, it will rapidly return since the response is dominated by the large eigenvalue. From the element E_{21} it is seen that the transient induced in x by a displacement in s is very small in magnitude. From the element E_{22}, it is seen that the response of cell concentration to a displacement from equilibrium is slow, being dominated by the small eigenvalue. From the element E_{12} it is seen that a displacement in x will induce a transient displacement in s of substantial magnitude and long response time.

In like manner, the capability of the control inputs D and s_1 can be determined by examining the elements of **F** as given in eq. (26). The earlier conclusions regarding D are confirmed by elements F_{11} and F_{21}. Element F_{22} reveals that while the sustained response of cell concentration to changes in feed composition is large, it is also slow, being dominated by the small eigenvalue. Furthermore, the tentative indication of noninteraction between s_1 and s as given by the influence coefficient matrix is not confirmed: changes in s_1 induce large, long, slow transients in s.

In summary, there is strong evidence for two qualitative conclusions. First, it will be relatively easy to achieve fast noninteracting control over substrate concentration using dilution factor. Second, care must be taken in the design of the cell concentration controller that must combat the strong natural interaction between feed concentration and substrate concentration and must speed up the process response by an order of magnitude.

MODERN CONTROL APPROACH

Modern control theory presents a formal method for obtaining control laws for systems described by linear equations. Here the presentation of Newell and Fisher [3] is followed since it handles the needs of process control well. The controller is to be of the discrete time quadratic optimal class. For this, it is necessary to produce a discrete time process model. It does not seem reasonable to do this on a basis other than that of the numerical example. The general form is

$$\mathbf{z}[(n + 1)T] = \mathbf{\Phi z}(nT) + \mathbf{\Delta u}(nT) \tag{27}$$

The matrices $\mathbf{\Phi}$ and $\mathbf{\Delta}$ are readily found from eq. (26). Taking $T = 0.1$,

$$\begin{bmatrix} s*[0.1(n+1)] \\ x*[0.1(n+1)] \end{bmatrix} = \begin{bmatrix} 0.003 & -1.48 \\ 0.0012 & 0.992 \end{bmatrix} \begin{bmatrix} s*(0.1n) \\ x*(0.1n) \end{bmatrix} + \begin{bmatrix} 1.49 & 1.48 \\ -0.00187 & 0.0081 \end{bmatrix} \begin{bmatrix} D*(0.1n) \\ s_1^*(0.1n) \end{bmatrix} \tag{28}$$

The controller is to be given as

$$\mathbf{u}(nT) = \mathbf{K}_{FB}\mathbf{z}(nT) + \mathbf{K}_{sp}\mathbf{y}_{sp}(nT) \tag{29}$$

where in our case

$$\mathbf{y}_{sp} = \begin{bmatrix} s_D^* \\ x_D^* \end{bmatrix}$$

$\mathbf{K}_{FB}$ and $\mathbf{K}_{sp}$ are constant matrices providing for state feedback and setpoint inputs to the process. Newell and Fisher provide results for a quite general objective function, but the simplest case in which only the squared error in the state variables is counted and they are equally weighted is suitable for the present purpose. In this case, their result reduces to

$$\mathbf{K}_{FB} = \mathbf{\Delta}^{-1}\mathbf{\Phi} \tag{30}$$

for the feedback matrix and for all practical purposes

$$K_{sp} = \Delta^{-1} \tag{31}$$

for the setpoint matrix. In terms of the numerical case,

$$\begin{bmatrix} D^* \\ s_1^* \end{bmatrix} = \begin{bmatrix} 0.121 & 99.7 \\ -0.124 & -99.4 \end{bmatrix} \begin{bmatrix} s^* \\ x^* \end{bmatrix} + \begin{bmatrix} 0.546 & -99.75 \\ 0.126 & 100.4 \end{bmatrix} \begin{bmatrix} s_D^* \\ x_D^* \end{bmatrix} \tag{32}$$

There are a number of reasons why it would not be advisable to attempt to control a fermentor using eq. (32). It can be shown that the control is of a type known as deadbeat, i.e., it trys to reduce the error to zero in exactly one sample period and hold it there for all time. Consider, for example, trying to move a fermentor from a substrate concentration of 0.1 and a cell concentration of 8 to $s = 0.05$ and $x = 10$ in 6 min! The high gain values in the matrices $\mathbf{K}_{FB}$ and $\mathbf{K}_{sp}$ are indicative of the difficulties.

The objective here is rather to extract from the modern control theory result insights that can aid in the design of practical controllers. To this end note that of the eight matrix elements in eqs. (32), four are much larger than the others. Furthermore, these four multiply x^* or x_D^*; all the coefficients relating to s^* and s_D^* are small. Since these equations are written in terms of normalized deviation variables, the difference in coefficient size cannot be attributed to variable scaling. With some precision it can be said that the optimal feedback controller bases its choice of both D and s_1 on cell concentration data rather than substrate concentration data. Also, when a change is called for in cell concentration it moves D and s_1 by equivalent amounts, but in opposite directions. When a change in substrate concentration is called for the predominant change is in dilution factor.

These broad generalizations from eq. (32) are consonant with the earlier generalizations from the transient response matrix, eq. (26). The optimal controller expects displacements in s to decay away rapidly on their own and therefore takes little action. It uses very high gain control action on errors in cell concentration to greatly speed the response. It decouples the cell concentration controller from substrate concentration.

DIRECT APPROACH TO CONTROLLER SYNTHESIS

Using these insights, consider a direct approach to the control problem beginning with eq. (16). Assume that the controller manipulates D and s_1 continuously to maintain $ds/dt = 0$ and $s^* = s_D^*$. In that case the first row of eq. (16) becomes a defining equation for D^*,

$$D^* = x^* - (Y\bar{s}_1/\bar{x})s_1^* + (Y\bar{s}/\overline{D}\bar{x})(\overline{D} + a\bar{x})s_D^* \tag{33}$$

Then define a "dimensional" control equation utilizing the error in x to manipulate s_1,

$$s_1 = K(x_D - x) + \bar{s}_1 \tag{34}$$

and express it in normal coordinates as

$$s_1^* = (\bar{x}K/\bar{s}_1)(x_D^* - x^*) \tag{35}$$

Using eq. (35), eq. (33) becomes

$$D^* = (1 + YK)x^* - YKx_D^* + (Y\bar{s}/\overline{D}\bar{x})(\overline{D} + a\bar{x})s_D^* \tag{36}$$

Equations (35) and (36) constitute a proposed control law for achieving QSS in fed-batch fermentation. They are analogous to eqs. (32) and can be expressed in the same form as

$$\begin{bmatrix} D^* \\ s_1^* \end{bmatrix} = \begin{bmatrix} 0 & 1 + KY \\ 0 & -\bar{x}K/\bar{s}_1 \end{bmatrix} \begin{bmatrix} s^* \\ x^* \end{bmatrix} + \begin{bmatrix} \dfrac{Y\bar{s}}{\bar{x}}\left(1 + \dfrac{a\bar{x}}{\overline{D}}\right) & -KY \\ 0 & \bar{x}K \\ & \bar{s}_1 \end{bmatrix} \begin{bmatrix} s_D^* \\ x_D^* \end{bmatrix} \tag{37}$$

The similarity can be underscored by evaluating at the base case with $K = 400$ (although such a high value is out of the question in practice).

$$\begin{bmatrix} D^* \\ s_1^* \end{bmatrix} = \begin{bmatrix} 0 & 101 \\ 0 & -99.9 \end{bmatrix} \begin{bmatrix} s^* \\ x^* \end{bmatrix} + \begin{bmatrix} 0.67 & -100 \\ 0 & 99.9 \end{bmatrix} \begin{bmatrix} s_D^* \\ x_D^* \end{bmatrix} \tag{38}$$

The striking similarity of eqs. (38) and (32) implies that the generalizations drawn from the form of the optimal control law were substantially correct.

The transient response of the system with this control can be found by substituting eq. (37) into eq. (16), which yields after some manipulation

$$\begin{bmatrix} s^{*\prime} \\ x^{*\prime} \end{bmatrix} = \begin{bmatrix} -(\overline{D} + a\overline{x}) & 0 \\ aY\overline{s} & -\overline{D}(1 + KY) \end{bmatrix} \begin{bmatrix} s^* \\ x^* \end{bmatrix}$$

$$+ \begin{bmatrix} \overline{D} + a\overline{x} & 0 \\ \dfrac{\overline{DY}\overline{s}}{\overline{x}}\left(1 + \dfrac{a\overline{x}}{\overline{D}}\right) & \overline{D}KY \end{bmatrix} \begin{bmatrix} s_D^* \\ x_D^* \end{bmatrix} \tag{39}$$

The behavior of the substrate concentration is ideal, and the cell concentration response time is speeded up by the factor $1 + KY$. However, there are interaction terms in the cell concentration response. They are small, as can be seen by using the numerical example with $K = 36$ (factor of 11 less than the "gain" of the optimal controller). The second row of eq. (39) gives

$$\frac{dx^*}{dt} + x^* = 0.9x_D - 0.000125s_D^* + 0.0067(s^* - s_D^*) \tag{40}$$

which is arranged to show that there is both a transient and a steady-state offset in the cell concentration induced by a change in desired substrate concentration: these are negligible.

The need for a control system was originally established in eq. (21), which showed that only about one time constant of the uncontrolled system response can be observed in a practical apparatus. The control system proposed in eq. (37) can be evaluated by the same procedure, giving

$$t_M/\tau_{LCL} = (1 + KY)\ln(V_M/V_0) \tag{41}$$

where τ_{LCL} is the largest time constant of the closed-loop response. With the same apparatus considered previously, $1 + KY$ time constants of the closed-loop response can be observed. There is a definite possibility that several distinct quasi-steady-states can be observed in a single run.

DISCUSSION

Equation (37) constitutes the proposed linear control law for making the QSS accessible. This proposal should be criticized on three main lines: first, by contrasting it with the optimal control law given by eq. (32); second, by qualitatively assessing its sensitivity to model errors; and third, by considering the problems of implementation.

Equation (37) has three advantages over eq. (32) as a control law. First, the assumptions that lead to it are set out explicitly and can be compared against reality in the application. Second, the gain coefficients are given directly in terms of the operating point. Third, it is tunable with a single adjustable gain. Against these advantages is the disadvantage that a small degree of interaction has been retained.

Equation (37) gives the controller parameters directly in terms of the process operating point and key parameters such as Y and a. These latter two are not apt to be known *a priori* and may in fact change with time during a fermentation. Provided the gain K is made reasonably large, the cell concentration response will be insensitive to errors or changes in these parameters: this is the general property of a feedback controller.

Since the substrate control is not based on a direct measurement of substrate concentration, errors in the process model will translate into response errors in the controlled system. The fast response of substrate mitigates this for some applications; a simple calibration of the setpoint input could solve the problem, for example. A second possibility is to retune the controller using formulas given in eq. (37); for example, the parameter Y at QSS may be estimated as $\bar{x}/\bar{s}_1$. A third approach would be to incorporate feedback into the dilution factor control equation. A good alternative would be

$$D^* = x^* - \frac{Y\bar{s}_1}{\bar{x}} s_1^* + \frac{Y\bar{s}}{\bar{D}\bar{x}} (\bar{D} + a\bar{x})s^* + \frac{K_2\bar{D}}{\bar{s}} (s_D^* - s^*) \qquad (42)$$

Here a new feedback gain K_2 is introduced.

The decision regarding measurement or not of substrate concentration is one implementation issue. The basic recommendation is to accept a bit of sensitivity if the measurement is difficult. On the other hand, it is essential that cell concentration be measured. Less obvious from the mathematical development is the need to measure fermentor volume. The physical manipulated variable is total feed rate to the fermentor that is obtained from the required dilution factor and the current volume.

The second manipulated variable used here is also somewhat indirect in physical terms. The final control elements must produce a specified feed rate at a specified concentration. The concept is that the actual feed will be found as a blend of a concentrated and a dilute feed stream having the requisite feed rate and composition.

Nomenclature

A	matrix defined by eq. (17)
a	see eq. (12)
B	matrix defined by eq. (17)
b	see eq. (12)
D	dilution factor (hr^{-1})
E	state transition matrix [eq. (24)]
F	state response matrix [eq. (24)]

K	scaler controller gain (dimensionless)
K	controller gain matrix (dimensionless)
n	time series index
s	substrate concentration (g/liter)
s_1	concentration of substrate in the feed (g/liter)
t	time (hr)
u	general forcing function vector
V	volume (liter)
x	cell concentration (g/liter)
Y	yield (dimensionless)
y	setpoint vector
z	general state vector
Δ	discrete response matrix [eq. (27)]
λ	eigenvalue (hr^{-1})
μ	specific cell growth rate (dimensionless)
σ	specific substrate uptake rate (dimensionless)
τ	time constant (hr)
Φ	discrete state transition matrix [eq. (27)]
$\bar{}$	steady-state value, e.g., $\bar{x}$
$*$	normalized incremental variable, e.g., $x^* = (x - x)^*/\bar{x}$

boldface lower case variable denotes vector
boldface upper case variable denotes matrix

References

[1] H. C. Lim, B. J. Chen, and C. C. Creagan, *Biotechnol. Bioeng.*, 19, 425 (1977).
[2] I. Dunn and J.-R. Mor, *Biotechnol. Bioeng.*, 17, 1803 (1975).
[3] R. B. Newell and D. G. Fisher, *Automatica*, 8, 247 (1972).

Computer Control of Glucose Feed to a Continuous Aerobic Culture of *Saccharomyces cerevisiae* Using the Respiratory Quotient

R. SPRUYTENBURG, I. J. DUNN, and J. R. BOURNE

Chemical Engineering Laboratory, Swiss Federal Institute of Technology, CH-8092 Zurich, Switzerland

INTRODUCTION

Under aerobic conditions the growth characteristics of *Saccharomyces cerevisiae* on glucose depend on the glucose concentration in the fermentation medium. Above a critical level (50–100 mg/liter [1,2]), metabolism will be fermentative even in the presence of excess oxygen [3], whereby glucose is partially converted to ethanol and carbon dioxide at the expense of biomass production. In a batch fermentation of *S. cerevisiae* on glucose, this behavior manifests itself in two growth phases [4]: due to its high initial concentration glucose is converted to ethanol, carbon dioxide, and biomass by fermentative metabolism in a first phase, during which the specific oxygen uptake rate Q_{O_2} [ml/hr g dry weight (DW)] is low compared to the carbon dioxide production rate Q_{CO_2}, and thus the respiratory quotient (RQ) ($RQ = Q_{CO_2}/Q_{O_2}$) is greater than 1. After complete conversion of glucose, ethanol is used in a second phase, during which metabolism is oxidative, and RQ theoretically equals 0.67.

In a glucose-limited continuous culture of *S. cerevisiae* metabolism depends on the growth rate [4], which at steady state equals the dilution rate D. At low dilution rates, growth is oxidative, the specific gas conversions Q_{O_2} and Q_{CO_2} are proportional to the dilution rate, and RQ theoretically equals 1. Above a critical dilution rate, metabolism switches to the fermentative type, causing RQ to increase and the yield of biomass to drop.

For bakers' yeast production, a loss of productivity due to fermentative growth is to be avoided. For this purpose glucose should be fed to the culture at a rate that will maintain oxidative conversion. Control of the feed rate is in principle possible by the feedback of a metabolism-related process variable. Simplest would be to directly control glucose concentration below the critical level; this, however, cannot be realized due to the lack of sterilizable on-line measurement equipment. The detection of fermentative growth is possible by measuring ethanol in the liquid or gas phase [5], or by noting changes in the RQ [1,6]. Aiba et al. [1] and Wang et al. [6] examined the use of RQ for feed rate control of a fed-batch fermentation of *S. cerevisiae*. In both these investigations RQ was not used primarily as a control variable in a closed feedback loop. In the present work a digital process computer has been used to compute the RQ from a chromatographic exit-gas analysis and to implement digital feed rate control of a glucose-limited continuous culture of *S. cerevisiae* [7].

CONCEPT FOR FEED RATE CONTROL

The biomass needed to start a continuous fermentation has to be produced in a preliminary batch process. The problem investigated in this work then consisted of defining an end-point criterion of the batch process to allow the computer to initialize the continuous culture, determining a start-up procedure to reach the dilution rate of maximum oxidative growth, and maintaining this operating point.

The end of the batch process is given by the desired oxidative metabolism of the organisms, as it occurs during the batch's second growth phase. The criterion used to determine the end-point of the batch growth is based on the fact that the carbon dioxide production rate drastically decreases twice in the course of the batch fermentation: first after complete glucose conversion, then after complete ethanol conversion.

Adaptation of the culture takes place when it is switched from ethanol-limited batch growth to glucose-limited continuous growth. For this reason and to maintain the desired low glucose concentration, the initial feed rate had to be determined empirically.

For productivity reasons, the feed rate should be increased as rapidly as possible after the preliminary adaptation phase, but as the dynamics of biological systems are poorly understood, this maximum rate of change has to be determined empirically from the RQ values. The feed rate was increased either rampwise or stepwise. The computer stopped the ramp change either at a prescribed dilution rate or at the detection of fermentative growth. Using step changes with small amplitudes (up to 0.01 hr^{-1}), a change took place every time a given number of RQ values indicated no fermentative growth.

In practice, measurement errors blur the theoretical RQ limit value of 1.0 and as a consequence filtering of the gas analysis results by mean-value computation was necessary. A limit value of RQ somewhat above 1.0 was defined to account for the fact that a distribution of cells with either one of the two possible metabolisms is to be expected in the neighborhood of the actual switch-over point. Both these precautions caused an increased time lag in the control loop. Five RQ values were needed for a mean value, and a step change in dilution rate took place only after two consecutive mean RQ values under the predefined limit RQ value.

Fermentative growth can occur because of two reasons: first in a dynamic state if the feed rate is changed too fast; secondly in a steady state at dilution rates above the critical one. If no further increase is possible after a long settling time, the critical dilution rate has been reached. To guarantee a stable operation at the optimum dilution rate, the feed rate was then decreased slightly.

As RQ is a complex process variable, precautions have to be taken to assure constancy of all operating variables. Temperature, pH, pO$_2$, stirring speed, and gas flow rate were therefore provided with a computer connection for limit supervision. Each of these variables was associated with a tolerance range fixed by experience, within which deviations from the respective set-points were assumed to have no influence on RQ. Violation of a limit value of one variable during a 5-min period resulted in the rejection of the RQ value for this period. The cumulative sum method was applied for trend detection of the process variables (pH, pO$_2$, T, stirring speed, and gas flow rate). The cumulative sum is computed as follows [8]:

$$S_{i+1} = S_i + x_{i+1} - \bar{x} \qquad (1)$$

where S is the recurrent cumulative sum of a given variable, x_{i+1} is the current value of that variable, and $\bar{x}$ is its mean value. The cumulative sum changes monotonically if the operating level of the observed variable is subjected to a step change. The new operating level can then be computed as follows:

$$\bar{x}_{new} = \bar{x}_{old} + \Delta S / \Delta n \qquad (2)$$

with $\bar{x}_{new}$ being the new operating level, $\bar{x}_{old}$ is the old level, and ΔS is the change of the cumulative sum over Δn samples. From eq. (2) it can then be inferred if the operating point still lies within the tolerance limit.

The five process variables were sampled by the computer every 15 sec, and 20 probes of each variable were needed to correct the stored operating points, which were thus updated every 5 min to be available at every 5-min output of the gas analysis.

INSTRUMENTATION

Experiments were carried out in a 50-liter double-jacketed stirred-tank fermentor [9], interfaced with a time-sharing PDP 11/45 process computer. The location of the computer and process in different buildings about half a mile apart required a special interface for serial digital data transmission on a single twisted pair of telephone lines. The interface B-3000 (Borer AG, Switzerland) consisted of a control unit placed in the computer, and of a terminal on the process location. AD and DA conversion thus took place at the process; the terminal was further provided with relay outputs and facilities for operator communication, such as alarm lamps, decimal display, decimal, and function keyboards.

Instrumentation allowed measurement of pH (Ingold probe, Knick 47 amplifier), temperature (Pt 100 resistance thermometer), pO$_2$ (IL 531 electrode and amplifier), stirrer speed, gas flowrate (Brooks mass flowmeter 5812), and fermentor weight (strain gauge Philips PR 6040/05). pH was controlled by a new computer algorithm [10], temperature with a conventional PI controller, and fermentor weight with a limit controller actuating a magnetic valve; the sensitivity of this control system was about 400 mg, and its long-term stability such that after 10 days of continuous fermentation the actual fermentor contents amounted to 39.7 kg at a 40 kg setpoint. A magnetic flowmeter (Eckardt) was used for measurement of nutrient feed rate; a PI controller provided control under conventional operation, whereas under computer operation the feed pump was under digital control. Controllable peristaltic pumps (Watson Marlow) provided for flow of liquids (neutralizing agent and nutrient).

Exhaust gas was analyzed with a gas chromatograph separating oxygen, carbon dioxide, and nitrogen with helium being the carrier. Samples could be taken every 4.5 min automatically. Peak values in 5-digit BCD form were serialized and sent to a standard teletype interface in the computer over a separate pair of telephone lines. The data were read from the register and decoded by a dedicated cyclic program.

The oxygen consumption and carbon dioxide production rates were computed from the gas analysis by means of steady-state material balances as follows:

$$Q_{O_2}X = [G_{in}y(O_2)_{in} - G_{out}y(O_2)_{out}]/V \qquad (3)$$

$$Q_{CO_2}X = [G_{out}y(CO_2)_{out} - G_{in}y(CO_2)_{in}/V \qquad (4)$$

with G being the gas flow rate (mmol/hr) and y is the mole fractions of oxygen, carbon dioxide, and nitrogen. Only G_{in} was measured, as G_{out} results from a nitrogen balance:

$$G_{out} = G_{in}y(N_2)_{in}/y(N_2)_{out} \qquad (5)$$

RQ follows from eqs. (3) and (4):

$$RQ = Q_{CO_2}X/Q_{O_2}X \qquad (6)$$

Temperature and humidity effects are neglected in eqs. (3)–(5) because they cancel out in eq. (6).

The gas chromatograph was extremely sensitive to changes in atmospheric pressure and ambient temperature, and its close supervision by the computer was necessary in order to detect when recalibration was necessary. For this, the sum of the measured partial pressures was computed after each analysis and the total pressure thus obtained compared with a tolerance limit based on the recurrently computed standard deviation of the total pressure.

The software system developed to carry out the experiments consisted of separate programs for initialization, operator communication; limit supervision of pH, temperature, pO$_2$, gas flow rate, and stirrer speed; sampling data from the gas chromatograph, and for gas analysis and feed rate control. All programs were written in FORTRAN under the RXS11D operating system.

EXPERIMENTAL CONDITIONS

Continuous growth experiments of bakers' yeast (LBG H 1022) were carried out with 1 and 5% glucose-limited synthetic feed medium, without the addition of yeast extract. The actual glucose concentrations in the feed were 7.7 and 43 g/liter, respectively. The experiments were carried out under sterile conditions at pH 5, 30°C, and 1000 rpm stirrer speed. To allow the exact detection of oxygen consumption and carbon dioxide production rates with the gas chromatograph, aeration was kept low throughout the experiments, but special care was taken to avoid oxygen limitation. Fermentor contents were controlled at 40 kg. The sampling interval of the gas chromatograph was 4.5 min; thus a step change in dilution rate took place approximately every 40 min, after two mean RQ values under the limit value.

RESULTS AND DISCUSSION

Figures 1–4 show the development of dilution rate, oxygen consumption rate ($Q_{O_2}X$), carbon dioxide production rate ($Q_{CO_2}X$), and RQ during the course of selected experiments. Experiments with a 5% glucose feed were carried out first.

Experiment 1

The first continuous experiment (Fig. 1) was started with a dilution rate of 0.1 hr^{-1}. As is evident from Figure 1, the gas conversions ($Q_{O_2}X$ and $Q_{CO_2}X$) exhibit erratic oscillations during the first 14 hr. An initial rise of RQ demonstrates the formation of ethanol, which, however, eventually disappeared again. There seemed to be no indication of fermentative growth afterwards, so a ramp change of the dilution rate was initialized (increase in dilution rate 0.05 hr^{-1} in 6 hr). However, after 1.5 hr, RQ reached a value of 1.6 and as a consequence the feed had to be interrupted manually.

This experiment showed that the biological system was severely disturbed

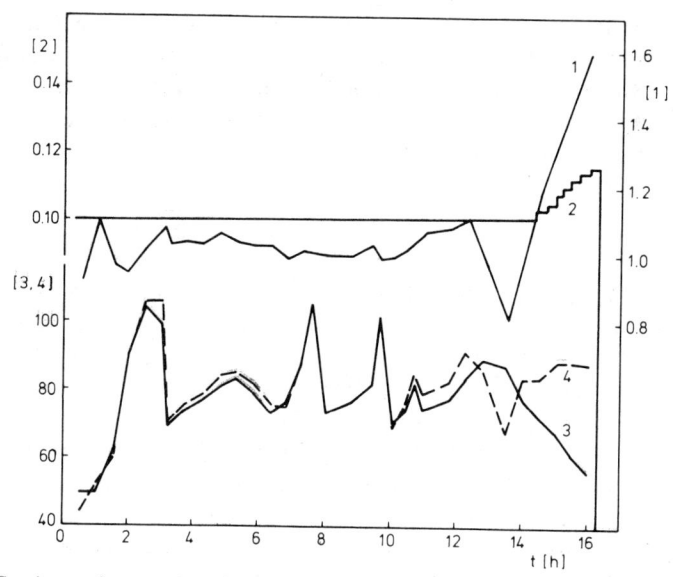

Fig. 1. Continuous fermentation of S. cerevisiae with 5% glucose-limited synthetic feed medium. 1 = RQ, 2 = D (hr^{-1}), 3 = $Q_{O_2}X$ (mmol/liter hr), 4 = $Q_{CO_2}X$ (mmol/liter hr).

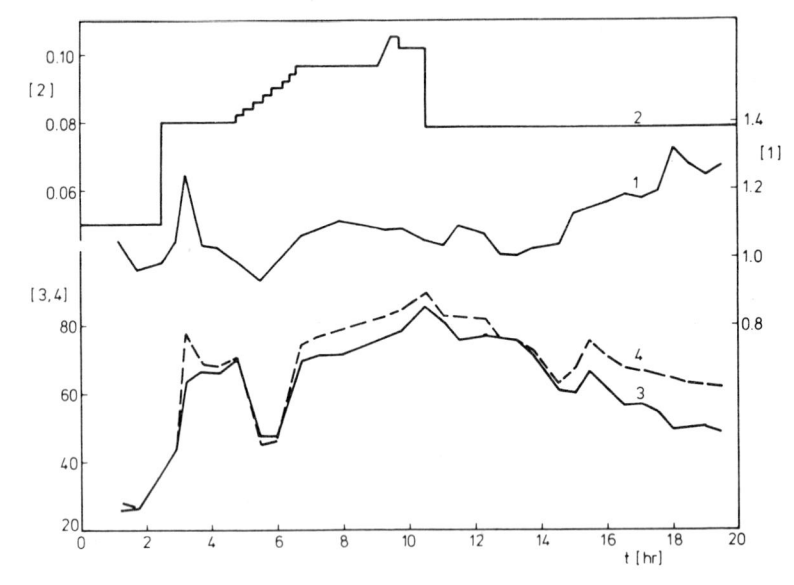

Fig. 2. Continuous fermentation of *S. cerevisiae* with 5% glucose-limited synthetic feed medium. 1 = RQ, 2 = D (hr^{-1}), 3 = $Q_{O_2}X$ (mmol/liter hr), 4 = $Q_{CO_2}X$ (mmol/liter hr).

by the initial dilution rate of 0.1 hr^{-1}, as indicated by the changing gas conversions. Since the RQ values varied between 0.95 and 1.05, the limit RQ value for control was subsequently fixed at 1.05.

Following the shut-down of this first experiment, start-up with a lower initial dilution rate (0.05 hr^{-1}) was tried after an 8-hr pause, but the organisms returned to fermentative metabolism at the slightest disturbance. From this it was concluded that the change of metabolism exhibits a large hysteresis, and that a mere feed rate reduction does not guarantee rapid reversal to oxidative growth. As a consequence, the computer program was modified to stop the nutrient feed automatically after the occurrence of three consecutive RQ values above 1.2.

Experiment 2

The second experiment (Fig. 2) was started with a dilution rate of 0.05 hr^{-1}, which was increased to 0.08 hr^{-1} after 2.5 hr of operation, because the RQ values stayed mainly under the limit value 1.05. Following this step change, RQ reflects ethanol formation during a short period, but after one more hour conditions seemed to be steady enough to manually initialize a ramp change of the dilution rate (increase in dilution rate 0.04 hr^{-1} in 5.5 hr). This increase apparently was too fast, and the computer stopped the ramp change after three RQ values above 1.05. After three more hours at constant feed rate, a new ramp change of dilution rate was started, with the same effect as previously; as a consequence the ramp change was stopped and the dilution rate manually defined at a safe value of 0.08 hr^{-1}. However, in spite of this precaution, irreversible fermentative growth

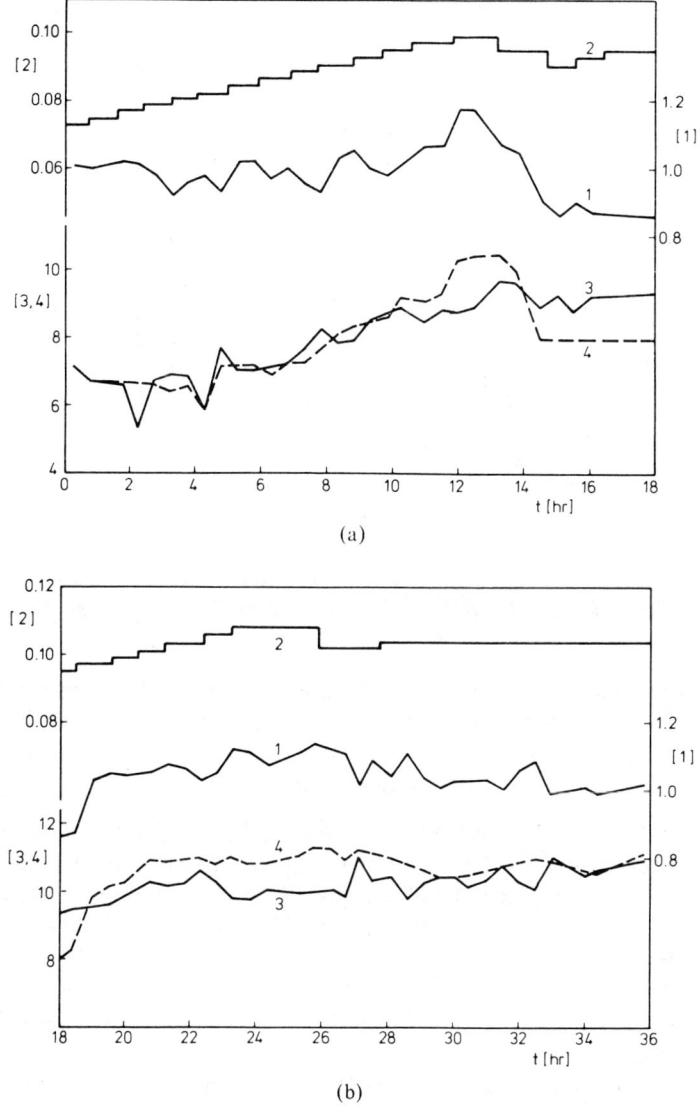

Fig. 3. (a) Continuous fermentation of *S. cerevisiae* with 1% glucose-limited synthetic feed medium. 1 = RQ, 2 = D (hr^{-1}), 3 = $Q_{O_2}X$ (mmol/liter hr), 4 = $Q_{CO_2}X$ (mmol/liter hr). (b) Continuous fermentation of *S. cerevisiae* with 1% glucose-limited synthetic feed medium. 1 = RQ, 2 = D (hr^{-1}), 3 = $Q_{O_2}X$ (mmol/liter hr), 4 = $Q_{CO_2}X$ (mmol/liter hr).

slowly started 4.5 hr after the last change in feed rate. Under such conditions the feasibility of a feed rate control based on RQ is questionable.

At higher feed concentrations the critical level of glucose concentration is reached faster, and faster controller response is required. The results of the above experiments seem to indicate that the response time was not adequate. The next

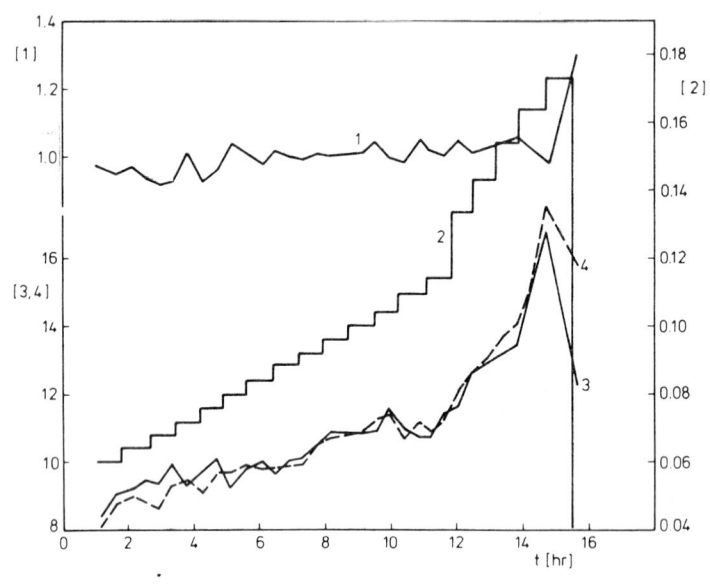

Fig. 4. Continuous fermentation of *S. cerevisiae* with 1% glucose-limited synthetic feed medium and yeast extract. 1 = RQ, 2 = D (hr^{-1}), 3 = $Q_{O_2}X$ (mmol/liter hr), 4 = $Q_{CO_2}X$ (mmol/liter hr).

experiments were carried out with a 1% glucose feed in order to check the feasibility of the proposed control method under more favorable conditions.

Experiment 3

Figures 3(a) and 3(b) show an experiment run by the computer alone, without manual interference. The fermentation was run with a dilution rate of 0.07 hr^{-1} for the first 10 hr (not shown). Subsequently the computer increased the dilution rate stepwise by 0.002 hr^{-1} whenever two consecutive mean RQ values of five probes each were below the limit value of 1.05. After 12 hr of small step changes, RQ rose to a value of 1.15, and the computer decreased the dilution rate in two steps, each 0.004 hr^{-1} in magnitude. This precaution sufficed to avoid fermentative growth. The dilution rate at this time was 0.1 hr^{-1}.

To make sure that oxidative growth at a higher dilution rate was really not possible, the control response was delayed by raising the RQ limit value for metabolism distinction from 1.05 to 1.07. As RQ meanwhile had again dropped to safe values, due to the previous decrease of dilution rate, the computer increased the feed rate progressively, but after 12 more hours RQ rose to 1.2 at a dilution of 0.11 hr^{-1}. The subsequent feed rate reduction was sufficient to avoid fermentative metabolism, and the experiment was run during another 10 hr as a stability test for operation near the determined optimal dilution rate of 0.1 hr^{-1}.

In this experiment the proposed scheme for feed rate control was successful.

Experiment 4

The next experiment (Fig. 4) shows that difficulties can also arise with growth on a 1% glucose feed if the dilution rate is increased too fast. This experiment was run with an addition of yeast extract to the 1% medium, in order to enable oxidative growth at higher feed rates and thus to extend the start-up period. The program parameters were changed several times in the course of the experiment, until finally a step change of 0.01 hr^{-1} in dilution rate took place every 40 min. This rate of increase in feed rate eventually was too fast to allow oxidative growth any longer, and RQ rose so fast that fermentative growth could no longer be reversed by a simple decrease of the dilution rate, and the computer stopped the feed after three consecutive RQ values above 1.2.

CONCLUSIONS

A low initial dilution rate at start-up of continuous operation with a subsequent stepwise increase of the feed rate allowed the optimal dilution rate to be reached without fermentative growth.

The success of feed rate control based on the feedback of RQ is mainly impaired by the delays in the control loop. These arise from the sampling of the gas chromatograph and from the unavoidable statistical treatment of the gas analysis. These could possibly be reduced by use of a sufficiently accurate continuous gas analyzer. A delay inherent to the feedback nature of the control loop is due to the well-mixed tank hydrodynamics: when the rate of change of feed rate exceeds the rate of change of growth rate, fermentative metabolism will not occur until after glucose concentration in the reactor rises above the critical level. With 1% feed concentrations, the glucose concentration in the reactor increases relatively slowly with a small overshoot; this possibly outweighs the control delays, and feed rate control is feasible. At higher feed concentrations, a limit RQ value for control closer to 1.0 than the value 1.05 used here would be required, which, however, does not seem realisable.

Whether the culture can return to oxidative growth depends on the hysteresis characteristics of the biological response; the switch from fermentative to oxidative growth apparently requires more time if the glucose concentration overshoot has been severe.

Most feedback control systems operate successfully with a process output variable such as RQ or ethanol concentration in this case, but the controllability of the present system is dictated by the hysteresis and irreversibilities of the biological culture, which are largely unknown. The ideal control would involve direct glucose measurement to control the feed rate, and all overshoot would be eliminated. However, it is conceivable that a satisfactory RQ control system could be designed if the biological dynamics were well understood.

Nomenclature

D	dilution rate (hr^{-1}) (reciprocal residence time)
G	gas flow rate (mmol/hr)
n	number of probes
$Q_{O_2}X$	O_2 uptake rate (mmol/liter hr)
$Q_{CO_2}X$	CO_2 production rate (mmol/liter hr)
RQ	respiratory quotient (Q_{CO_2}/Q_{O_2})
S	cumulative sum
V	volume of reacting medium (liter)
x	value of a state variable
$\bar{x}$	mean value of a state variable
X	cell concentration (g/liter)
$y(CO_2)$	mole fraction of CO_2 in gas
$y(N_2)$	mole fraction of N_2 in gas
$y(O_2)$	mole fraction of O_2 in gas

The authors are indebted to Professor A. Fiechter, Microbiology Institute, for the use of the 50-liter fermentor, and to D. Karrer for his valuable help with the experiments.

References

[1] S. Aiba, S. Nagai, and Y. Nishisawa, *Biotechnol. Bioeng., 18,* 1001 (1976).
[2] E. Oura, *Biotechnol. Bioeng., 16,* 1197 (1974).
[3] H. P. Knoepfel, Ph.D. thesis No. 4905, ETH Zurich, 1972.
[4] K. Von Meyenburg, Ph.D. thesis No. 4279, ETH Zurich, 1969.
[5] Austrian Patent No. 3194-70.
[6] H. Y. Wang, C. L. Cooney, and D. I. C. Wang, *Biotechnol. Bioeng., 19,* 69 (1977).
[7] R. Spruytenburg, Ph.D. thesis No. 6080, ETH Zurich, 1977.
[8] R. H. Woodward and P. L. Goldsmith, *Mathematical and Statistical Techniques for Industry* (Oliver and Boyd, Edinburgh, 1964), Monograph No. 3.
[9] D. Karrer, A. Einsele, and A. Fiechter, *Proceedings of the Vth International Fermentation Symposium* (Springer, Berlin, 1976).
[10] R. Spruytenburg, N. D. P. Dang, I. J. Dunn, J. R. Mor, A. Einsele, A. Fiechter, and J. R. Bourne, *Chem. Eng., 310,* 447 (1976).

Computer Control of the Pekilo Protein Process

AARNE HALME

Department of Process Engineering, University of Oulu, Oulu, Finland

INTRODUCTION

The Pekilo process is a new fungal-type single-cell protein (SCP) process that utilizes suitable carbohydrates as the substrate, such as sulfite spent liquor, and as the microbe a certain strain of *Paecilomyces varioti* microfungus. The process was developed in the Central Laboratory, Finland, by the so-called side-product group of pulp-and-paper industry in the late 1960s. A full-scaled 10,000 ton/year plant that utilizes sulfite spent liquor, has been in operation since 1975 in Jämsänkoski, Finland. The plant was designed and delivered by Tampella Ltd., which has been responsible for further development and marketing of the Pekilo process. The process and its basic properties have been described before in more detail [1, 2].

The development of the computer control system for the Pekilo process was started at the beginning of 1977 as a cooperative project between a research group from Tampere University of Technology, Tampella, and Rintekno, a company that was developing a computer system for fermentation processes. The project was preceded by thorough preliminary studies on process dynamics, control principles, and argumentation of the computer system, which were done by the university group and Tampella [3, 4]. At the present stage, the system is tested in the inoculation line of the Jämsänkoski plant.

SHORT DESCRIPTION OF THE PROCESS AND ITS OPERATION

Process

A simplified flow scheme of the Pekilo process utilizing sulfite spent liquor is shown in Figure 1. In the pretreatment, the SO_2 content of the liquor is lowered by steam stripping. In the stripper, potassium and phosphorus are added as potassium chloride and phosphoric acid, respectively. The liquor is then cooled and fed to the fermentor. Fermentation is aerobic and takes place continuously with a 3–5 hr detention time, pH 4.5–4.7, and temperature 38–39°C. Nitrogen is taken from ammonia, which is added for pH control of this process. Foam

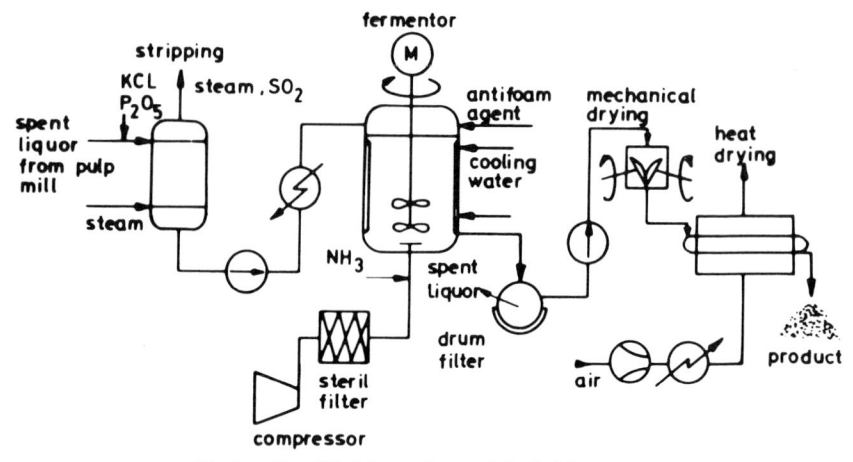

Fig. 1. Simplified flow schema of the Pekilo process.

control is accomplished by chemical antifoams. The metabolic heat makes the fermentation quite exothermic and water cooling is needed to maintain temperature control. In the fermentation, the microfungus utilizes mainly acetic acid, monosaccharides, and aldonic acids as the carbon source, which are consumed in this order. Because the acetic acid content of the spent liquor is usually much lower than the monosaccharide content, and the consumed part of the aldonic acids is small, the monosaccharides are, in practice, the main substrate. The main component of the monosaccharides in soft-wood cooking is mannose and in hard-wood cooking is xylose. The biomass growth rate is limited partly by the carbon source and partly by oxygen, which is due to the highly viscous fermentation emulsion. Nevertheless the attained growth rates are quite high, about 3.5–4 g/liter/hr. The cell mycelium is separated from the outlet flow of the fermentor by filtration and is dried first by mechanical pressing and then by heat. The product is finally pressed into pills for easy handling and transportation. The protein content ($6.25N$) in the product is 50–60% (w/w) and the dry matter content is 92–95% [2]. The production facilities in the Jämsänkoski plant include two stripping, fermentation, and filtering units, and one drying unit. The working volume of the production fermentors is 360 m^3 each. For start-up of the plant, there is an inoculation line, which includes 15-m^3 and 150-liter fermentors. The plant has its own laboratory for production supervision and research purposes.

Instrumentation and Process Operation

The Pekilo process can be satisfactorily operated with quite conventional instrumentation. In the fermentors, the usual variables such as emulsion level, foam level, temperature, pressure, pH, agitator power, air flow rate, and liquor feeding rate are directly measured and controlled. The O_2 and CO_2 concentrations in the exhaust gases are measured by automatic analyzers; these analyzers being the only ones used in the process. The measurement of the dissolved oxygen (DO) concentration is not very useful in the Pekilo process, because in normal operation it has a very low value that cannot be reliably measured. Measurable values give only the indications of severe disturbances in the growth process, which can also be observed from exhaust gas measurements. Such variables as the biomass concentration, substrate and nutrients concentrations, and the protein content of the product are measured by sample analysis. In the stripper units, the variables measured and controlled are pressure, temperature, column level, and steam flow to the column. The SO_2 concentrations before and after stripping are measured by sample analysis. In the filter unit (a drum), the rotating speed and basin level are controlled. In the heat drier (a multiple-belt drier), the most important variable controlled is the product moisture. This moisture cannot be reliably measured directly. Instead, the moisture of the inlet material and temperatures and moistures in the air flow can be measured. Based on this information, the product moisture can be controlled indirectly.

Production control is done primarily by controlling the fermentation units. Pretreatment and secondary treatment units are adapted to this functioning. The biomass growth rate can be affected by several variables. The most important control variable is the dilution rate. In addition, the aeration rate and agitator power can be used. Changes of temperature may also be used under certain exceptional conditions. As to the availability of the substrate liquor, the process is dependent on the functioning of the pulp plant. Furthermore, the amount of liquor and its sugar content may vary. When looking at the most important cost variables in the plant operation, it can be noted that the consumption of electrical energy is mainly determined by aeration and agitation in the fermentation units, and the consumption of heat (steam) energy in the drying unit. The most valuable chemical is ammonia, and also phosphoric acid and antifoam agent mean significant costs. Instead, the consumption of potassium is of secondary importance. Because the spent liquor is burned after the Pekilo process, the value of the substrate is calculated according to the decreasing burning value of the liquor. The sugar consumption does not decrease the burning value notably, but dilution waters from the process (mainly waters used for washing the biomass) increase the evaporation costs.

ARGUMENTATION OF THE COMPUTER SYSTEM

The computer system is one, in many cases optional, part of the production equipment. Besides technical points related to hardware and software, its economic justification is as important when defining starting points for the developmental work. In the case of the Pekilo process, the following points formed the ground for the decision to develop the system:

i) Accurate operation of the process requires many repeated measurements, storing the results, and calculations.

ii) Typically long time constants in the growth process make accurate process control difficult by conventional methods.

iii) By more accurate process control, energy and chemicals (per ton) can be saved.

iv) It is hoped that the residual concentrations of the nutrients remain at low and steady levels also because it helps in further treatment of the spent liquor (evaporation).

v) Average production rate and protein contents can be increased by maintaining optimal conditions and eliminating disturbances in the growth process.

vi) Reliability of the process may be improved by more efficient supervision. By avoiding some shutdowns, yearly production can be increased easily, because the start-up of the process takes a relatively long time (in certain cases 2-3 days).

When justifying the profitability of the computer system it was estimated that a 15-20% saving in energy consumption, 25-30% saving in phosphorus acid, and 5-10% saving in ammonia consumption per ton produced would be realized along with a 10-15% increase in production. Most of the energy saving is from steam in the drier due to more accurate moisture control. Savings in nutrient (K,P) consumption is quite high, because accurate dosage control is difficult to obtain by conventional methods and a certain overdosage is necessary in practice. A small amount of ammonia is saved by improved pH control. The increase in production is due to the growth control in the fermentor and decreasing the shutdown time of the process. Also due to improved moisture control, the mean dry matter content of the product may be decreased 2-3%, which increases production. From these factors, the pay-back time of an appropriate size minicomputer system (hardware + software) was estimated to be between 1.5 and 2 years, which corresponds to the values generally accepted in the process industry.

PROCESS ANALYSIS AND ESTIMATION TECHNIQUES

Fermentor Model

The main emphasis was given to the fermentor and drying parts of the process. For both sections, a dynamic model was developed and identified. The fermentor model is more interesting in this context. It is based on the usual Monod kinetics with oxygen limitation and describes the following phenomena: mycelium growth rate, substrate and nutrient consumption, oxygen transfer and consumption, cell age distribution of the biomass, and changes in the protein content.

The model has been described in more detail elsewhere [3]. A number of nonstandard features can, however, be mentioned here. Experiments show that part of the oxygen transfer may take place by direct absorption from air bubbles. This phenomenon has been modeled by dividing the total oxygen consumption into two parts, one corresponds to uptake from the liquid phase and the other from the gas-liquid film (Fig. 2). Cell age distribution is used to indicate the "age state" of the biomass in the fermentor. Changes in oxygen yield coefficient and protein content can be correlated to age distribution. In Figure 3 there is

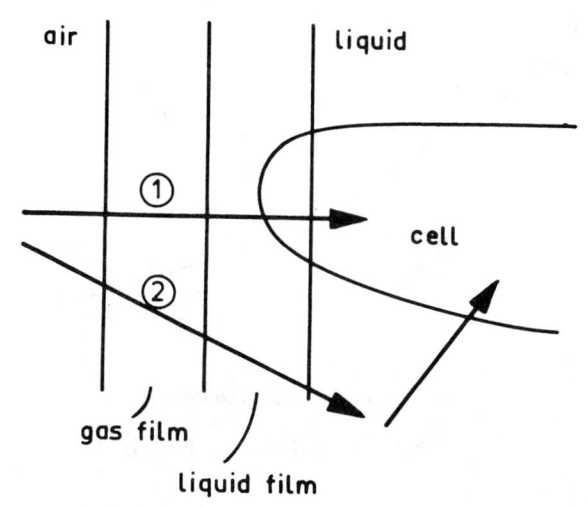

Fig. 2. Oxygen transfer in the Pekilo process. 1, Direct transfer path, 2, path via the liquid.

an example of a dynamic situation where the process is disturbed by changing the dilution rate. The changes in the protein content and oxygen yield coefficients can be observed clearly and can be compared to changes in the age distribution. The distribution is described simply by an index that gives the maximum age for 80% of the biomass (Fig. 4, not used to obtain practical correlations). The principal dynamic equations used for fermentor modeling are given below.
Biomass:
$$\dot{X} = [\mu(S,\tilde{C}) - k_D]X - DX \qquad (1)$$
Substrate:

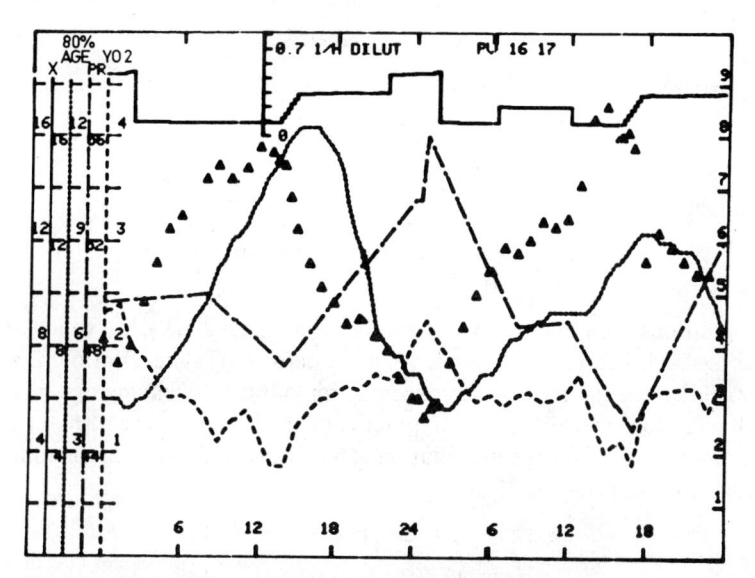

Fig. 3. Variations in the 80% age, protein content (PR) and the oxygen yield coefficient (Y_{O_2}) in an experiment in which the dilution rate was disturbed (15 m³ fermentor).

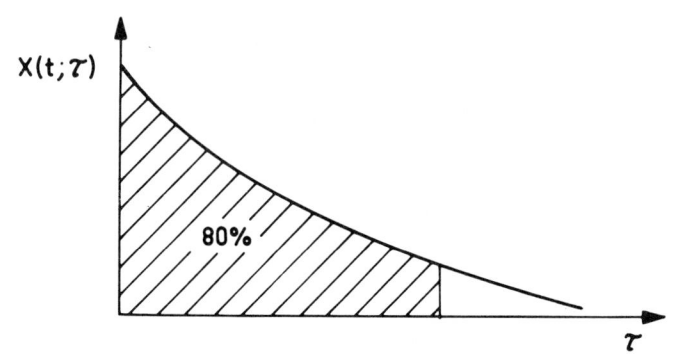

[Figure: Cell age distribution curve X(t;τ) vs τ, with shaded area labeled 80%]

Fig. 4 Cell age distribution in the steady state. Striped area illustrates the distribution of 80% of total biomass. Upper limit of the age interval is called 80% age.

$$\dot{S} = D(S_0 - S) - (1/Y_{X/S})\mu(S,\tilde{C})X \tag{2}$$

Nutrients (P,K):

$$\dot{C}_N = D(C_{N0} - C_N) - (1/Y_{X/N})\mu(S,\tilde{C})X \tag{3}$$

Dissolved oxygen:

$$\dot{C} = K_L a(C^* - C) - DC - \rho O_2 CON \tag{4}$$

$$\mu(S,\tilde{C}) = \hat{\mu}[S/(S + K_s)][\tilde{C}/(\tilde{C} + K_c)] \tag{5}$$

$$\tilde{C} = \frac{K(1 + X/K_x)^{-1}C^* + C}{K(1 + X/K_x)^{-1} + 1} \tag{6}$$

$$\rho = \frac{C}{K(1 + X/K_x)^{-1} + 1} \tag{7}$$

$$K = K_1(K_x K_2)^{-1} \tag{8}$$

Age distribution:

$$\frac{\partial}{\partial t}X(\tau;t) + \frac{\partial}{\partial \tau}X(\tau;t) = -DX(\tau;t) \tag{9}$$

Boundary conditions:

$$\begin{aligned} X(\tau;0) &= X_0(\tau) \quad \text{(given)} \\ X(0,t) &= \mu X \end{aligned} \tag{10}$$

The notations are standard except for the parameters K_1, K_2, K_x, which are related to the direct oxygen transfer (in more detail [3]). The notation $X(\tau;t)$ in the age distribution means the concentration of biomass having age τ at time t. All equations are discretized in practical calculations. For eqs. (1)–(4) this is a standard job. The age distribution can be best discretized to the following recursive state form.

$$X_1(k + 1) = \Delta t G(k)[1 - \Delta t D(k)]$$

$$X_2(k + 1) = [1 - \Delta t D(k)]X_1(k)$$

$$\vdots \qquad (11)$$
$$X_n(k+1) = [1 - \Delta t D(k)] X_{n-1}(k)$$
$$X_{n+1}(k+1) = [1 - \Delta t D(k)][X_{n+1}(k) + X_n(k)]$$

where $G(k)$ means the growth rate at interval $[k\Delta t, (k+1)\Delta t]$ and $X_i(k)$ the following age classes:

$$X_1(k) = \int_0^{\Delta \tau} X(\tau; k\Delta t)\, d\tau$$
$$X_2(k) = \int_{\Delta \tau}^{2\Delta \tau} X(\tau; k\Delta t)\, d\tau$$
$$\vdots \qquad (12)$$
$$X_n(k) = \int_{(n-1)\Delta \tau}^{n\Delta \tau} X(\tau; k\Delta t)\, d\tau$$
$$X_{n+1}(k) = \int_{n\Delta \tau}^{\infty} X(\tau, k\Delta t)\, d\tau$$

The model fits the experimental data quite well; examples are given in Figures 5 and 6. Changes in parameters, such as in yield coefficients, have been studied and several useful correlations were found. The effects of aeration and agitation were also studied extensively. Figure 7 shows an example of an experimental set of curves, which give the maximum growth rate as a function of aeration rate

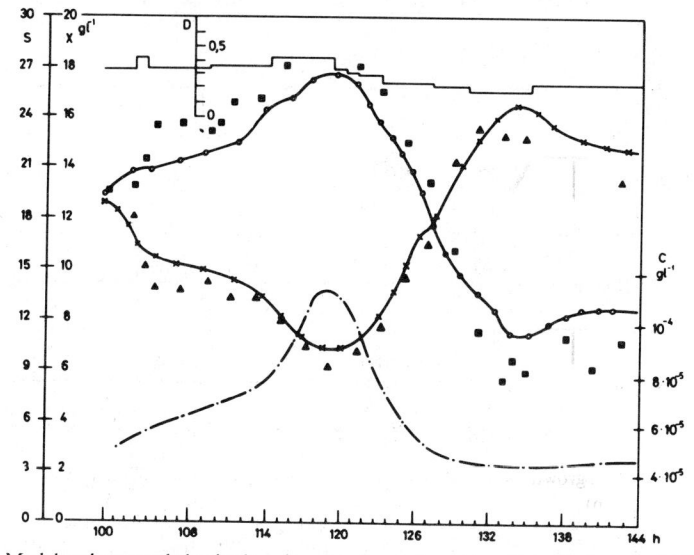

Fig. 5. Model and process behavior in a dynamical experiment in which dilution rate is disturbed: (a) biomass (X); (b) substrate (S) and dissolved oxygen (C). (■) S_{meas}; (O) S_{sim}; (▲) X_{meas}; (*) X_{sim}; (- — -) C_{sim}.

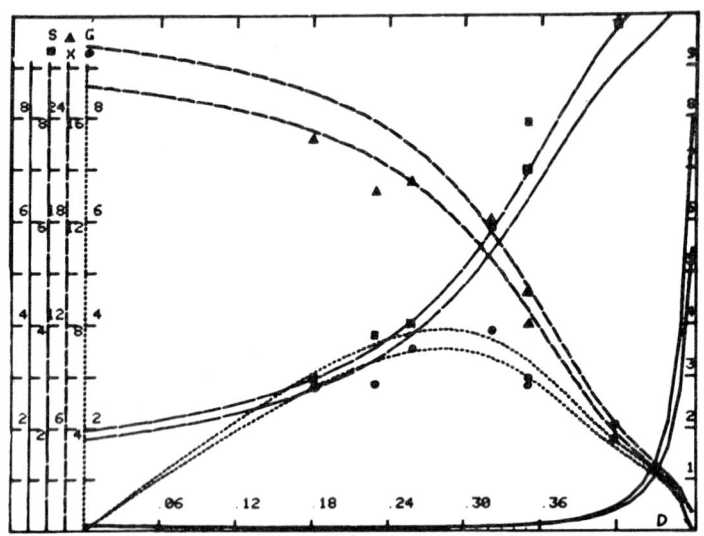

Fig. 6. Model and process behavior in the steady state. (■) S; (▲) X; (●) G. $K = 0.06$; $K_L a = 80$; $A = 0.4$; $B = 0.1$; $K_X = 14$; $K_C = 0.0001$; $K_S = 4$; $\mu_M = 0.55$; $Y = 0.57$; $C_T = 0.008$; $S_0 = 33, \ldots, 36$.

(AP) and impeller power (IP). This set of curves (strongly dependent on the fermentor geometry) can be used to select the optimum power combination between the compressor and impeller.

The fermentor model is useful for understanding the process behavior, studying different alternatives, and process optimization. The fermentor model has been programmed for the process control computer so that it can be run on-line.

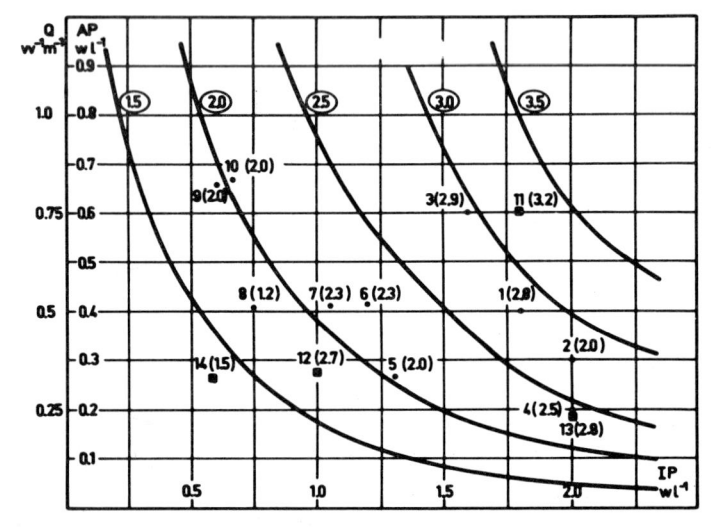

Fig. 7. Maximum growth rates as a function of aeration (AP) and impeller (IP) powers in a 15-m³ fermentor (Cemap).

Applications of Estimation Techniques

To obtain more exact information from the process between samples and to reduce sampling rates, a method was developed to estimate the growth rate, and the biomass, substrate, and nutrient concentrations in the fermentor. The basis of the method lies in the use of the cell age distribution. Consider the usual oxygen consumption equation

$$O_2CON = a(1/Y)\mu X + bX \tag{13}$$

where especially the coefficient b may vary in practice. Taking into account the age distribution, this equation may be written in the form

$$O_2CON = \frac{a}{Y}\mu X + \sum_{i=1}^{n} b_i X_i \tag{14}$$

where $X_1, \ldots, X_n$ are the age classes defined in eq. (11) and $b_1, \ldots, b_n$ are constants that are not expected to vary. The oxygen yield coefficient can now be written in the form

$$Y_{O_2} \triangleq \frac{\mu X}{O_2CON} = \left(\frac{a}{Y} + \frac{1}{\mu}\sum_{i=1}^{n} b_i \frac{X_i}{X}\right)^{-1} \tag{15}$$

It can thus be expected that the variables X_i/X explain the variations in the yield coefficient. In practice, 4–5 1-hr age classes have turned out to be sufficient to obtain a good correlation. The growth rate can be estimated from oxygen consumption when the yield coefficient is known. Then, eqs. (1)–(3) are used to estimate the biomass, substrate, and nutrient concentrations. As an example, Figure 8 shows the real and estimated oxygen yield coefficients and the estimated and measured biomass in a dynamic fermentation experiment.

MAIN FUNCTIONS IN THE COMPUTER SYSTEM

Hardware and Software

The computer system has been realized as a multicomputer system, where the main computer is a PDP 11/34 with 48k core memory, two disk units, and a floppy disk. The operating system is Digital Equipment Corporation RSX 11-M real-time operating system. As the operator communication units, there are two CRT terminals, two typewriters, and a specially designed keyboard. The main computer has been connected via substations to the process (only one in use in the test system). The substations include a PDP 11/04 computer with 16k core memory and the process interface.

The software has been generally organized so that substations perform data acquisition and prehandling, alarm checking, and stabilizing control (lower-level DDC loops). The main computer operates the data file system, operator communication, process calculations including higher level control, and reporting. A commercial control software package (PROSCON 102) was used as the basic

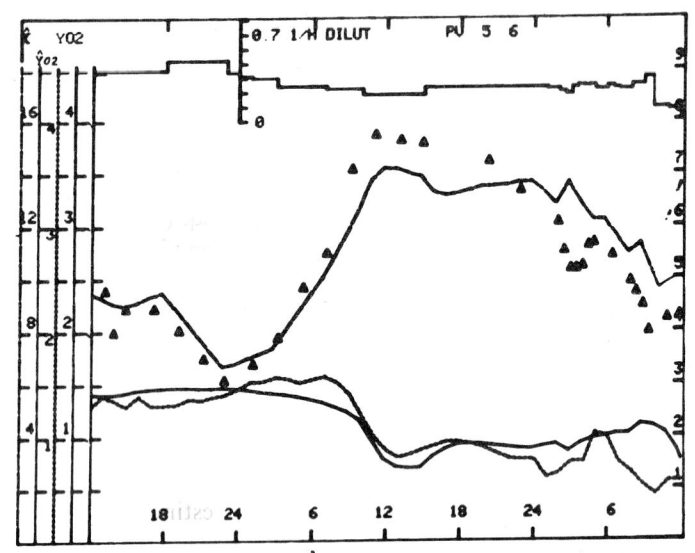

Fig. 8. Functioning of the growth estimator. Real (Y_{O_2}) and estimated ($\hat{Y}_{O_2}$) yield coefficients and estimated biomass ($\hat{X}$) together with measured values (Δ) as a function of time (in hr).

program in the substation. Programming language is mainly assembler. In the main computer, the software is divided into several quite independent modules, which take care of different tasks. The programming language is FORTRAN. The operator communication, data file, and general reporting systems are about the same as in the PFCS system described in the preceding symposium [5]. This set of programs includes only a few features that are characteristic for the Pekilo process. The features special to the process are included in the calculations, higher level control, and special reporting programs.

Data Acquisition and Storing

As a special feature, there is quite a large number of sample analysis results that are regularly given to the computer. The time interval between samples varies from 1 to 6 hr and the laboratory results are delayed $\frac{1}{2}$–5 hr from the sample time. Because of this, special features are needed in the preliminary treatment of the analysis results and the data bank organization to ensure that these results are available in the right form for calculation. In the automatic measurements, the sample interval in the substations is 5–10 sec. However, only 1- or 2-min mean values are transferred to the main computer for calculation and 10-min mean values are stored in the data bank.

Process Calculations

The set of calculation programs includes several groups of variables that are related to the handling of measurement data, mass and energy balances, estimation, on-line dynamic modeling, controlling, and reporting. For example, one fermentor consists of 80–100 such variables. Most of these programs are specific to the Pekilo process and include the "intelligence" of the system. Calculations are performed every 1 or 2 min when new information is received from the substations so that all groups are scheduled at least once within 10 min.

Controls

In stabilizing control, the computer does not perform anything new if compared to the conventional control system. In the reference system, basic controls have been realized mainly as supervisory control, i.e., the setpoints of analog controllers are controlled by the computer. On the higher level, there exist objects for optimization controls, all of which have not yet been proved. These are as follows. In the fermentor: nutrient dosage control, aeration and agitation power control, and growth and protein content optimization. In the drier: product moisture control. In the strippers: steam feed control.

In the nutrient dosage control, the concentrations of the nutrients (K,P) in the fermentor are kept at minimum acceptable values. The control is based on the estimated values of the growth rate and the nutrient concentrations. In aeration and agitation power control, the aim is to set the electrical energy consumption to its optimal value while taking into account, e.g., changes in the production rate. The principle is to have a suitable balance between oxygen and substrate limitation and an optimal balance between aeration and agitation powers. The growth can be maximized by controlling the dilution rate within given limits. The protein content can be maximized by maintaining a high value of the specific growth rate.

The product moisture is controlled near its lowest permitted value. This automatically sets the steam consumption in the drier to its minimum value. In the stripper, steam is also saved when overstripping is avoided by calculating continuously optimal values for steam feeding.

Operator Communication and Reporting

The operator communication takes place by the aid of two CRT terminals and a special keyboard. The system also includes separate alarm and reporting typewriters. The basic functions and operations are similar to those of the PFCS system described in Ref. 5. Shift and day reports are given as standard reports.

CONCLUSION AND DISCUSSION

The computer system considered is one of a few computer applications in industrial fermentation processes. The system is not yet ready, but the first experience with the test system shows that it acts quite satisfactorily and gives a relevant contribution to the process operation. The feeling is that most of the goals set up in the argumentation stage, mentioned earlier, can be attained. Estimation of fermentor variables, such as the growth rate, concentration of the biomass, and the substrate and nutrient concentrations, are in key position in process calculation and optimization. The methods and algorithms used for estimation have turned out to be sufficiently accurate and reliable that process supervision and control can be based on the estimates. A reason may be that the process is a simple biomass production process, which is quite stable in the normal operating state. The computer system is quite similar to usual control computer systems in the process industry. The need for process analysis and calculation programs is, however, greater than usual and flexible operation and alteration of these programs is necessary. The basic software system (MFCS) used has turned out to be practical in this respect.

References

[1] H. Romantchuk, "The Pekilo Process," in *Single Cell Protein*, S. R. Tannenbaum and D.I.C. Wang, Eds. (MIT Press, Cambridge MA, 1974).

[2] K. Forss, K. Passinen, and E. Sjöström, "Utilization of the spent sulfite liquor component in the Pekilo protein process and influence of the process on environmental problems of the sulfite mill," in *Proceedings of the Symposium on Wood Chemistry, Pure and Applied* (ACS, Los Angeles, 1974).

[3] A. Halme, A. Holmberg, and E. Tiussa, "Modeling and control of a protein fermentation process utilizing the spent sulfite liquor," in *Proceedings of IFAC Symposium on Environmental Systems Planning, Design and Control* (Pergamon, London, 1978).

[4] A. Halme, H. Kiviranta, and M. Kiviranta, "Study of a single-cell protein fermentation process for computer control," in *Proceedings of 5th IFAC Conference on Digital Computer Applications to Process Control* (North-Holland, The Hague, 1977).

[5] A. Meskanen, "Design of the man-machine interface for computer coupled fermentation," in *Proceedings of the Workshop on Computer Applications in Fermentation Technology 1976*, R. P. Jefferis III, Ed. (Verlag Chemie, Berlin, 1977).

Computer Control of Fermentation Plants

RALF LUNDELL

Rintekno Oy, Kotkapolku 2, 02620 Espoo 62, Finland

INTRODUCTION

A fermentation plant is an industrial enterprise that manufactures one or several products by means of microbes, then recovers, purifies, and packages these products in bulk or formulated form, and markets them to yield a profit.

The fermentation plant typically has a yearly turnover in the range of 3 million (M) to 30 M dollars. The raw materials and utilities account for the largest part of the production costs. Since many plants are run in three shifts and normally are very sparsely instrumented and automated, the labor costs are quite substantial. Most plants are forced to maintain quite a large organization for product control and process development.

Today there are four main groups of industrial fermentation processes, as indicated in Figure 1 [1]. The products of groups 1 and 2 are mainly produced in plants designed for a single product, whereas groups 3 and 4 include processes run in multiproduct plants, i.e., a penicillin plant often produces Pen G, Pen V, and erythromycin, as well as semisynthetic penicillins.

As can also be seen from Figure 1, there are only a few continuous fermentation processes. The majority of the batch fermentations are of the fed-batch-type, i.e., nutrients or precursors are added to the fermentation vessel during the batch to optimize the product yield. This adds a continuous feature to the batch operation requiring a more strict control approach. In addition, process operations such as washing, sterilization, filling, and emptying of the fermentation equipment create a need for sequence control. Many batch fermentations are followed by continuous recovery operations that utilize solvent distillation units. These are similar to normal chemical plant unit operations.

The fermentation industry has, during the past years, benefitted strongly from the development of stable microbial mutants with very high productivity. Process developments, such as the introduction of the fed-batch operation mode, the increase of mass transfer capacity in the fermentor vessels by increased power input, and the improved aseptic design lowering the batch contamination frequencies, have also positively affected the fermentation plant economy.

These improvements and the increased understanding of the biochemistry have created a need for a more sophisticated control approach in the fermentation plants. This can be achieved by the use of computer technology.

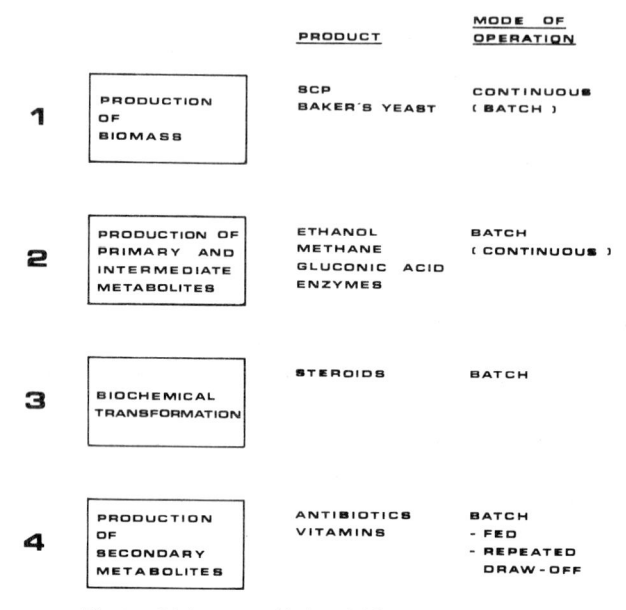

		Product	Mode of Operation
1	PRODUCTION OF BIOMASS	SCP BAKER'S YEAST	CONTINUOUS (BATCH)
2	PRODUCTION OF PRIMARY AND INTERMEDIATE METABOLITES	ETHANOL METHANE GLUCONIC ACID ENZYMES	BATCH (CONTINUOUS)
3	BIOCHEMICAL TRANSFORMATION	STEROIDS	BATCH
4	PRODUCTION OF SECONDARY METABOLITES	ANTIBIOTICS VITAMINS	BATCH - FED - REPEATED DRAW-OFF

Fig. 1. Main types of industrial fermentation processes.

COMPUTER CONTROL IN THE PROCESS INDUSTRY

Today, the use of computers and digital electronics for automatic control and rationalization of the process industry is a reality. The forerunners in this field have been the chemical and petrochemical industries as well as the power stations and the woodworking industry [2].

In general, the driving forces behind such a development in the process industry are the following:

1) To operate the plant as economically as possible, e.g., to optimize the use of raw materials and utilities, increase production and improve quality.

2) To operate the plant on a flexible mode, e.g., to easily adapt the production program to market fluctuations.

3) To operate the plant as safely as possible.

4) To operate the plant below permissible pollution limits.

5) To remove or improve low-quality jobs.

6) To obtain full documentation.

These objectives concern management of a fermentation plant as well as the management of every processing unit. However, apart from the general trend in the process industry, there are only a few examples of computerization in the fermentation industry. The reasons for the relatively slow introduction of computer systems in this branch of the process industry are several. As mentioned earlier, the productivity of fermentation plants has been greatly increased by strain and process improvements. Furthermore, the microbial processes are of a very complex nature, and industrially reliable instruments and analyzers capable of directly measuring metabolic variables are lacking. A further obstacle has been that existing plants typically have only a few measuring points (temperature, pressure, and pH). Consequently very few data-acquisition and direct-control possibilities are available.

Today, with improved instrumentation, more reliable and less expensive computer hardware, and a better understanding of fermentation processes, computer automation of the fermentation industry has begun [3-6].

PROCESS CONTROL HIERARCHY

A fermentation plant is distributed in many ways in nature. The process equipment is spread over the plant site while the control and decision functions are distributed on several levels of the plant organization. This can be seen in Figure 2, which schematically shows the hierarchy of process control in a fermentation plant [7].

The highest process control level is devoted to production planning on a long-range basis, i.e., weeks to months. It considers, e.g., fluctuations in prices of raw materials and products, and governmental restrictions on the production scheduling.

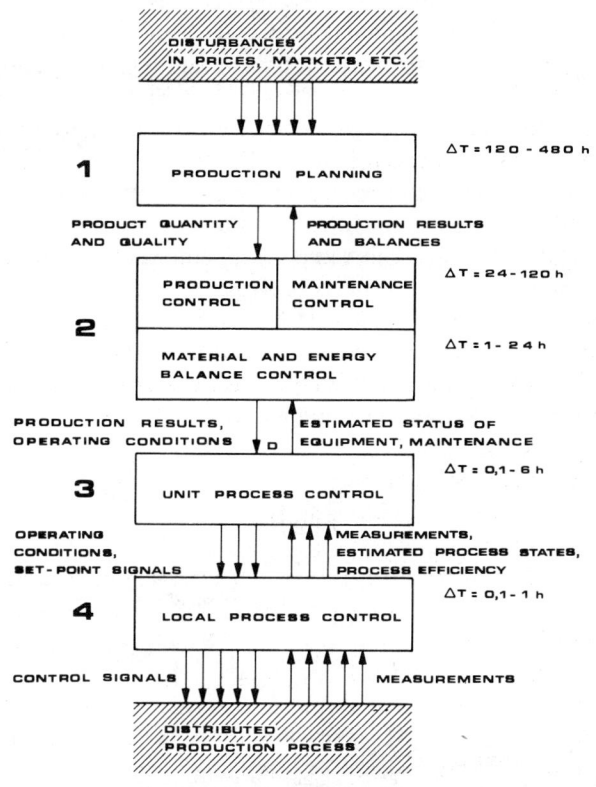

Fig. 2. Hierarchy of process control.

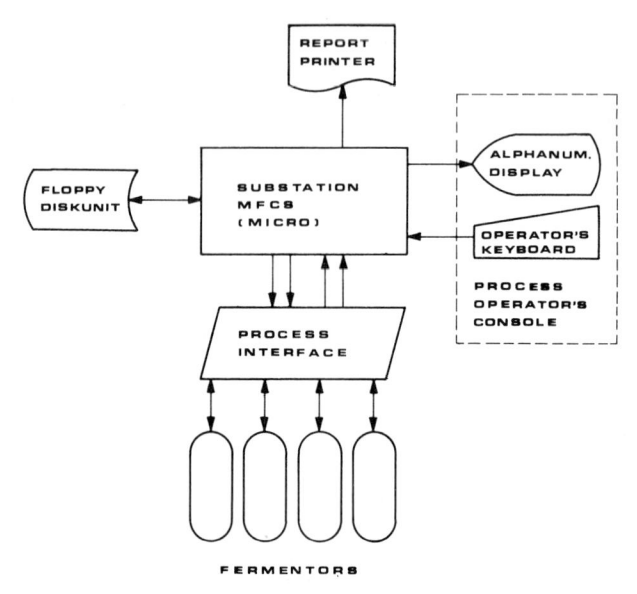

Fig. 3. Multifermentor control system.

The next level views production control tasks with a time span from days to weeks. This level is concerned with equipment failures, storage problems, and quality irregularities, as well as with start-up and shut-down of different processing steps in the fermentation plant. Typically the disturbances at this level are normally stochastic in nature.

The third level can be referred to as the unit process control. Fermentors, extraction units, distillation units, and cooling systems belong to this level of control functions. The time scale ranges from minutes to hours. This level involves fairly complicated models of the process dynamics and requires expensive analysis efforts.

The lowest level in the hierarchy comprises the traditional process control tasks, controlling heat exchangers, storage tanks, condensers, etc., by local loops. The time scale runs from seconds to minutes. No complicated mathematical treatment is involved.

TABLE I
Estimated Cost of a MFCS Substation System

Objective:	Simultaneous control of four fermentors
Tasks:	Sterilization sequencing, nutrient feeding, control of four parameters, data logging, basic operator communication
Costs:	Hardware $45 000, software $25 000

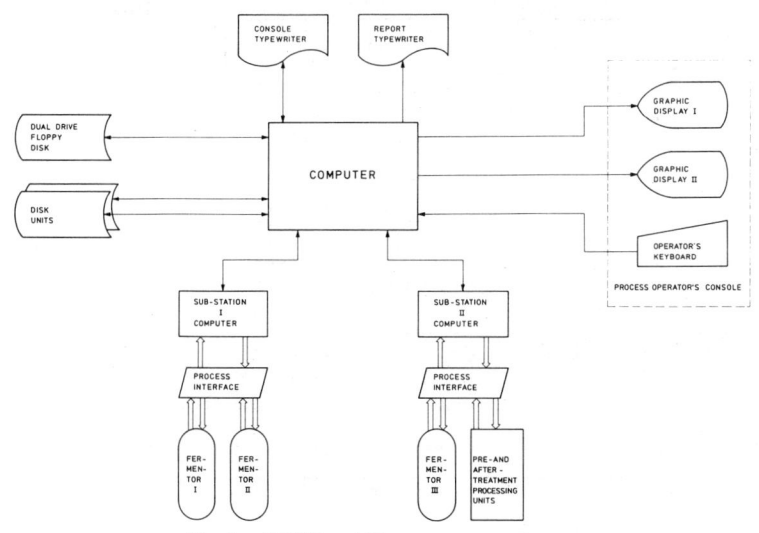

Fig. 4. MFCS multifermentor control system.

Today, there are efforts in the process industry to master each of these levels separately and as an integrated whole to gain a fast and reliable tool for economy-based decision making on every level of the plant organization.

The realization of a computer control hierarchy in the process industry has proceeded stepwise. It has been related to the development of instrumentation and analyzers. The approach has been to move toward increased system complexity starting from the lowest levels of the hierarchy. In this way, attention is focused on every individual part of the process entity at a time, giving a thorough understanding of the separate pieces of the puzzle.

TABLE II

Estimated Costs of a MFCS System with One Master Computer and Two Substations

Objective:	Control of a continuous fermentation plant, three fermentors, pretreatment, product recovery
Tasks:	Process monitoring, stabilizing control, optimizing control, on-line processing of data, data collection and reduction, advanced operator communication
Costs:	Hardware: $150 000
	Software: $125 000

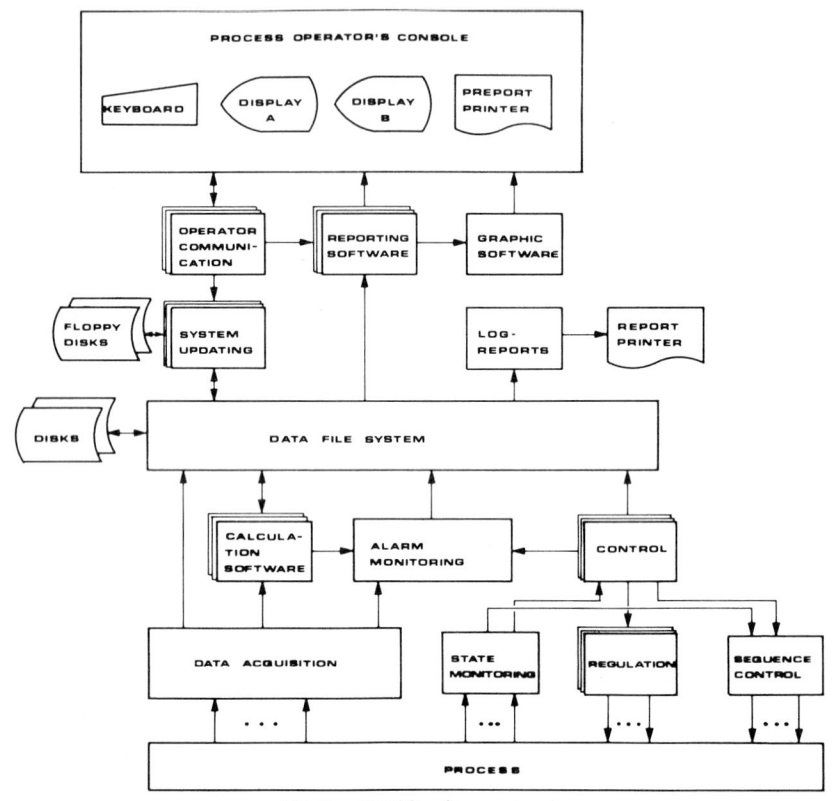

Fig. 5. MFCS software structure.

MFCS CONCEPT

The MFCS is a multifermentor computer control system developed for data processing and control of several fermentors simultaneously [8-10]. The system is a distributed hierarchic control entity, comprising one master computer to which can be connected a twin back-up computer and several substation computers, as well as all the necessary communication devices. The system is very

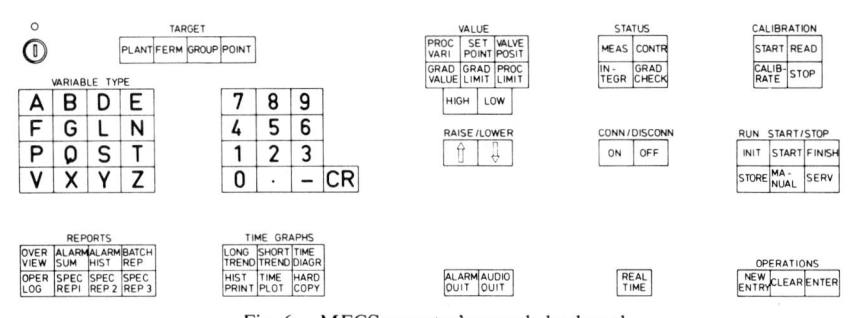

Fig. 6. MFCS operator's console keyboard.

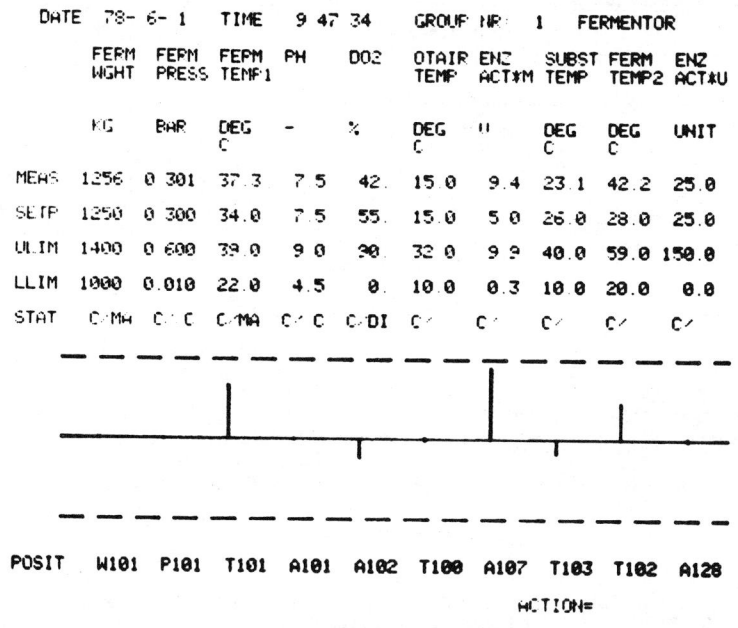

```
DATE  78- 6- 1   TIME   9 47 34   GROUP NR   1   FERMENTOR

      FERM  FERM  FERM  PH   DO2   OTAIR ENZ    SUBST FERM  ENZ
      WGHT  PRESS TEMP1              TEMP  ACT*M TEMP  TEMP2 ACT*U

      KG    BAR   DEG   -    %     DEG   U      DEG   DEG   UNIT
                  C                 C            C     C

MEAS  1256  0.301 37.3  7.5  42.   15.0  9.4    23.1  42.2  25.0
SETP  1250  0.300 34.0  7.5  55.   15.0  5.0    26.0  28.0  25.0
ULIM  1400  0.600 39.0  9.0  90.   32.0  9.9    40.0  59.0  150.0
LLIM  1000  0.010 22.0  4.5  0.    10.0  0.3    10.0  20.0  0.0
STAT  C/MA  C/C   C/MA  C/C  C/DI  C/    C/     C/    C/    C/
```

POSIT W101 P101 T101 A101 A102 T100 A107 T103 T102 A128

ACTION=

Fig. 7. MFCS fermentor report.

modular in structure enabling a straightforward expansion. It can be used for DDC or supervisory control. The substations can be used as a first step on the way toward integrated computer control of the fermentation plant or pilot plant.

MFCS Substation System

A substation MFCS system is shown in Figure 3. It can be connected to several fermentors, distillation columns, and processing units for data acquisition and handling, as well as for direct digital control purposes. The system is sturdy and inexpensive, as can be seen from Table I.

The main drawback of such a system is that the use of more complicated mathematical models and, thus, control functions are quite limited. However, it is an excellent tool for data gathering and straightforward control tasks and represents a good first step into the computer age.

MFCS Distributed System

The MFCS hierarchic computer system is shown in the form of a two-substation system in Figure 4. More substations can easily be connected to the master computer. The master computer handles the calculation and communication tasks. The primary data acquisition and control tasks of the computer system are performed at the substation level. In a larger plant this concept will save cabling costs and reduce the load on the master computer. It also increases

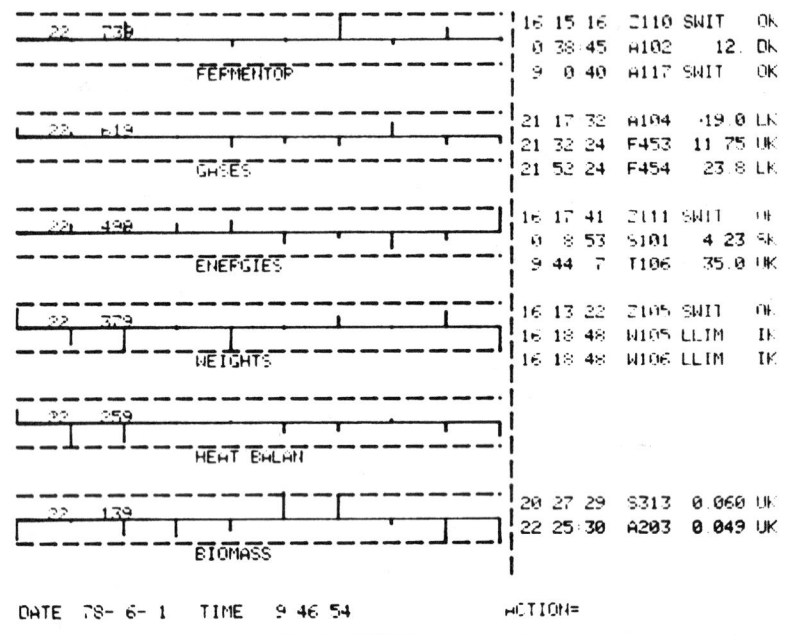

Fig. 8. MFCS group report.

the reliability of the system and makes it possible to use DDC control, thereby reducing the instrumentation costs.

The first installation of this hierarchic system was at the PEKILO factory in 1978. The computer system controls the fermentation processes, the pretreatment of the substrates, and the aftertreatment stages, dewatering, and drying of the cell mass.

The costs of installing such a system are indicated in Table II. The decision to invest in a computer system also brought with it an increase in the level of instrumentation and a recruitment of personnel with process computer experience, e.g., a control engineer and a programming specialist. The personnel recruitment was necessary to take full advantage of the installed computer system.

THE MFCS SOFTWARE

The structure of the MFCS system is shown in Figure 5. The software consists of the following main blocks [10].

Data acquisition: involves reading analog measurements, linearizing and filtering the data, and converting the values to engineering units.

Digital state monitoring: Includes reading of contact states (from limit switches, computer/manual switches, etc.) as well as reacting to alarm inputs.

Alarm monitoring: Detection, reporting, and logging.

Calculation of derived variables: The most fermentation-oriented part of

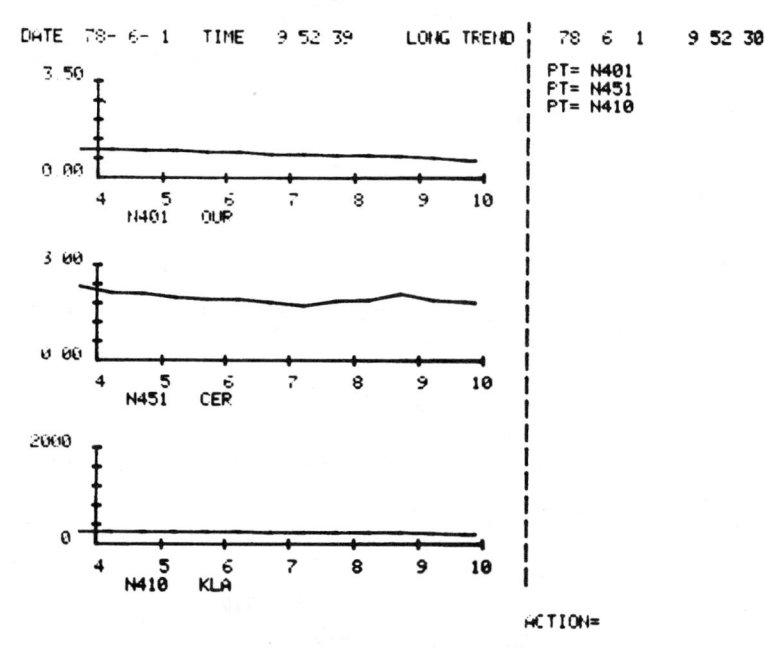

Fig. 9. MFCS long trend diagrams.

the system. A total of 120 derived variables have been included in the MFCS system. These variables are calculated using analog measured data, manually input data like laboratory analyses, as well as other derived variables. Each variable is calculated at an individual, suitable frequency. The structure of the software allows for addition of any number of new variables and calculation formulas.

Data file updating programs: Organize the measured, manually entered and derived variables into fermentor-related history files for use in calculation, operator communication, and reporting programs.

System updating programs: Perform the generation of process-dependent parameter files, addition of new measurement or control loops, changing the format of reports, etc.

Operator communication programs: Perform the interchange or information between the computer system and the process operator.

Reporting programs: Generate the various alarm reports, the shift and daily reports for continuous fermentation processes, and the batch reports for batch fermentations processes.

Graphical report programs: Generate the trend diagrams, the time history diagrams; and the monitoring deviation diagrams.

Control modules: Include programs for DDC and supervisory control using a choice of various control algorithms, as well as the programs for setpoint profiling.

Sequence control programs: Sterilization, filling and emptying, starting and stopping of pumps, etc.

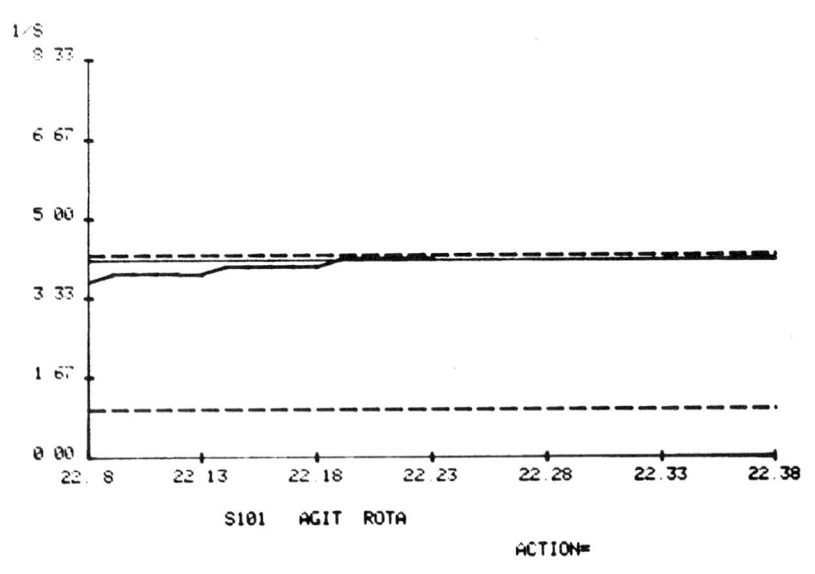

Fig. 10. MFCS short trend diagram.

OPERATOR COMMUNICATION WITH MFCS

The operator communication within a fermentation plant is of great importance. In order to achieve an easily operated communication link, graphic displays and keyboards are used in the MFCS systems. The operator graphic display of a MFCS installation has a special feature added to the standard display unit to facilitate reliable and fast response for the process operator. This feature consists of the sets of pushbuttons around the screen for direct addressing of a selected fermentor within a plant, a selected group of variables within a fermentor, or a variable within a group. Figure 6 shows the layout of a specially designed operator's console keyboard to activate the operator communication functions like requesting a report, entering an analytical result, entering a calibration value, changing a setpoint.

To illustrate some of the features of the operator communication with the MFCS-system, data from a batch fermentation of *Bacillus subtilis* has been chosen. The fermentation was performed in a 2000 liter pilot-fermentor at the Technical Research Center of Finland. The pilot-fermentor and its computer-control system are described in Refs. 9 and 11. Figure 7 shows a fermentor overview report from the protease fermentation. The fermentation variables are grouped according to their logical consistency to six groups of ten variables. If more variables or groups are needed, additional pages can be called by pushing a "next page" button at the display. The short, vertical line segments represent deviations of the variables from their setpoints, or in case of non-controlled variables, from their normal values. The dashed lines are upper and lower alarm

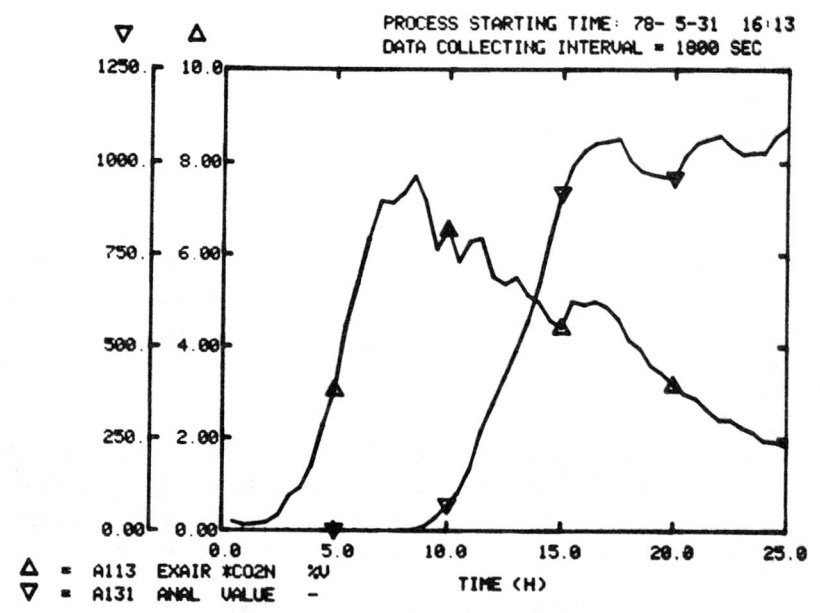

Fig. 11. MFCS batch report, sample 1.

limits. When an alarm or deviation is observed, a closer look at the situation is gained by pressing a button at the display beside the target fermentation variables group. A group report (Fig. 8) is now displayed, giving the numeric values of setpoints, measured or calculated values, alarm limits, etc.

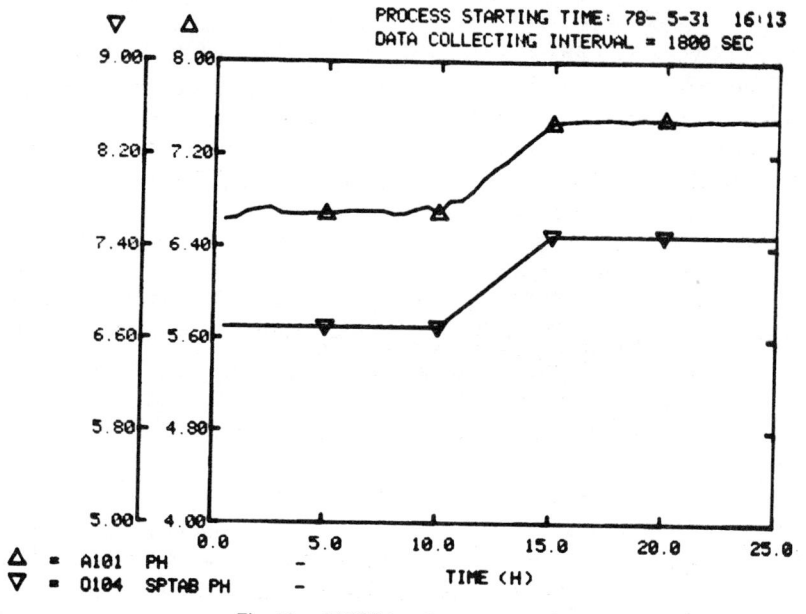

Fig. 12. MFCS batch report, sample 2.

Very often the trends of the variables are of interest. The past history of a chosen variable of the displayed group is displayed when the corresponding button below the variable is pushed (Fig. 9). Additional variables in the same group can be selected by pushing corresponding buttons. If a variable from another group is wanted for trend display, the selection is performed either by changing the group display and direct addressing, or by using the operator's keyboard to key-in the identification code and pressing the "long trend" button.

When performing dynamic experiments, e.g., for tuning control loops, a short-time trend of a variable is handy. This is produced by pushing the "short trend" button. Initially, the short-time history of the selected variable is displayed. Afterwards, the computer updates the picture at short intervals, e.g., 10 sec, which allows the continuous tracking of the variables behavior (Fig. 10).

An additional MFCS feature is the graphic display of batch reports at the end of a fermentation. Figures 11 and 12 show some features of the batch protease fermentation. The pH profile shown had been established based on earlier batch results and was initiated by the operator at the beginning of the batch.

MFCS PROJECT REALIZATION

The MFCS concept is a basic computer system for the fermentation industry that can be applied to any user's process. This eases the economic burden and definitely shortens the total system installation time. It also enables the user to concentrate on solving specific control problems.

The nature and complexity of fermentation plants vary widely. It is therefore difficult to make broad statements on the justification of computer control. Even though two plants produce the same product and have the same overall control objectives, their equipment, instrumentation, and even processes differ. Thus, each fermentation plant must be analyzed individually for sound economic decision making. It is obvious that the more complex a process is, the greater is the potential for computer automation. The trend of automation in the fermentation industry is the same as in the process industry as a whole, i.e., a rapidly increasing use of process computers.

References

[1] D. Herbert, in *6th International Symposium on Continuous Culture of Microorganisms* (Oxford U.P., London, 1975).
[2] P. Uronen and S. Heikkilä, in *Automation Symposium Society of Control Technology in Finland* (Helsinki, 1978).
[3] H. Metz and F. Wenzel, in *Fifth International Fermentation Symposium* (Verlag, Berlin, 1976).
[4] A. Halme, H. Kiviranta, and M. Kiviranta, *Digital Computer Applications to Process Control*, Van Nauta Lemke, Ed. (IFAC and North Holland, Amsterdam, 1977).
[5] R. P. Jefferis III, in *8th Meeting of the North West European Microbiological Group* (Helsinki, 1976).
[6] R. P. Jefferis III, *Process Biochem.*, 10 (1975).

[7] O. A. Asbjørnsen, in *European Federation of Chemical Engineering: 166th Event* (Firenze, 1976).
[8] R. Lundell, P. Laiho, and A. Meskanen, in *European Federation of Chemical Engineering: 166th Event* (Firenze, 1976).
[9] M. Nihtilä and J. Virkkunen, *Digital Computer Applications to Process Control,* Van Nauta Lemke, Ed. (IFAC and North Holland Publishing, Amsterdam, 1977).
[10] A. Meskanen, ASM meeting, Las Vegas, 1978.
[11] A. Meskanen, in *Workshop on Computer Applications in Fermentation Technology 1976* (Verlag Chemie, New York, 1977).

Author Index

Aiba, S., 269
Alvarez, J., 149
Ando, T., 103
Armiger, W. B., 215

Beaverstock, M. C., 241
Blachere, H. T., 205
Bošnjak, M., 155
Boulton, R., 167
Bourne, J. R., 359
Boyle, T. J., 349

Cadman, T. W., 25
Cheruy, A., 303
Cooney, C. L., 1; 13; 95
Cordonnier, M., 227
Creagan, C. C., 335

Dobry, D. D., 39
Drakeford, J., 231
Dunn, I. J., 359
Durand, A., 303

Endo, I., 321
Erickson, L. E., 49

Gerhardt, P., 137
Gray, P. P., 125

Halme, A., 369
Hatch, R. T., 25
Hennigan, P. J., 257
Hewetson, J. W., 125

Inoue, I., 321
Ishikawa, T., 103
Ishikawa, Y., 103

Jefferis III, R. P., 231
Joergensen, B. B., 85
Johanides, V., 155
Jong, T. H., 125
Jost, J. L., 39

Kamiyama, N., 103
Kernevez, J. P., 227

Kitabata, T., 283
Kitsunai, T., 103
Kjaergaard, L., 85
Klein, S. S., 231
Kobayashi, T., 73
Kryze, J., 227

Lebeault, J. M., 227
Lim, H. C., 257; 335
Lundell, R., 381

Miwa, K., 103
Mohler, R. D., 257; 335
Moo-Young, M., 179
Moran, D. M., 215
Moreira, A. R., 179
Mori, H., 73

Nagamune, T., 321

Ohashi, M., 103
Oikawa, Y., 103

Peringer, P., 205

Ricaño, J., 149

Schultz, J. S., 65
Shibata, T., 103
Shimizu, S., 73
Shiota, M., 283
Shioya, S., 283
Shoda, M., 103
Sims, G., 65
Spruytenburg, R., 359
Stieber, R. W., 137
Swartz, J. R., 95

Takamatsu, T., 283
Topolovec, V., 155
Trearchis, G. P., 241
Tsao, G. T., 257

Undén, Å., 61

Van Dedem, G., 179

AUTHOR INDEX

Wang, D. I. C., 13
Wang, H. Y., 13
Watabe, T., 103
Watanabe, Y., 103
Weigand, W. A., 257; 335

Wilder, C., 25

Yano, T., 73

Zabriskie, D. W., 117

Subject Index

Activity, 283
Affinity sensors, 65
Ammonium lactate, 137
Analyses, 61
ATP, 179
Available electron balance, 49

Bakers' yeast, 205, 359
Biomass, estimation, 61, 117
  high-density, 65
  measurement, 117
  productivity, 335

Carbon balance, 49
Carbon dioxide production, 39
Cell age distribution, 369
Cell population balance, 283
Cellulases, 179
Computer-aided fermentor, 269
Computer-coupled system design, 215
  function, 215
  specification, 215
Computer interfacing, 215
Control, 95, 227, 349, 359
  adaptive, 1
  computer, 1, 13, 25, 39, 257, 269, 335, 369, 381
  data logging, 227
  data treatment, 227
  feedback, 1
  feed forward, 1
  multivariable, 283
  noninteracting, 283
  on-line computer, 149
  optimal, 205
  PI, 283
  process, 39, 269
  rate, 167
  temperature, 167

Data, analysis, 49
  consistency, 49
  recording, 39
Determination of the culture metabolic activity, 117
Dialysis, 137

Diauxic growth, 179
Dissolved oxygen sensor, 103
DO-stat system, 65
Dual substrate, 167
Dummy state variable, 283
Dynamic programming, 149

Economics, 1
Erythromycin biosynthesis, 303
Ethanol-assimilating yeast, 283
Enzyme induction, 179
  regulation, 179

Feed rates, 61
Fermentation, 137, 335
  aerobic, 359
  alcoholic, 167, 321
  bakers' yeast, 13
  batch, 167, 257, 321
  continuous, 257
  control, 231, 241
  dissolved oxygen, 39
  fed-batch, 257, 335, 349
  monitoring, 61
  optimal, 349
  optimization, 269
  penicillin, 1
  repeated fed-batch, 205, 335
Fermentation modeling, 61, 125, 369
Fermentation models, 179
ff gas, 61
Fitting theory to experiment, 155
Fluorescent antibodies, 25
Fluorescence techniques, 117

Gas analysis, 13
Gas balances, 359
Glucose effect, 359
Growth estimation, 369

Identification, 149
Immobilized penicillinase, 125
Indirect measurement, 61, 95
Influence of efficiency, 349
Inhibition, 167
Instrumentation review, 61

SUBJECT INDEX

Intracellular monitoring, 117

Kinetics, 137

Lactobacillus, 137
Laser flow microphotometer, 25
Linear model, 349
Lipase, 179

Maintenance, 49
Material balancing, 1, 13
Mathematical model, 155, 283
Maximum principle, 335
Membranes, 137
Metabolite repression, 179
Metabolites, 65
Methanol concentration, 65, 149
Microbial population dynamics, 25
Microcomputer, system homogeneity, 227
   reproducibility, 227
Mixed cultures, 25
Model, 167
   identification, 269
Modeling, 137
   batch-culture, 179
   continuous-culture, 179
   tubular fermentor, 179
Monitoring, fermentation, 61

NADH, 117
Nonlinear programming statistical tests, 149

Objectives, pilot-plant system, 215
   plant-scale system, 215
On-line analysis, 39
Optimization, 1, 257, 303, 335
Oxygen uptake, 39

Parameter estimation, 269
Penicillin fermentation, 1

Penicillin-sensing electrode, 125
Permeases, 179
PI control, 283
Process control, 39, 269
Process kinetics, 155
Prognosis, 155

Quasi-steady state, 349

Redox potential, 85
Respiratory quotient, 13, 359

Sensors, 103
Silicone tubing, 65
Simulation, 155
Single-board microcomputer, 231
Single-cell protein, 335, 369
State observer, 283
State space, 349
Step disturbance, 283
Supervisory digital control, 257
   digital panel meter, 257
System hardware configuration, 215
   software configuration, 215
Systems engineering, 349

Three-dimensional growth concept, 155
Time optimal, 335
Tubing sensor, 65

Ultrafiltrate, 137

Viability, 167

Whey, 137
Wine, 167

Yeast, 149
Yield, 49

Published Symposia of Biotechnology & Bioengineering

1969   No. 1   Global Impacts of Applied Microbiology II
                        Edited by Elmer L. Gaden, Jr.
1971   No. 2   Biological Waste Treatment
                        Edited by Raymond P. Canale
1972   No. 3   Enzyme-Engineering
                        Edited by Lemuel B. Wingard, Jr.
1973   No. 4   Advances in Microbial Engineering (Part 1)
                        Edited by B. Sikyta, A. Prokop, and M. Novák
1974   No. 4   Advances in Microbial Engineering (Part 2)
                        Edited by B. Sikyta, A. Prokop, and M. Novák
1975   No. 5   Cellulose as a Chemical and Energy Resource
                        Edited by C. R. Wilke
1976   No. 6   Enzymatic Conversion of Cellulosic Materials: Technology and Applications
                        Edited by Elmer L. Gaden, Jr., Mary H. Mandels, Elwyn T. Reese, and
                        Leo A. Spano
1977   No. 7   Single Cell Protein from Renewable and Nonrenewable Resources
                        Edited by Arthur E. Humphrey and Elmer L. Gaden, Jr.
1978   No. 8   Biotechnology in Energy Production and Conservation
                        Edited by Charles D. Scott
1979   No. 9   Computer Applications in Fermentation Technology
                        Edited by William B. Armiger

All the above symposia can be individually purchased through the Subscription Department, John Wiley & Sons.